T4-AEB-144

CAMAC Instrumentation and Interface Standards:

- Description of CAMAC
- Definitions of CAMAC Terms
- Modular Instrumentation and Digital Interface System (CAMAC)
  ANSI/IEEE Std 583-1982
- Serial Highway Interface System (CAMAC)
  ANSI/IEEE Std 595-1982
- Parallel Highway Interface System (CAMAC)
  ANSI/IEEE Std 596-1982
- Multiple Controllers in a CAMAC Crate
  ANSI/IEEE Std 675-1982
- Block Transfers in CAMAC Systems
  ANSI/IEEE Std 683-1976 (R1981)
- Real-Time BASIC for CAMAC
  ANSI/IEEE Std 726-1982
- Subroutines for CAMAC
  ANSI/IEEE Std 758-1979 (R1981)

Published by
The Institute of Electrical and Electronics Engineers, Inc

Distributed in cooperation with
Wiley-Interscience, a division of John Wiley & Sons, Inc

Library of Congress Catalog Number 8185060

ISBN 0-471-89737-X

© Copyright 1982 by

The Institute of Electrical and Electronics Engineers, Inc
345 East 47th Street, New York, NY 10017

*No part of this publication may be reproduced in any form,
in an electronic retrieval system or otherwise
without the prior written permission of the publisher.*

May 24, 1982

SH08482

CAMAC Instrumentation and Interface Standards:

- Description of CAMAC

- Definitions of CAMAC Terms

- Modular Instrumentation and Digital Interface System (CAMAC)
  ANSI/IEEE Std 583-1982 — **583**

- Serial Highway Interface System (CAMAC)
  ANSI/IEEE Std 595-1982 — **595**

- Parallel Highway Interface System (CAMAC)
  ANSI/IEEE Std 596-1982 — **596**

- Multiple Controllers in a CAMAC Crate
  ANSI/IEEE Std 675-1982 — **675**

- Block Transfers in CAMAC Systems
  ANSI/IEEE Std 683-1976 (R1981) — **683**

- Real-Time BASIC for CAMAC
  ANSI/IEEE Std 726-1982 — **726**

- Subroutines for CAMAC
  ANSI/IEEE Std 758-1979 (R1981) — **758**

Foreword

This volume includes IEEE Standards defining the CAMAC modular instrumentation and interface system and associated digital highways, and recommended practices for CAMAC systems. They are based on reports of the NIM Committee* (National Instrumentation Methods Committee) of the US Department of Energy and the ESONE Committee** of European Laboratories.

The basic CAMAC Standard is ANSI/IEEE Std 583-1982, Modular Instrumentation and Digital Interface System, which specifies CAMAC modules and crates together with a data highway (Dataway) and "rules of the road" for the Dataway and Dataway signals. ANSI/IEEE Std 595-1982, Serial Highway Interface System (CAMAC), defines a serial highway interface system for interconnecting CAMAC crate assemblies and for communicating between crates and computers or other external controllers, using bit serial or byte serial data transmission. A parallel highway interface system, also for use with crate assemblies, but using bit parallel transmission of data, is defined in ANSI/IEEE Std 596-1982, Parallel Highway Interface System (CAMAC); ANSI/IEEE Std 675-1982, Multiple Controllers in a CAMAC Crate, makes provision for auxiliary controllers. The standard on Block Transfers in CAMAC Systems, ANSI/IEEE Std 683-1976 (R1981), Recommended Practice for Block Transfers in CAMAC Systems is supplementary to the basic CAMAC Standard (ANSI/IEEE Std 583-1982) and contributes to increased system compatibility by encouraging use of a limited number of block-transfer algorithims. CAMAC software standards include ANSI/IEEE Std 726-1982, Real-Time BASIC for CAMAC, and ANSI/IEEE Std 758-1979 (R1981), Subroutines for CAMAC. Also included in this volume is a compilation of definitions of terms, as used in the context of CAMAC, from Department of Energy Report DOE/ER-0104.

The development of the CAMAC system was first proposed by the Harwell Laboratory of the United Kingdom Atomic Energy Research Establishment. Harwell requested the ESONE Committee of European Laboratories to assume responsibility for the development and sought also the collaboration of the United States NIM Committee. The ESONE and NIM Committees collaborated actively throughout the CAMAC development, and the resulting standards reflect the needs of a wide body of users and potential users in a variety of disciplines throughout the world. Several of the CAMAC standards have been adopted in a number of countries and by the International Electrotechnical Commission (IEC).

The NIM/ESONE collaboration continues in the maintenance and extension of these documents and, where appropriate, generation of additional hardware and software standards and recommendations for CAMAC systems.

Requests for information should be addressed to Louis Costrell, National Bureau of Standards, Washington, D.C. 20234.

*NIM Committee[1]
L. Costrell, Chairman

C. Akerlof	N. W. Hill	L. B. Mortara
E. J. Barsotti	D. Horelick	V. C. Negro
B. Bertolucci	M. E. Johnson	L. Paffrath
J. A. Biggerstaff	C. Kerns	D. G. Perry
A. E. Brenner	F. A. Kirsten	I. Pizer
R. M. Brown	P. F. Kunz	E. Platner
E. Davey	R. S. Larsen	S. Rankowitz
W. K. Dawson	A. E. Larsh, Jr	S. J. Rudnick
S. R. Deiss	R. A. LaSalle	G. K. Schulze
S. Dhawan	N. Latner	W. P. Sims
R. Downing	F. R. Lenkszus	D. E. Stilwell
T. F. Droege	R. Leong	J. H. Trainor
C. D. Ethridge	C. Logg	K. J. Turner
C. E. L. Gingell	S. C. Loken	H. Verweij
A. Gjovig	D. R. Machen	B. F. Wadsworth
B. Gobbi	D. A. Mack	L. J. Wagner
D. B. Gustavson	J. L. McAlpine	H. V. Walz
D. R. Heywood		D. H. White

NIM Executive Committee for CAMAC
L. Costrell, Chairman

E. J. Barsotti   F. A. Kirsten
J. A. Biggerstaff   R. S. Larsen
S. Dhawan   D. A. Mack
J. H. Trainor

NIM Dataway Working Group
F. A. Kirsten, Chairman

E. J. Barsotti   C. Kerns
J. A. Biggerstaff   P. F. Kunz
L. Costrell   R. S. Larsen
S. Dhawan   D. R. Machen
A. Gjovig   L. Paffrath
D. R. Heywood   S. Rankowitz
D. Horelick   S. J. Rudnick

NIM Software Working Group
D. B. Gustavson, Jr, Chairman
W. K. Dawson, Secretary

A. E. Brenner   P. A. LaSalle
R. M. Brown   F. B. Lenkszus
S. R. Deiss   C. A. Logg
S. Dhawan   J. McAlpine
C. D. Ethridge   L. B. Mortara
M. E. Johnson   D. G. Perry

NIM Mechanical and Power Supplies Working Group
L. J. Wagner, Chairman

L. Costrell
C. Kerns
S. J. Rudnick
W. P. Sims

NIM Analog Signals Working Group
D. I. Porat, Chairman

L. Costrell
C. E. L. Gingell
N. W. Hill
S. Rankowitz

NIM Serial System Subgroup
D. R. Machen, Chairman

E. J. Barsotti
D. Horelick
F. A. Kirsten
R. G. Martin
L. Paffrath
S. J. Rudnick

NIM Block Transfer Subgroup
E. J. Barsotti, Chairman

W. K. Dawson
F. A. Kirsten
R. A. LaSalle
F. R. Lenkszus
R. G. Martin
R. F. Thomas, Jr

NIM Multiple Controllers Subgroup
P. F. Kunz, Chairman

E. J. Barsotti   D. R. Machen
F. A. Kirsten   R. G. Martin

---

[1] National Instrumentation Methods Committee of the US Department of Energy.

**ESONE Committee[2]

W. Attwenger, Austria, *Chairman 1980–81*; H. Meyer, Belgium, *Secretary*

Representatives of ESONE Member Laboratories

W. Attwenger, Austria	E. Kwakkel, Netherlands	A. C. Peatfield, England
R. Biancastelli, Italy	J. L. Lecomte, France	I. C. Pyle, England
L. Binard, Belgium	J. Lingertat, D. R. Germany	B. Rispoli, Italy
J. Biri, Hungary	M. Lombardi, Italy	M. Sarquiz, France
B. Bjarland, Finland	M. Maccioni, Italy	W. Schoeps, Switzerland
D. A. Boyce, England	P. Maranesi, Italy	R. Schule, F. R. Germany
B. A. Brandt, F. R. Germany	C. H. Mantakas, Greece	P. G. Sjolin, Sweden
F. Cesaroni, Italy	D. Marino, Italy	L. Stanchi, Italy
P. Christensen, Denmark	H. Meyer, Belgium	R. Trechcinski, Poland
W. K. Dawson, Canada	K. D. Muller, F. R. Germany	M. Truong, France
M. DeMarsico, Italy	J. G. Ottes, F. R. Germany	P. Uuspaa, Finland
C. A. DeVries, Netherlands	A. D. Overtoom, Netherlands	H. Verweij, Switzerland
H. Dilcher, F. R. Germany	E. C. G. Owen, England	A. J. Vickers, England
B. V. Fefilov, USSR	L. Panaccione, Italy	S. Vitale, Italy
R. A. Hunt, England	M. Patrutescu, Roumania	M. Vojinovic, Yugoslavia
W. Kessel, F. R. Germany	R. Patzelt, Austria	K. Zander, F. R. Germany
R. Klesse, France		D. Zimmermann, F. R. Germany

ESONE Executive Group (XG)

W. Attwenger, Austria, XG, *Chairman 1980–81*
H. Meyer, Belgium, *Secretary*

P. Christensen, Denmark	M. Sarquiz, France
M. Dilcher, F. R. Germany	R. Trechcinski, Poland
B. Rispoli, Italy	A. Vickers, England
H. Verweij, Switzerland	

ESONE Technical Coordination Committee (TCC)

A. C. Peatfield, England, *TCC Chairman*
P. Christensen, Denmark, *TCC Secretary*

W. Attwenger, Austria	W. Kessel, F. R. Germany
R. Biancastelli, Italy	J. Lukacs, Hungary
G. Bianchi, France	R. Patzelt, Austria
H. Dilcher, F. R. Germany	P. J. Ponting, Switzerland
P. Gallice, France	W. Schoeps, Switzerland
S. Vitale, Italy	

ECA/ESONE CAMAC Document Maintenance Study Group (DMSG)

P. Gallice, France, *Chairman*

R. C. M. Barnes, England	F. Iselin, Switzerland
L. Besse, Switzerland	H. Meyer, Belgium
J. Davis, England	H. J. Trebst, F. R. Germany

When this publication was approved on September 17, 1981, by the IEEE Standards Board it had the following membership:

I. N. Howell, Jr, *Chairman* Irving Kolodny, *Vice Chairman*

Sava I. Sherr, *Secretary*

G. Y. R. Allen	Jay Forster	F. Rosa
J. J. Archambault	Kurt Greene	R. W. Seelbach
J. H. Beall	Loering M. Johnson	J. S. Stewart
J. T. Boettger	Joseph L. Koepfinger	W. E. Vannah
Edward Chelotti	J. E. May	Virginius N. Vaughan, Jr
Edward J. Cohen	Donald T. Michael*	Art Wall
Len S. Corey	J. P. Riganati	Robert E. Weiler

*Member emeritus

---

[2]European Standards on Nuclear Electronics Committee.

CAMAC Instrumentation and Interface Standards:

Description of CAMAC

1. Introduction

Incompatible instruments have long been the rule in laboratories and industrial organizations. The problems they pose are enormous and expensive. The concern is with interfaces, mechanical, electrical, and from a signal standpoint. Besides the inefficiency inherent in a multiplicity of different interfaces, the man hours of design effort is staggering, and this effort is needlessly repeated for installation after installation.

ANSI/IEEE Std 583-1982, Modular Instrumentation and Digital Interface System (CAMAC), defines a standardized instrumentation system designated CAMAC, for Computer Automated Measurement and Control, that has been developed to alleviate these problems and that is gaining wide international acceptance. The system features a fully specified data highway (Dataway) together with modular functional units that are completely compatible and that are available from diverse sources.

Additional levels of compatibility are achieved through the use of standardized parallel and serial highways that have been developed for use with the basic CAMAC system. A number of general articles describing the basic system and the highways have appeared in the literature.

2. Interfacing

Fig 1[1] shows a typical configuration in which various instruments are interfaced to each other and to a computer. Large numbers and large varieties of interfaces are involved. Though this uneconomical arrangement predates current technology, it is none the less in very widespread use. The money and effort expended on such interfacing is immense, and that is the problem that has been alleviated by the CAMAC system.

Fig 2 shows the CAMAC arrangement which involves standardization of the mechanical, electrical, and signal interfaces. The standardization of the mechanical and electrical interfaces permits installation of a multiplicity of modular instruments, of various designs and from diverse sources, in a common enclosure (or crate) and operating from a common power supply. The signal interface is provided by a bused "Dataway" at the rear of the crate with clearly spelled out signal standards and "rules of the road" for the Dataway and for the connections to the Dataway.

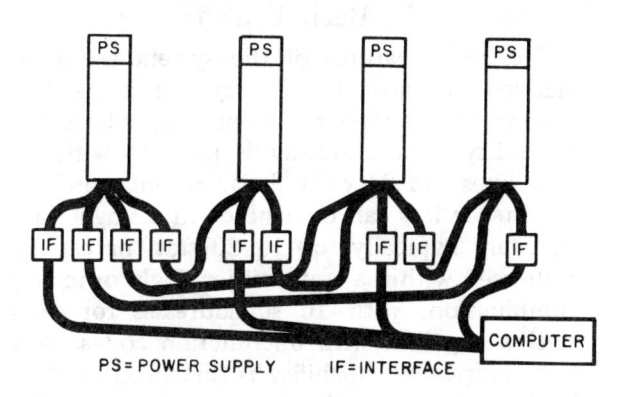

Fig 1
Typical Signal Interfacing

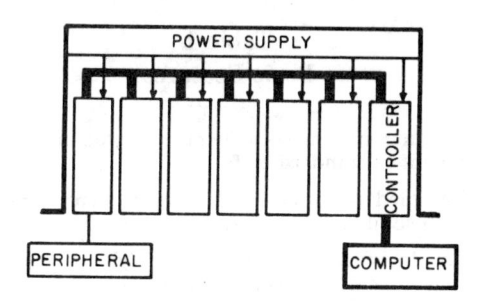

Fig 2
CAMAC Arrangement
[Data Highway (Dataway) is Heavy Black Line Going to All Modules]

Fig 3
Typical Computer CAMAC System

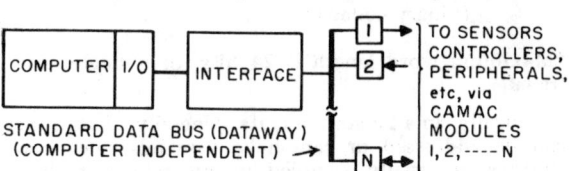

---
[1] Figs 1, 2, 3, 10, and 11 are from ANSI/IEEE Std 583-1982 (Figs K1A-K1E).

Fig 3 shows in a general way how sensors, peripherals, controls, etc, communicate with a computer via CAMAC modules through a common Dataway and a single interface.

3. Basic Features

The basic features of the system are summarized in Table 1. The system has a $1\mu s$ Dataway cycle time and a very high addressing capability. The Dataway is parallel, with 24 read lines and 24 write lines, so that words of 24 bits or less can be handled in a single data transfer. Typically, up to 23 stations can be addressed within a crate, either singly or in any combination, with 16 subaddresses for each station together with 32 function codes. This high addressing capability is further multiplied by 7 in the case of the standard 7 crate parallel highway system and by 62 in the case of the standard serial highway system, both of which are discussed later. Table 2 summarizes the data buses and the addressing and function capability.

4. CAMAC Instruments or Plug-In Units

CAMAC instruments are modular functional units that are accomodated in a CAMAC crate as in Fig 4. A typical CAMAC instrument (or module) is shown in Fig 5. The module is based on the standard 8 3/4 in (~222 mm) high panel and has a basic single width of 17 mm (~0.7 in). The modules can come in any multiple of this basic single width. The specifications allow complete flexibility with regard to the internal design of the modular instruments. The instrument is treated as a "black box" with the standardization confined to the external mechanical configuration, the electrical power requirements, and the signal inputs and outputs. This permits instrument compatibility and interchangeability without restricting the internal functioning of the instruments or the design details.

Table 1
Basic Features

(1) Modular system with functional plug-in units that mount in standard crate

(2) Designed to exploit high packing density possible with solid-state devices

(3) Plug-in units connect to data highway (Dataway) that is part of crate and carries data, control signals, and power

(4) Can connect to on-line computer though use of computer is optional

(5) Assemblies of crates may be interconnected by means of parallel or serial highways

Table 2
Data Buses and Addressing and Function Capability

Read Lines (Buses)	24*
Write Lines (Buses)	24*
Station Address (dedicated lines)	24†
Station Demand (dedicated lines)	24†
Subaddresses (per Station Address) (binary coded)	16
Functions (binary coded)	32

*Maximum word length is 24 bits for a single data transfer.

†Typically. In a 25 station crate, generally 23 stations can be addressed or can make demands since two stations are usually occupied by the crate controller.

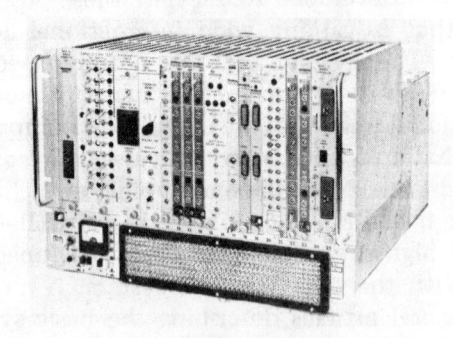

Fig 4
CAMAC Crate with Modules from Several Manufacturers
(Crate Controller Is at Extreme Right)

5. Crates

The CAMAC crate serves as the common housing for CAMAC instruments and also provides the Dataway for all the instruments that are inserted in the crate. The instruments receive digital inputs from and transmit their digital outputs along the Dataway.

All stations into which the modular instruments can be inserted are identical except for the control station which is located at the extreme right. This station is reserved for the crate controller which handles all communications between the instruments and the Dataway

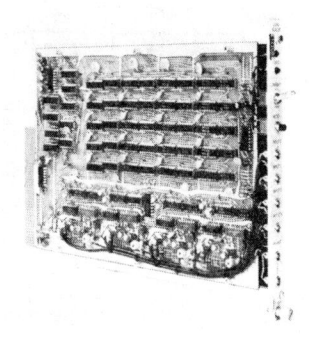

**Fig 5
Typical CAMAC Instrument**

and also between the crates and the computer or other external controller. The crate controller seen at the extreme right in Fig 4 occupies the control station plus the adjacent normal station.

A typical CAMAC crate can accomodate up to 25 single-width modules (though a minimum of two of these are required for the crate controller), or a lesser number of wider modules. Single-width modules and modules of any multiple width can be inserted in any combination.

6. Dataway

The Dataway at the rear of a CAMAC crate consists primarily of bused lines that interconnect the stations and that are utilized to transmit read, write, controls and addressing information. The Dataway includes also power supply buses as well as a limited number of dedicated lines. The controller addresses specific stations, singly or in any combination, by means of dedicated station lines. A second set of dedicated lines is used by the instruments to alert the controller when they want attention. This interrupt capability is an important feature of CAMAC that has proved to be of immense value. The Dataway consists of buses and dedicated lines as indicated in Fig 6.

**Fig 6
Dataway Wiring**

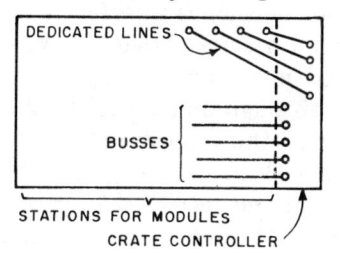

In a 25 station crate, 24 are designated "normal" stations and are wired identically. The wiring of the control station is different from that of the normal stations. Since a crate controller must have access to the Read R and Write W bus lines, it must occupy at least one normal station in addition to the control station.

The Dataway lines fall into four categories as follows:

(1) *All Station Bus Lines* linking corresponding Dataway connector contacts at all stations, including the control station.

(2) *Normal Station Bus Lines* linking corresponding Dataway connector contacts at all normal stations, but not connected to the control station. (These include the Read and Write lines.)

(3) *Individual Lines* between the control station and each normal station. There are two such lines for each normal station:
 (a) the N line (Station Number line) by which the control station addresses specific normal stations, and
 (b) the L line (Look-at-Me line) which is a *demand* line used by the station to advise the control station that it wishes attention.

(4) *Patch Contacts* (for each station) that are not connected to the Dataway and are available for patch connections.

7. CAMAC Systems, Crate Controllers and Branch Drivers

CAMAC systems vary from simple single-crate systems to complex and extensive systems with a vast number of crates distributed over a considerable area. Fig 7 shows a self-contained single-crate system with blind counters and a visual display unit operating through a crate controller. The instruments within the crate all speak a common language to the Dataway. The controller, while also speaking the common Dataway language, serves as a translator if the crate is connected to a computer as in Fig 8.

**Fig 7
Self-Contained CAMAC System**

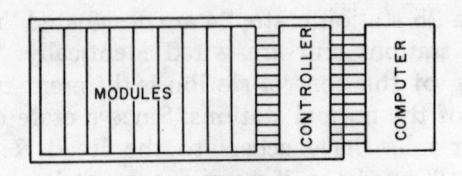

Fig 8
CAMAC System with Controller Communicating with Computer

The controller is then the one unit in the crate that must be able to communicate directly with the computer. Controllers of this sort, dedicated to specific computers, have been produced for a number of computers.

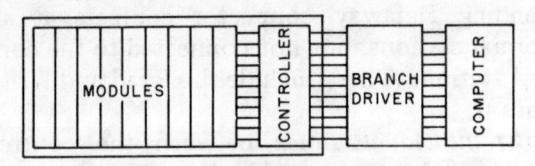

Fig 9
CAMAC System with Branch Driver

In many instances, it is advantageous to make even the crate controller computer independent. This is done by interposing a branch driver between the crate controller and the computer as in Fig 9. The crate controller then speaks the Dataway language and a standardized highway language. The Branch Driver also speaks the highway language, but it assumes the role of dedicated translator and must speak also the computer language. A single branch driver, capable of serving a multiplicity of crates, now becomes the one computer dependent device in the installation.

8. Highways

The interconnection between crates and between the crates and a computer or other controller is called the Highway. With dedicated crate controllers designed to interface with specific computers, the Highway becomes an extension of the computer I/O bus. Where branch drivers are used, that portion of the Highway that interconnects the crates and branch driver can be computer independent. Such highways can be either parallel or serial, single-ended or balanced.

9. Parallel Highways

Parallel highways are inherently capable of the highest data transfer rate and have maximum transparency to the Dataway. They are especially attractive where the highway length

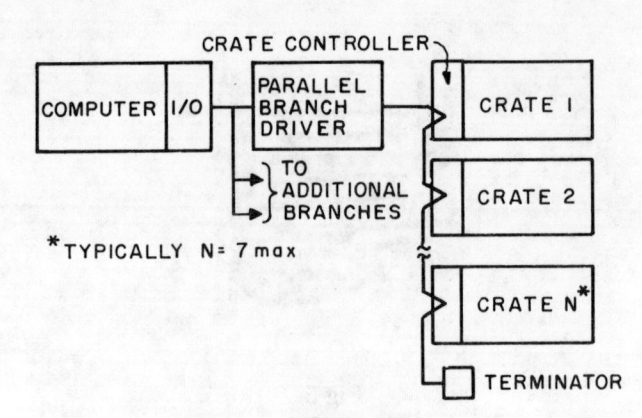

Fig 10
Parallel Highway System

*TYPICALLY N = 7 max

is not excessive so that a single-ended highway can be used [see A2.1 of ANSI/IEEE Std 596-1982, Parallel Highway Interface System (CAMAC)]. Parallel highways of greater length are feasible but this requires a balanced transmission mode, using either balanced transmitters and receivers or balanced-to-single-ended converters. Such converters permit use of single ended operation at the crates with balanced transmission over long-span portions of the highway. The standardized CAMAC parallel highway of ANSI/IEEE Std 596-1982 using branch drivers and well defined crate controllers, as shown in Fig 10, is in widespread use. The crate controllers are available from a number of sources, and branch drivers are available for a wide variety of computers.

10. Serial Highways

Serial highways using a limited number of lines are particularly attractive for installations where very long highways are required and where the highway wiring cost can be of considerable significance. Fig 11 shows diagramatically the standardized CAMAC Serial System of ANSI/IEEE Std 595-1982, Serial Highway Interface System (CAMAC), that is capable of handling up to 62 CAMAC crates. This serial system can be either bit serial or byte serial with 8 bits per byte.

The serial highway is interfaced to the computer through a serial driver or, where an unsophisticated interface of limited capability is acceptable, through an adaptor to the teletypewriter port of the computer. The highway is organized in a loop that begins at the defined output port of the serial driver, goes through a serial crate controller in each CAMAC crate,

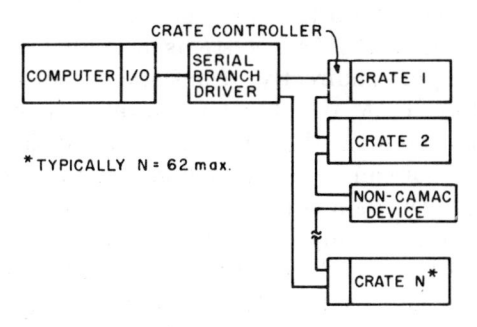

**Fig 11
Serial Highway System**

and then returns to the defined input port of the serial driver. The serial driver communicates with the computer through a port that is specific to the computer.

11. Use of Microprocessors

The capability of CAMAC systems can be further enhanced by the use of microprocessors within the modules as in Fig 12A. This permits 'local' processing and thus reduces the load on the Dataway. Alternatively, the conventional crate controller can be replaced or supplemented by microprocessor-type controllers that can reduce the load on the computer, as in Fig 12B or eliminate the need for a computer, as in Fig 12C.

12. Interfacing to Other Devices and Buses

Non-CAMAC devices can be connected into CAMAC systems either in the highway as in Fig 11 or via the Dataway by means of an interfacing module as in Fig 13. As shown in Fig 13 the measuring instrument interface bus of ANSI/IEEE Std 488-1978, Digital Interface for Programmable Instrumentation and Related System Components can be connected to the CAMAC Dataway of ANSI/IEEE Std 583-1982, by means of a simple interface modules, to an intelligent module with processing capability, or to an auxiliary controller. or to an auxiliary controller.

13. Auxiliary Controllers

The use of auxiliary controllers is provided for in ANSI/IEEE Std 675-1982, Multiple Controllers in a CAMAC Crate. Though Fig 12A shows a single auxiliary controller (in addition to the usual crate controller at the right), numerous auxiliary controllers can be installed in a single crate. An *Auxiliary Controller Bus* (ACB) is specified that provides for communication between the auxiliary controllers and the crate controller, and between the auxiliary controllers and the plug in units via the crate controller. A *Grant-in, Grant-out* line serves to award bus control in accordance with priorities allocated to the various auxiliary controllers.

14. Software

There is an increasing awareness of the importance of software economy in installations involving computers. The standardization and transportability of software becomes feasible with CAMAC systems because of the standardized Dataway operations and the high degree of computer independence. Subroutines for CAMAC, based largely on compatibility with FORTRAN (but not restricted to FORTRAN), are described in ANSI/IEEE Std 758-1979 (R1981). The use of BASIC is provided for in ANSI/IEEE Std 726-1982, Real-Time BASIC for CAMAC. ANSI/IEEE Std 683-1976 (R1981), Block Transfers in CAMAC Systems, recommends a limited family of block transfer algorithms for CAMAC use.

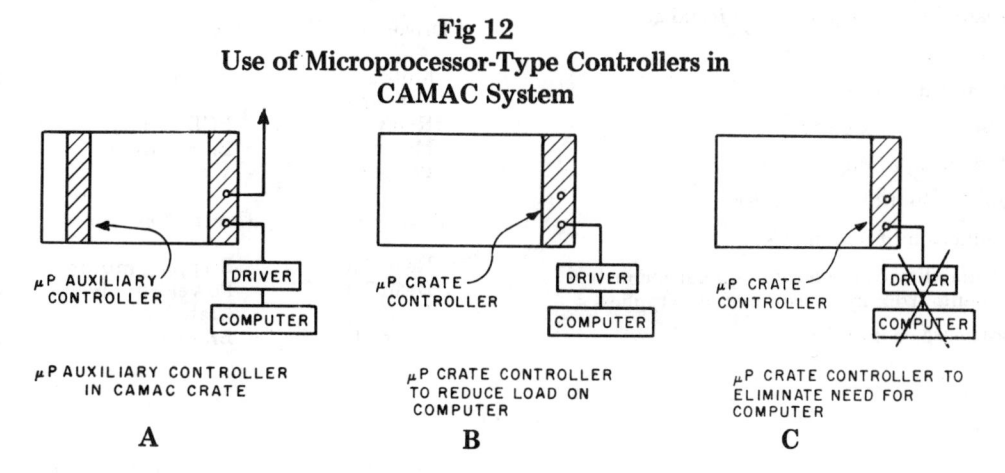

**Fig 12
Use of Microprocessor-Type Controllers in CAMAC System**

A — μP AUXILIARY CONTROLLER IN CAMAC CRATE

B — μP CRATE CONTROLLER TO REDUCE LOAD ON COMPUTER

C — μP CRATE CONTROLLER TO ELIMINATE NEED FOR COMPUTER

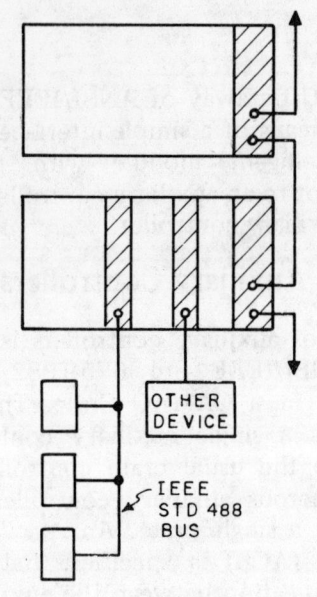

Fig 13
Connecting Other Devices and Buses (Including Bus of ANSI/IEEE Std 488-1978) Into CAMAC System Via Interfacing Modules

15. Advantages, Cost Effectiveness, Applications

A significant number of users in the industrial and medical areas and in a variety of laboratories in America, in Europe, and in other parts of the world have independently arrived at the conclusion that CAMAC is cost effective. As a result CAMAC is in widespread production and use throughout the industrialized world.

Table 3
Advantages of CAMAC

(1) Flexibility — interchangeability
(2) Optimization of systems
(3) Ease of restructuring
(4) Deferred obsolescence, updating capability
(5) High degree of computer independence
(6) Reduction of interfaces
(7) Ready interchange between installations
(8) Reduction of inventories
(9) Increased utilization
(10) Ease of serviceability
(11) Reduction of down time
(12) On-the-shelf blank modules
(13) Reduction of design effort
(14) Availability of numerous compatible instruments from many commercial suppliers
(15) Software economies

Table 3 lists numerous advantages that have been important factors in the decisions to implement CAMAC systems and that are directly or indirectly translatable into cost savings. CAMAC is a user-developed system which permits the configuration of installations that can be easily upgraded with time and which can be made practically independent of the control computer except for small sections of hardware and software. Updating of installations, such as industrial control systems, has traditionally been very difficult. Since these difficulties are not present in CAMAC systems with standardized multisource functional units, economical modernizing of CAMAC industrial control systems is feasible.

The use of CAMAC in various industrial, laboratory, medical, and aerospace applications is described in the literature. Also, application and tutorial information on CAMAC and ANSI/IEEE Std 583-1982, has been presented in IEEE short courses in numerous cities throughout the United States. The US Energy Research and Development Administration (ERDA) Report TID-26618, CAMAC *Tutorial Articles*, is a particularly informative document.

Table 4
Countries in Which CAMAC Instruments are Manufactured (Incomplete Listing)

Belgium	Netherlands
Canada	Norway
China	Poland
Denmark	Roumania
France	Switzerland
Germany	United Kingdom
Hungary	USA
Italy	USSR
Japan	

Table 5
Some of the Types of CAMAC Instruments Commercially Available

Drivers	Converters	
Motor	A to D	Encoders
Synchro	D to A	Registers
Relay	BCD-Binary	Triac units
Meter	Binary-BCD	Generators
Peripheral	V to F	Multiplexers
		Digitizers
Counters	Controllers	DVMs
		Buffers
Real-Time	Stepping motor	Clock
Dead-Time	Power Supply	Fan outs
Up-Down	Crate	Latchs
Interval	Branch	Gates

Definitions of CAMAC Terms

1. Introduction

The definitions herein are intended to provide concise explanations of the main aspects of terms used in the CAMAC documents. The definitions refer to the use of the terms in the context of CAMAC through some of the terms are also used in other contexts. They do not modify or supersede the more formal and comprehensive definitions contained in the individual CAMAC documents in this volume.

2. Categories

The definitions are listed in Section 3 in categories 3.1 through 3.8 as below.
3.1 Names
3.2 Crates, Bins, Assemblies, Systems
3.3 Modules, Plug-In Units
3.4 Controllers
3.5 Highways, Drivers
3.6 General Terms in the Context of CAMAC
3.7 Specialized Terms for CAMAC Equipment
3.8 Specialized Terms for Messages on the CAMAC Serial Highway
Index

3. Definitions

An asterisk indicates that a term is defined formally in the CAMAC standards (see also Section 4, Index).

3.1 NAMES

3.1.1 *CAMAC. A standardized modular instrumentation and digital system as defined in ANSI/IEEE Std 583-1982 (often treated as an acronym for Computer Automated Measurement and Control.)

3.1.2 ESONE. A multi-national committee representing European nuclear laboratories. It produced the initial CAMAC specification and collaborated with NIM in the maintenance and extension of CAMAC.

3.1.3 NIM. (1) A committee sponsored by the US Department of Energy and associated with the US National Bureau of Standards. It produced the NIM instrumentation system specifications, endorsed the use of CAMAC, and collaborated with ESONE in the extension and maintenance of CAMAC.
(2) A standardized modular instrumentation system consisting of NIM modules and NIM BINS as defined in US Department of Energy Report TID-20893.

3.2 CRANES, BINS, ASSEMBLIES, SYSTEMS

3.2.1 crate. A general term referring to either a CAMAC crate or a CAMAC compatible crate.

3.2.2 *CAMAC crate. A mounting unit or housing for plug-in units that includes a dataway and conforms to the requirements of ANSI/IEEE Std 583-1982.

3.2.3 CAMAC compatible crate. A mounting unit or housing for plug-in units in which CAMAC modules can be mounted and operated in accordance with the dataway requirements of ANSI/IEEE Std 583-1982 but that does not conform to the full requirements for a CAMAC crate.

3.2.4 CAMAC crate assembly. An assembly of a crate controller and one or more CAMAC modules mounted in a CAMAC crate (or CAMAC compatible crate) and operable in conformity with the dataway requirements of ANSI/IEEE Std 583-1982.

3.2.5 CAMAC system. A system including at least one CAMAC crate assembly.

3.2.6 NIM bin. A mounting unit or housing for NIM modules that includes bussed connectors at the rear for mating with connectors on the modules to provide power at the modules, and that conforms to the requirements of US Department of Energy Report TID-20893.

3.3 MODULES, PLUG-IN UNITS

3.3.1 module. A general term referring to CAMAC modules and NIM modules. The type (CAMAC or NIM) is made clear by the context in which the term is used.

3.3.2 *CAMAC module. A modular functional plug-in unit that mounts in one or more normal stations of a CAMAC crate and conforms to the requirements of ANSI/IEEE Std 583-1982 including the use of dataway lines as defined in Section 5, therein (the term thus excludes crate controllers that occupy the control station and auxiliary controllers that occupy normal stations).

3.3.3 *plug-in unit. General term for modular units, including CAMAC modules and crate controllers, that mount in CAMAC crates and conform to the requirements of ANSI/IEEE Std 583-1982.

3.3.4 NIM module, NIM instrument. A modular functional unit or instrument that mounts in a NIM bin and conforms to the requirements of US Department of Energy Report TID-20893.

3.4 CONTROLLERS

3.4.1 *(crate) controller. A functional unit that mounts in the control station and one or more normal stations of a CAMAC crate (or CAMAC compatible crate) and that controls dataway operations. It communicates with CAMAC modules via the dataway in accordance with ANSI/IEEE Std 583-1982 receiving or generating dataway signals in accordance with Section 5, therein. In many instances it links the dataway with external highways and computers.

3.4.2 parallel crate controller. A crate controller that serves as the communication link between the dataway and a parallel highway.

3.4.3 *serial crate controller. A crate controller that serves as the communication link between the dataway and a serial highway.

3.4.4 *auxiliary crate controller. A controller that mounts in one or more normal stations of a crate and that can control dataway operations in order to communicate via the dataway, utilizing the auxiliary controller bus.

3.4.5 *crate controller Type A1. A crate controller, defined by ANSI/IEEE Std 596-1982 for use with the CAMAC branch highway.

3.4.6 *crate controller Type A2. A crate controller, defined by ANSI/IEEE Std 675-1982, for use with the CAMAC branch-highway and containing features that permit operation with one or more auxiliary controllers within a crate via an auxiliary controller bus.

3.4.7 *serial crate controller Type L2. A serial crate controller defined by ANSI/IEEE Std 595-1982 for use with the CAMAC serial highway.

3.5 HIGHWAYS, DRIVERS

3.5.1 highway. General term referring to parallel highway and serial highway.

3.5.2 highway for a CAMAC system. An interconnection between CAMAC crate assemblies or between one or more CAMAC crate assemblies and an external controller.

3.5.3 CAMAC highway. A highway for a CAMAC system that conforms to the requirements of the CAMAC branch-highway of ANSI/IEEE Std 596-1982 or the CAMAC serial highway of ANSI/IEEE Std 595-1982.

3.5.4 parallel highway. A highway for a CAMAC System in which bits comprising a data word, a command or other information are transmitted simultaneously on multiple lines.

3.5.5 *CAMAC branch-highway. A parallel

highway that conforms to the requirements of ANSI/IEEE Std 596-1982. It consists of a multi-wire digital highway interconnecting up to 7 crate controllers and a branch driver. Also referred to as a CAMAC parallel highway.

3.5.6 CAMAC parallel highway. Synonymous with CAMAC branch-highway.

3.5.7 serial highway. A highway for a CAMAC System in which data, commands and other information are transmitted in bit-serial or byte-serial mode (see also bit-serial highway and byte-serial highway).

3.5.8 *CAMAC serial highway. A serial highway that conforms to the requirements of ANSI/IEEE Std 595-1982. It may be either bit-serial or byte-serial (in 8-bit bytes), and can accommodate up to 62 CAMAC crates or other controlled devices in a loop configuration.

3.5.9 bit-serial highway. A serial highway in which data, commands and other information are transmitted in bit-serial mode.

3.5.10 byte-serial highway. A serial highway in which data, commands and other information are transmitted in byte-serial mode.

3.5.11 *branch. An assembly of up to seven CAMAC crates and a branch driver conforming to the requirements of ANSI/IEEE Std 596-1982.

3.5.12 highway driver. A unit that communicates with a CAMAC system via a highway and that in many instances interfaces to a computer or other external controller.

3.5.13 *CAMAC branch driver. A highway driver for the CAMAC branch-highway that can control branch operations in conformity with the requirements of ANSI/IEEE Std 596-1982.

3.5.14 *CAMAC serial driver. A highway driver for the CAMAC serial highway, that can control operations in conformity with the requirements of ANSI/IEEE Std 595-1982.

3.5.15 *port. A defined interface between a highway and a crate controller or highway driver. (The CAMAC branch and serial highways are defined primarily in terms of their characteristics at their ports.)

3.5.16 *D-port. A defined port for the CAMAC serial highway, conforming fully with ANSI/IEEE Std 595-1982. Each CAMAC serial driver and serial crate controller has two D-Ports, one for input and one for output.

3.5.17 *U-port. A port for the CAMAC serial highway, conforming with the message structure defined in ANSI/IEEE Std 595-1982, but using signals that are undefined in IEEE CAMAC publications.

3.5.18 system crate. A CAMAC crate assembly in which specialized plug-in units are used to link one or more sources of commands to one or more highway drivers.

3.6 GENERAL TERMS, IN THE CONTEXT OF CAMAC.

3.6.1 bit. A binary digit. A signal or item of information with only two permitted values, 0 and 1.

3.6.2 byte. A group of bits, typically 8 bits, usually constituting a fraction of a word or message.

3.6.3 word. A group of bits, typically the maximum group that is processed as a unit within a computer or other controller. Hence a group of 24 bits in CAMAC. In general, the grouping of bits into a word does not necessarily imply their relative numerical values.

3.6.4 *field. A subdivision of a CAMAC serial highway message, consisting of a group of bits in one or more bytes, and containing a unit of information, for example, crate address, column-parity, read data.

3.6.5 *bit serial. The mode of transmission on the CAMAC serial highway in which all the bits constituting a byte or message are transmitted through the port in time sequence on one pair of lines.

3.6.6 *byte serial. The mode of transmission on the CAMAC serial highway in which the bits constituting a byte are transmitted through the port simultaneously on eight pairs of lines, and the bytes constituting a message are transmitted in time sequence.

3.6.7 *data. Information carried by the read and write lines of the CAMAC dataway or branch-highway or by data fields of CAMAC serial highway messages.

3.6.8 *read. (1) The direction of data transfer from controlled devices toward controllers. (In CAMAC, specifically from modules to crate controllers or auxiliary crate controllers and from crate controllers to highway drivers and external controllers).

(2) Dataway and highway lines assigned to read data.

3.6.9 *write. (1) The direction of data transfer from controller towards controlled devices. (On the CAMAC dataway, from the crate controller or auxiliary crate controller to CAMAC modules; on highways, from an external controller to highway drivers and so to crate controllers.)

(2) Dataway and highway lines assigned to write data.

3.6.10 *message. A group of bytes transmitted through a CAMAC serial highway port, and forming a syntactical unit whose beginning and end are indicated by delimiter bytes.

3.6.11 handshake. An interlocked exchange of signals between a data source and data acceptor, controlling the transfer of data. (Used on the CAMAC branch-highway and on the external connections to many CAMAC modules.)

3.6.12 *start bit. A synchronizing bit, indicating the beginning of a byte in bit-serial transmission.

3.6.13 *stop bit. A synchronizing bit, indicating the end of a byte in bit-serial transmission.

3.6.14 *byte frame. An 8-bit byte, together with a start bit and one or more stop bits, and transmitted in bit-serial mode through the D-ports of the CAMAC serial highway.

3.7 SPECIALIZED TERMS FOR CAMAC EQUIPMENT

3.7.1 *dataway. A multi-wire assembly at the rear of a CAMAC Crate that:
(1) Interconnects the stations
(2) Serves as a communication link between plug-in units within a CAMAC crate
(3) Provides power to the plug-in units via power busses that are part of the dataway
(4) Conforms to the requirements for a dataway as specified in ANSI/IEEE Std 583-1982.

3.7.2 *station. A general term referring to normal stations and control stations of a CAMAC crate.

3.7.3 *normal station. One of the mounting positions for plug-in units in a CAMAC crate, providing access to the dataway. Accessed lines of the dataway include the read and write lines and two individual lines that connect to the control station. One of the individual lines is for address (station number) and the other for demands (look-at-me).

3.7.4 *control station. The single mounting position in a CAMAC crate reserved for the crate controller and giving access to all the station number and look-at-me lines, but not the data lines. The control station is the right-most position in a CAMAC crate.

3.7.5 *bus-line. A line in the dataway that joins corresponding contacts at all normal stations and, in certain cases, the control station. All dataway lines except station number and look-at-me are bus-lines.

3.7.6 *individual line. A line in the dataway that joins one contact at a normal station to one contact at the control station. Individual lines are used for station number and look-at-me.

3.7.7 *CAMAC operation. A dataway operation or branch operation or serial highway command-reply transaction.

3.7.8 *dataway operation. A CAMAC data transfer or control operation on the dataway, characterized by the generation of busy and

CAMAC TERMS

strobe signals (includes command operations and unaddressed operations).

3.7.9 *command operation. A CAMAC operation or branch operation characterized by the presence of a command consisting of station number, subaddress and function.

3.7.10 *unaddressed operation. A dataway operation characterized by one of the common control signals, initialize or clear, without a command.

3.7.11 *graded-L operation. A special form of command operation on the CAMAC branch-highway, in which the read-write lines of the highway are used to transfer a composite graded-L word from all crate controllers to the branch driver.

3.7.12 *command. Signals on the dataway or highway specifying one or more crates, one or more stations, a subaddress, and a function.

3.7.13 *station number. (1) Identification assigned to a station in a crate ($1 \leq N \leq 25$). The station number is part of the command.

(2) Individual dataway lines or associated highway lines or message fields addressing one or more stations (and hence addressing the modules occuping the stations).

(3) The signals on these lines, or the contents of these fields.

(4) Identification assigned in an internal feature of a crate controller ($26 \leq N \leq 32$).

3.7.14 *subaddress. (1) Identification assigned to a specific subsection of a CAMAC module ($0 \leq A \leq 15$). The subaddress is part of the command.

(2) Dataway bus lines or associated highway lines or message fields carrying information which, when decoded in a CAMAC module, addresses a subsection of the module.

(3) The signals on these lines, or the contents of these fields.

3.7.15 *function. (1) Part of the command ($0 \leq F \leq 31$), specifying the action to be taken by a CAMAC module and cate controller during a command operation.

(2) Dataway bus lines or associated highway lines or message fields carrying information which when decoded in a CAMAC module, specify the action to be taken during a command operation.

(3) The signals on these lines, or the contents of these fields.

3.7.16 *crate address. (1) Identificiation assigned to a CAMAC crate assembly in a multi-crate CAMAC system. The crate address is part of the command in such systems.

(2) Individual lines on the CAMAC branch-highway, or a field of the command message on the CAMAC serial highway, carrying the crate address information.

3.7.17 *common control signals. Dataway signals initialize, clear, and inhibit, which are used without an accompanying command.

3.7.18 *initialize. (1) One of the common control signals associated with an unaddressed operation, and typically used at switch-on to set a CAMAC system or crate assembly to a defined state.

(2) The dataway bus line and any corresponding highway line carrying the Initialize signal.

3.7.19 *inhibit. (1) A common control signal used to prevent action (such as data taking) in CAMAC modules.

(2) The dataway bus line carrying the Inhibit signal.

3.7.20 *clear. (1) A common control signal, associated with an unaddressed operation, that serves to set data registers to zero.

(2) The dataway bus line carrying the clear signal.

3.7.21 *strobe. (1) Specific timing signals (S1 and S2) that occur on the dataway during command operations and during unaddressed operations, and that must be present for actions to be initiated in the module.

(2) Dataway bus lines (S1 and S2) on which the strobe signals are carried.

3.7.22 *busy. (1) A signal indicating that a dataway operation is in progress.

(2) The dataway bus line on which the Busy signal is carried.

3.7.23 *command accepted. (1) A binary indication whether an addressed module has recognized the command.

(2) The dataway bus-line (X), branch-highway line (BX), and serial highway bit (SX), carrying this information.

(3) The signals on these lines, or the value of the SX bit.

3.7.24 *response. (1) A binary indication of the state of an internal feature of an addressed module.

(2) The dataway bus line (Q), branch-highway line (BQ), and serial highway bit (SQ) carrying this information.

(3) The signals on these lines, or the value of the SQ bit.

3.7.25 *demand. An unsolicited request for service (typically for a program interrupt or for a data transfer to or from memory) originating from a CAMAC module. Demands are transmitted on the look-at-me lines to the crate controller. They may be processed by a lam grader or SGL encoder, and transmitted by a graded-L operation on a branch-highway, or a demand message on a serial highway.

3.7.26 *demand handling. The transmission of look-at-me demands from modules via the dataway and, as appropriate, via the dataway and, as appropriate, via crate controllers, lam graders, SGL encoders, highways, highway drivers, and system crates.

3.7.27 *look-at-me. A common term for the means by which CAMAC modules generate demands for service (see look-at-me line signal and look-at-me request).

3.7.28 *look-at-me line signal (L). An individual line in the dataway, on which a CAMAC module can generate demands (L signals) for interrupts or data transfers. There is one L line from each normal station to the control station.

3.7.29 *look-at-me request (LAM). An individual demand within a CAMAC module. One or more lam requests within a module may be associated with the same dataway L line.

3.7.30 *lam grader. A unit that selects, rearranges or combines the dataway look-at-me (L) signals within one CAMAC Crate to form a set of graded-L signals. Typically used as an ancillary to a crate controller in a CAMAC branch-highway system.

3.7.31 *graded-L signals (GL). A selection, rearrangement, or combination of the dataway look-at-me (L) signals, forming the graded-L word.

3.7.32 *SGL encoder. A unit that selects, rearranges or combines the dataway look-at-me (L) signals within one CAMAC crate to form the serial graded-L field of a demand message on a CAMAC serial highway. Typically used as an ancillary to a serial crate controller.

3.7.33 *block transfer. The sequential transfer of multiple words. (In CAMAC, specifically a sequence of single CAMAC operations in response to a single special command).

3.7.34 *auxiliary controller bus (ACB). A bus linking auxiliary crate controllers to a crate controller. It allows the auxiliary controllers to address all normal stations and to receive look-at-me signals from all normal stations.

3.8 SPECIALIZED TERMS FOR MESSAGES ON THE CAMAC SERIAL HIGHWAY

3.8.1 *command message. A message on the CAMAC serial highway, from the serial driver to a serial crate controller, that conveys a command and, if appropriate, write data.

3.8.2 *reply message. A message on the CAMAC serial highway, from a serial crate controller to the serial driver, in response to a command message. It may convey read data.

3.8.3 *command-reply transaction. The implementation of a command operation on a CAMAC serial highway by means of a command message and the resulting reply message.

3.8.4 *demand message. An unsolicited message on the CAMAC serial highway, from a serial crate controller to the serial driver, in response to a dataway look-at-me demand. It includes a serial graded-L-field identifying the demand.

3.8.5 *serial graded-L field. A group of 5 bits in a demand message on the CAMAC serial highway, carrying information identifying a demand.

3.8.6 *delimiter byte. A byte that identifies the end of a message on the CAMAC serial highway (see end byte, endsum byte, and wait byte).

3.8.7 *nondelimiter byte. Any byte that is not a delimiter byte.

3.8.8 *delimiter bit. A bit in a byte on the CAMAC serial highway that, when set (bit 7 = 1) identifies the byte as a delimiter byte.

3.8.9 *END byte. A delimiter byte that terminates command messages on the CAMAC serial highway.

3.8.10 *WAIT byte. One of a sequence of delimiter bytes generated between messages on the CAMAC serial highway to permit the insertion and propagation of demand messages.

3.8.11 *ENDSUM byte. A delimiter byte that terminates each reply message or demand message on the CAMAC serial highway. The ENDSUM byte carries the column-parity field of the geometric error detection code.

3.8.12 *SUM byte. A nondelimiter byte in a command message on the CAMAC serial highway, carrying the column-parity field of the geometric error detection code.

3.8.13 *column-parity field. A field in bits 1-6 of the SUM and ENDSUM bytes on the CAMAC serial highway that provides the column-parity component of the geometric error-detection scheme.

3.8.14 *header byte. The first byte of a command message, reply message, or demand message on the CAMAC serial highway, that includes the crate address field.

3.8.15 *SPACE byte. One of a sequence of nondelimiter bytes in a command message on the CAMAC serial highway, space bytes are generated by the serial driver and, in normal operation, are subsequently replaced by a reply message.

4. Index

Term	Term No	Relevant ANSI/IEEE publications[1].		
auxiliary controller bus (ACB)	3.7.34	675*		
auxiliary crate controller	3.4.4	675*	595	
bin, NIM	3.2.6	†*		
bit	3.6.1	583	595	596
		675	683	
bit serial	3.6.5	595*		
bit-serial highway	3.5.9	595		
block transfer	3.7.33	583*	683*	
branch	3.5.11	596*		
branch driver, CAMAC	3.5.13	596*		
bus-line	3.7.5	583*		
busy	3.7.22	583*	595	596
byte	3.6.2	595		
byte frame	3.6.14	595*		
byte serial	3.6.6	595*		
byte-serial highway	3.5.10	595		
CAMAC	3.1.1	583*		

[1] See Section 2 for corresponding documents.
† US Department of Energy Report TID-20893.
*See the specific standard for definition.

Index (*continued*)

DEFINITIONS OF

Term	Term No	Relevant ANSI/IEEE publications[1].		
CAMAC branch driver	3.5.13	596*		
CAMAC branch highway	3.5.5	596*		
CAMAC compatible crate	3.2.3			
CAMAC crate	3.2.2	583*		
CAMAC crate assembly	3.2.4			
CAMAC highway	3.5.3			
CAMAC module	3.3.2	583*		
CAMAC operation	3.7.7	583	595	596
CAMAC parallel highway	3.5.6	596		
CAMAC serial driver	3.5.14	595*		
CAMAC serial highway	3.5.8	595*		
CAMAC system	3.2.5			
clear	3.7.20	583*	595	596
column-parity field	3.8.13	595*		
command	3.7.12	583*	595	596
command accepted	3.7.23	583*	595	596
command message	3.8.1	595*		
command operation	3.7.9	583*	596*	
command-reply transaction	3.8.3	595*		
common control signals	3.7.17	583*		
control station	3.4.1	583*		
crate	3.2.1	583		
crate address	3.7.16	596*	595*	
crate, CAMAC	3.2.2	583*		
crate controller	3.4.1	583*	595	596*
crate controller, auxiliary	3.4.4	675*	595	
crate controller, parallel	3.4.2	583	596	
crate controller, serial	3.4.3	595*		
crate controller, serial Type L2	3.4.7	595*		
crate controller Type A1	3.4.5	596*		
crate controller Type A2	3.4.6	675*		
D-port	3.5.16	595*		
data	3.6.7	583*		
dataway	3.7.1	583*		
dataway operation	3.7.8	583*		
delimiter bit	3.8.8	595*		
delimiter byte	3.8.6	595*		
demand	3.7.25			
demand handling	3.7.26	596		
demand message	3.8.4	595*		
END byte	3.8.9	595*		
ENDSUM byte	3.8.11	595*		
ESONE	3.1.2			
field	3.6.4			
function	3.7.15	583*	595	596
graded-L operation	3.7.11	596*		
graded-L signals (GL)	3.7.31	596*		
handshake	3.6.11	596		

[1] See Section 2 for corresponding documents.
†US Department of Energy Report TID-20893.
*See the specific standard for definition.

CAMAC TERMS   Index (*continued*)

Term	Term No	Relevant ANSI/IEEE publications[1].
header byte	3.8.14	595*
highway	3.5.1	
highway driver	3.5.12	
highway for a CAMAC system	3.5.2	
individual line	3.7.6	583*
inhibit	3.7.19	583* 595 596
initialize	3.7.18	583* 595 596
instrument, NIM	3.3.4	†*
lam grader	3.7.30	596*
look-at-me	3.7.27	583
look-at-me line signal	3.7.28	583* 595
look-at-me request	3.7.29	583
message	3.6.10	595
module	3.3.3	583
module, CAMAC	3.3.2	583*
module, NIM	3.3.4	†* 583
NIM	3.1.3	†*
NIM bin	3.2.6	†*
NIM instrument	3.3.4	†*
NIM module	3.3.4	†* 583
nondelimiter byte	3.8.7	595*
normal station	3.7.3	583*
parallel crate controller	3.4.2	596
parallel highway	3.5.4	596
plug-in unit	3.3.3	583*
port	3.5.15	595* 596*
read	3.6.8	583* 595 596
reply message	3.8.2	595*
response	3.7.24	583* 595 596
serial crate controller	3.4.3	595*
serial crate controller Type L2	3.4.7	595*
serial graded-L field	3.8.5	595*
serial highway	3.5.7	595
SGL encoder	3.7.32	595*
space byte	3.8.15	595*
start bit	3.6.12	595*
station	3.7.2	583*
station number	3.7.13	583* 595 596
stop bit	3.6.13	595*
strobe	3.7.21	583* 595 596
subaddress	3.7.14	583* 595 596
SUM byte	3.8.12	595*
system crate	3.5.18	
U-port	3.5.17	595*
unaddressed operation	3.7.10	583*
WAIT byte	3.8.10	595*
word	3.6.3	
write	3.6.9	583* 595 596

[1] See Section 2 for corresponding documents.
† US Department of Energy Report TID-20893.
* See the specific standard for definition.

ANSI/IEEE
Std 583-1982
(Revision of ANSI/IEEE
Std 583-1975)

An American National Standard

IEEE Standard Modular Instrumentation and Digital Interface System (CAMAC*)

Sponsor

Instruments and Detectors Committee of the IEEE Nuclear and Plasma Sciences Society

Approved September 17, 1981

IEEE Standards Board

Approved December 15, 1981

American National Standards Institute

*Computer Automated Measurement and Control

© Copyright 1982 by

The Institute of Electrical and Electronics Engineers, Inc
345 E. 47th Street, New York, NY 10017

No part of this publication may be reproduced in any form, in an electronic retrieval system or otherwise, without the prior written permission of the publisher.

IEEE Standards documents are developed within the Technical Committees of the IEEE Societies and the Standards Coordinating Committees of the IEEE Standards Board. Members of the committees serve voluntarily and without compensation. They are not necessarily members of the Institute. The standards developed within IEEE represent a consensus of the broad expertise on the subject within the Institute as well as those activities outside of IEEE which have expressed an interest in participating in the development of the standard.

Use of an IEEE Standard is wholly voluntary. The existence of an IEEE Standard does not imply that there are no other ways to produce, test, measure, purchase, market, or provide other goods and services related to the scope of the IEEE Standard. Furthermore, the viewpoint expressed at the time a standard is approved and issued is subject to change brought about through developments in the state of the art and comments received from users of the standard. Every IEEE Standard is subjected to review at least once every five years for revision or reaffirmation. When a document is more than five years old, and has not been reaffirmed, it is reasonable to conclude that its contents, although still of some value, do not wholly reflect the present state of the art. Users are cautioned to check to determine that they have the latest edition of any IEEE Standard.

Comments for revision of IEEE Standards are welcome from any interested party, regardless of membership affiliation with IEEE. Suggestions for changes in documents should be in the form of a proposed change of text, together with appropriate supporting comments.

Interpretations: Occasionally questions may arise regarding the meaning of portions of standards as they relate to specific applications. When the need for interpretations is brought to the attention of IEEE, the Institute will initiate action to prepare appropriate responses. Since IEEE Standards represent a consensus of all concerned interests, it is important to ensure that any interpretation has also received the concurrence of a balance of interests. For this reason IEEE and the members of its technical committees are not able to provide an instant response to interpretation requests except in those cases where the matter has previously received formal consideration.

Comments on standards and requests for interpretations should be addressed to:

> Secretary, IEEE Standards Board
> 345 East 47th Street
> New York, NY 10017
> USA

Foreword

[This Forward is not a part of ANSI/IEEE Std 583-1982, IEEE Standard Modular Instrumentation and Digital Interface System (CAMAC).]

The interface system on which this standard is based was developed by the ESONE Committee of European Laboratories** with the collaboration of the NIM Committee of the US Department of Energy (DOE).* This standard is based on ERDA† Reports TID-25875, July 1972 (corresponding to ESONE Report EUR 4100e) and TID-25877. The mandatory features and dimensions in this document are identical (see note) to those in the reports mentioned above and in Publications 482 and 516 of the International Electrotechnical Commission (IEC).

This standard was reviewed and balloted by the Nuclear Instruments and Detectors Committee of the IEEE Nuclear and Plasma Sciences Society. Because of the broad applicability of this instrumentation and interface system, coordination was established with numerous IEEE Societies, Groups, and Committees, including the Communications Society, Control Systems Society, Industry Applications Society, Industrial Electronics and Control Instrumentation Group, Instrumentation and Measurement Group, and the Power Generation and Nuclear Power Engineering Committees of the Power Engineering Society. This review and coordination resulted in valuable suggestions that have been incorporated into the standard.

The revision of this standard was in conjunction with the 1981 review (1982 issue) of the entire family of IEEE CAMAC standards undertaken to incorporate existing addenda and corrections into the standards.

At the time of approval of this standard, the membership of the Nuclear Instruments and Detectors Committee of the IEEE Nuclear and Plasma Sciences Society was as follows:

D. C. Cook, *Chairman* **Louis Costrell,** *Secretary*

J. A. Coleman	T. R. Kohler	P. L. Phelps
J. F. Detko	H. W. Kraner	J. H. Trainor
F. S. Goulding	W. W. Managan	F. J. Walter
F. A. Kirsten	G. L. Miller	S. Wagner
	D. E. Persyk	

At the time it approved this standard, the American National Standards Committee on Nuclear Instruments, N42, had the following representatives:

Louis Costrell, *Chairman* **D. C. Cook,** *Secretary*

Organization Represented	*Name of Representative*
American Conference of Governmental Industrial Hygienists	Jesse Lieberman
American Nuclear Society	Frank W. Manning
Health Physics Society	J. B. Horner Kuper
	J. M. Selby *(Alt)*
Institute of Electrical and Electronics Engineers	Louis Costrell
	Julian Forster *(Alt)*
	David C. Cook *(Alt)*
	A. J. Spurgin *(Alt)*
Instrument Society of America	M. T. Slind
	J. E. Kaveckis *(Alt)*
Lawrence Berkeley National Laboratory	Lee J. Wagner
Oak Ridge National Laboratory	Frank W. Manning
	D. J. Knowles *(Alt)*
US Department of the Army, Materiel Command	Abraham E. Cohen
US Department of Commerce, National Bureau of Standards	Louis Costrell
US Department of Energy	Gerald Goldstein
US Federal Emergency Management Agency	Carl R. Siebentritt, Jr
US Nuclear Regulatory Commission	Robert E. Alexander
US Naval Research Laboratory	David C. Cook
Members-at-Large	J. G. Bellian
	O. W. Bilharz, Jr
	John M. Gallagher, Jr
	Voss A. Moore
	R. F. Shea
	E. J. Vallario

†Department of Energy Research and Development Administration (now part of US Department of Energy).

*NIM Committee[1]
L. Costrell, Chairman

C. Akerlof	N. W. Hill	L. B. Mortara
E. J. Barsotti	D. Horelick	V. C. Negro
B. Bertolucci	M. E. Johnson	L. Paffrath
J. A. Biggerstaff	C. Kerns	D. G. Perry
A. E. Brenner	F. A. Kirsten	I. Pizer
R. M. Brown	P. F. Kunz	E. Platner
E. Davey	R. S. Larsen	S. Rankowitz
W. K. Dawson	A. E. Larsh, Jr	S. J. Rudnick
S. R. Deiss	R. A. LaSalle	G. K. Schulze
S. Dhawan	N. Latner	W. P. Sims
R. Downing	F. R. Lenkszus	D. E. Stilwell
T. F. Droege	R. Leong	J. H. Trainor
C. D. Ethridge	C. Logg	K. J. Turner
C. E. L. Gingell	S. C. Loken	H. Verweij
A. Gjovig	D. R. Machen	B. F. Wadsworth
B. Gobbi	D. A. Mack	L. J. Wagner
D. B. Gustavson	J. L. McAlpine	H. V. Walz
D. R. Heywood		D. H. White

NIM Executive Committee for CAMAC
L. Costrell, Chairman

E. J. Barsotti	F. A. Kirsten
J. A. Biggerstaff	R. S. Larsen
S. Dhawan	D. A. Mack

J. H. Trainor

NIM Dataway Working Group
F. A. Kirsten, Chairman

E. J. Barsotti	C. Kerns
J. A. Biggerstaff	P. F. Kunz
L. Costrell	R. S. Larsen
S. Dhawan	D. R. Machen
A. Gjovig	L. Paffrath
D. R. Heywood	S. Rankowitz
D. Horelick	S. J. Rudnick

NIM Software Working Group
D. B. Gustavson, Jr, Chairman
W. K. Dawson, Secretary

A. E. Brenner	P. A. LaSalle
R. M. Brown	F. B. Lenkszus
S. R. Deiss	C. A. Logg
S. Dhawan	J. McAlpine
C. D. Ethridge	L. B. Mortara
M. E. Johnson	D. G. Perry

NIM Mechanical and Power Supplies Working Group
L. J. Wagner, Chairman

- L. Costrell
- C. Kerns
- S. J. Rudnick
- W. P. Sims

NIM Analog Signals Working Group
D. I. Porat, Chairman

- L. Costrell
- C. E. L. Gingell
- N. W. Hill
- S. Rankowitz

NIM Serial System Subgroup
D. R. Machen, Chairman

- E. J. Barsotti
- D. Horelick
- F. A. Kirsten
- R. G. Martin
- L. Paffrath
- S. J. Rudnick

NIM Block Transfer Subgroup
E. J. Barsotti, Chairman

- W. K. Dawson
- F. A. Kirsten
- R. A. LaSalle
- F. R. Lenkszus
- R. G. Martin
- R. F. Thomas, Jr

NIM Multiple Controllers Subgroup
P. F. Kunz, Chairman

E. J. Barsotti	D. R. Machen
F. A. Kirsten	R. G. Martin

---

[1] National Instrumentation Methods Committee of the US Department of Energy.

**ESONE Committee[2]

W. Attwenger, Austria, *Chairman* 1980–81; H. Meyer, Belgium, *Secretary*

Representatives of ESONE Member Laboratories

W. Attwenger, Austria	E. Kwakkel, Netherlands	A. C. Peatfield, England
R. Biancastelli, Italy	J. L. Lecomte, France	I. C. Pyle, England
L. Binard, Belgium	J. Lingertat, D. R. Germany	B. Rispoli, Italy
J. Biri, Hungary	M. Lombardi, Italy	M. Sarquiz, France
B. Bjarland, Finland	M. Maccioni, Italy	W. Schoeps, Switzerland
D. A. Boyce, England	P. Maranesi, Italy	R. Schule, F. R. Germany
B. A. Brandt, F. R. Germany	C. H. Mantakas, Greece	P. G. Sjolin, Sweden
F. Cesaroni, Italy	D. Marino, Italy	L. Stanchi, Italy
P. Christensen, Denmark	H. Meyer, Belgium	R. Trechcinski, Poland
W. K. Dawson, Canada	K. D. Muller, F. R. Germany	M. Truong, France
M. DeMarsico, Italy	J. G. Ottes, F. R. Germany	P. Uuspaa, Finland
C. A. DeVries, Netherlands	A. D. Overtoom, Netherlands	H. Verweij, Switzerland
H. Dilcher, F. R. Germany	E. C. G. Owen, England	A. J. Vickers, England
B. V. Fefilov, USSR	L. Panaccione, Italy	S. Vitale, Italy
R. A. Hunt, England	M. Patrutescu, Roumania	M. Vojinovic, Yugoslavia
W. Kessel, F. R. Germany	R. Patzelt, Austria	K. Zander, F. R. Germany
R. Klesse, France		D. Zimmermann, F. R. Germany

ESONE Executive Group (XG)

W. Attwenger, Austria, XG, *Chairman* 1980–81
H. Meyer, Belgium, *Secretary*

P. Christensen, Denmark	M. Sarquiz, France
M. Dilcher, F. R. Germany	R. Trechcinski, Poland
B. Rispoli, Italy	A. Vickers, England
H. Verweij, Switzerland	

ESONE Technical Coordination Committee (TCC)

A. C. Peatfield, England, *TCC Chairman*
P. Christensen, Denmark, *TCC Secretary*

W. Attwenger, Austria	W. Kessel, F. R. Germany
R. Biancastelli, Italy	J. Lukacs, Hungary
G. Bianchi, France	R. Patzelt, Austria
H. Dilcher, F. R. Germany	P. J. Ponting, Switzerland
P. Gallice, France	W. Schoeps, Switzerland
S. Vitale, Italy	

ECA/ESONE CAMAC Document Maintenance Study Group (DMSG)

P. Gallice, France, *Chairman*

R. C. M. Barnes, England	F. Iselin, Switzerland
L. Besse, Switzerland	H. Meyer, Belgium
J. Davis, England	H. J. Trebst, F. R. Germany

When this standard was approved on September 17, 1981, by the IEEE Standards Board it had the following membership:

I. N. Howell, Jr, *Chairman*   **Irving Kolodny,** *Vice Chairman*

Sava I. Sherr, *Secretary*

G. Y. R. Allen	Jay Forster	F. Rosa
J. J. Archambault	Kurt Greene	R. W. Seelbach
J. H. Beall	Loering M. Johnson	J. S. Stewart
J. T. Boettger	Joseph L. Koepfinger	W. E. Vannah
Edward Chelotti	J. E. May	Virginius N. Vaughan, Jr
Edward J. Cohen	Donald T. Michael*	Art Wall
Len S. Corey	J. P. Riganati	Robert E. Weiler

*Member emeritus

Note: Figures prefixed by the letter K are supplementary illustrations.

---

[2]European Standards on Nuclear Electronics Committee.

Contents

SECTION				PAGE
	CAMAC and NIM Standards and Reports			9
1.	Introduction			11
2.	Interpretation			11
3.	Basic Features			11
4.	Mechanical Characteristics			13
	4.1	The Crate		13
		4.1.1	Dimensions	13
		4.1.2	Dataway Connector Sockets	14
		4.1.3	Optional Features of the Crate	14
	4.2	Plug-in Units		15
		4.2.1	Dimensions	15
		4.2.2	Dataway Connector Plug	15
		4.2.3	Insertion of the Plug-in Unit into the Crate	16
		4.2.4	Printed-Wiring Card	17
		4.2.5	Other Connectors	17
	4.3	Adaptor for NIM Units		17
	4.4	The Dataway		17
5.	Use of the Dataway Lines			22
	5.1	Commands		23
		5.1.1	Station Number N	23
		5.1.2	Subaddress $A8, A4, A2, A1$	23
		5.1.3	Function $F16, F8, F4, F2, F1$	23
	5.2	Strobe Signals $S1$ and $S2$		24
	5.3	Data		24
		5.3.1	Write Lines $W1-W24$	24
		5.3.2	Read Lines $R1-R24$	24
	5.4	Status Information		26
		5.4.1	Look-at-Me L	26
			5.4.1.1 Look-at-Me: Clear, Disable, and Test	26
			5.4.1.2 Look-at-Me: Commands for Access	29
			5.4.1.3 Look-at-Me: Gating	29
			5.4.1.4 Look-at-Me: Use for Operations Synchronization in Block Transfers, Supplementary Information.	29
		5.4.2	Busy B	29
		5.4.3	Response Q	30
			5.4.3.1 Use of Q for Block Transfers: Address Scan Mode	30
			5.4.3.2 Use of Q for Block Transfers: Repeat Mode	30
			5.4.3.3 Use of Q for Block Transfers: Stop Mode	32
			5.4.3.4 Use of Q for Block Transfers: Stop-on-Word mode, Supplementary Information	32
			5.4.3.5 Use of Q for Block Transfers, Supplementary Information	32
		5.4.4	Command Accepted X	32
	5.5	Common Controls Z, C, I		33
		5.5.1	Initialize Z	33
		5.5.2	Inhibit I	33
		5.5.3	Clear C	33
	5.6	Nonstandard Connections $P1-P7$		33
		5.6.1	Free Bus-lines $P1-P2$	33
		5.6.2	Patch Contacts $P3-P7$	34
	5.7	Power Lines		34

SECTION				PAGE
6.	Dataway Commands			34
	6.1	Read Commands: Function Codes $F(0)$ to $F(7)$		35
		6.1.1	Read Group 1 Register, Code $F(0)$	35
		6.1.2	Read Group 2 Register, Code $F(1)$	35
		6.1.3	Read and Clear Group 1 Register, Code $F(2)$	35
		6.1.4	Read Complement of Group 1 Register, Code $F(3)$	35
		6.1.5	Other Read Commands, Codes $F(4)-F(7)$	35
	6.2	Control Commands: Function Codes $F(8)-F(15)$		35
		6.2.1	Test Look-at-Me, Code $F(8)$	35
		6.2.2	Clear Group 1 Register, Code $F(9)$	35
		6.2.3	Clear Look-at-Me, Code $F(10)$	35
		6.2.4	Clear Group 2 Register, Code $F(11)$	36
		6.2.5	Other Control Commands, Codes $F(12)-F(15)$	36
	6.3	Write Commands: Function Codes $F(16)-F(23)$		36
		6.3.1	Overwrite Group 1 Register, Code $F(16)$	36
		6.3.2	Overwrite Group 2 Register, Code $F(17)$	36
		6.3.3	Selective Set Group 1 Register, Code $F(18)$	36
		6.3.4	Selective Set Group 2 Register, Code $F(19)$	36
		6.3.5	Selective Clear Group 1 Register, Code $F(21)$	36
		6.3.6	Selective Clear Group 2 Register, Code $F(23)$	37
		6.3.7	Other Write Commands, Codes $F(20)$ and $F(22)$	37
	6.4	Control Commands: Function Codes $F(24)-F(31)$		37
		6.4.1	Disable, Code $F(24)$	37
		6.4.2	Execute, Code $F(25)$	37
		6.4.3	Enable, Code $F(26)$	37
		6.4.4	Test Status, Code $F(27)$	37
		6.4.5	Other Control Commands, Codes $F(28)-F(31)$	38
	6.5	External Representation of the Command		38
7.	Signal Standards			38
	7.1	Digital Signals on the Dataway		38
		7.1.1	Voltage Standards for Dataway Signals	38
		7.1.2	Current Standards for Dataway Signals	38
		7.1.3	Timing of Dataway Signals	41
			7.1.3.1 Timing of Dataway Command Operations	41
			7.1.3.2 Timing of Unaddressed Operations	42
		7.1.4	Digital Signals on Non Standard Connections	43
	7.2	Other Digital Signals		43
		7.2.1	Unterminated Signals	43
		7.2.2	Terminated Signals	43
	7.3	Analog Signals		43
8.	Power Line Standards			44
9.	Forced-Air Ventilation (Supplementary Information)			45
10.	Use of Supplementary 6 V Power (Bus Lines Y1, Y2)			45

TABLES

Table 1	Standard Dataway Usage	18
Table 2	Contact Allocation at a Normal Station	19
Table 3	Contact Allocation at the Control Station	20
Table 4	The Function Codes	25
Table 5	Voltage Levels of Dataway Signals	38
Table 6	Standards for Signal Currents Through Dataway Connectors and for Pull-Up Current Sources	39

TABLES		PAGE
Table 7	Current Standard for Patch Contacts	42
Table 8	Unterminated Signals	43
Table 9	Terminated Signals	43
Table 10	Power Line Standards	44
Table K5.4.3.5	Single Module, Single Address Block Transfers	31

FIGURES

Fig 1	Unventilated Crate: Front View	46
Fig 2	Plan View of Lower Guides in Crate	47
Fig 3	Crate Side View: Section d-d (Fig 1)	48
Fig 4	Plug-In Unit: Side and Rear Views	49
Fig 5	Dataway Connector: (5.1)-(5.3) Plug: (5.4)-(5.8) Socket	50
Fig 6	Ventilated Crate: Front View	51
Fig 7	Adaptor for NIM Units	52
Fig 8	Typical Printed Wiring Card	53
Fig 9	Timing of a Dataway Command Operation	54
Fig 10	Timing of a Dataway Unaddressed Operation	54
Fig 11	Some LAM Structure Options	55
Fig 12	Typical Current Sharing with Supplementary 6 V Power	55
Fig 13	Coaxial Connector, Type 50CM	56
Fig K1A[3]	Standard Arrangement.	12
Fig K1B	Multiplicity of Interfaces Required by Nonstandard Arrangements.	12
Fig K1C	Typical System With Standard Data Bus (Dataway) and Dedicated Interface	12
Fig K1D	Parallel Highway System	12
Fig K1E	Serial Highway System	12
Fig K4.2.2	Typical Dataway Connector Plug	16
Fig K4.4	Dataway Wiring, Front View of Twenty-Five Station Crate	21
Fig K5.4.1A	Example of Single-Source Look-at-Me Logic	27
Fig K5.4.1B	Example of Multiple-Source Look-at-Me Logic Showing Operations on LAMs via Subaddresses	27
Fig K5.4.1C	Example of Multiple-Source Look-at-Me Logic Showing Operations on LAMs via Databits	28
Fig K7.1.2	Current Distribution, CAMAC Dataway	40

APPENDIXES

Appendix A	Definitions of Module and Controller	58
Appendix B	Digital Signal Classes and Standard Markings for Coaxial Connectors (Supplementary Information)	59
Appendix C	Preferred Connectors and Contact Assignments	60
Appendix D	Typical Crate Mounted Power Supply and Ventilation Unit With Crate/Power Supply Interface Housing	66
Appendix E	CAMAC Categories	75
Appendix F	Bibliography	76
INDEX		78

---

[3] Note: Figures prefixed by the letter K are supplementary illustrations.

CAMAC and NIM Standards and Reports

Title	IEEE, ANSI Std No	IEC No	DOE No	EURATOM (EUR) or ESONE No
CAMAC Instrumentation and Interface Standards*	SH08482* (Library of Congress No 8185060)	—	—	—
Modular Instrumentation and Digital Interface System (CAMAC)	ANSI/IEEE Std 583-1982	516	TID-25875† and TID-25877†	EUR 4100e
Serial Highway Interface System (CAMAC)	ANSI/IEEE Std 595-1982	640	TID-26488†	EUR 6100e
Parallel Highway Interface System (CAMAC)	ANSI/IEEE Std 596-1982	552	TID-25876† and TID-25877†	EUR 4600e
Multiple Controllers in a CAMAC Crate	ANSI/IEEE Std 675-1982	729	DOE/EV-0007	EUR 6500e
Block Transfers in CAMAC Systems	ANSI/IEEE Std 683-1976 (Reaff 1981)	677	TID-26616†	EUR 4100 suppl
Amplitude Analog Signals within a 50 Ω System	—	—	TID-26614	EUR 5100e
The Definition of IML A Language for Use in CAMAC Systems	—	—	TID-26615	ESONE/IML/01
CAMAC Tutorial Articles	—	—	TID-26618	—
Real-Time BASIC for CAMAC	ANSI/IEEE Std 726-1982	§	TID-26619†	ESONE/RTB/03
Subroutines for CAMAC	ANSI/IEEE Std 758-1979 (Reaff 1981)	713	DOE/EV-0016†	ESONE/SR/01
Recommendations for CAMAC Serial Highway Drivers and LAM Graders for the SCC-L2	—	—	DOE/EV-0006	ESONE/SD/02
Definitions of CAMAC Terms	Included in SH08482	678	DOE/ER-0104	ESONE/GEN/01
Standard Nuclear Instrument Modules NIM	—	547**	TID-20893 (Rev 4)	—

†Superseded by corresponding IEEE Standard listed.

*This is a hard cover book that contains ANSI/IEEE Std 583-1982, ANSI/IEEE Std 595-1982, ANSI/IEEE Std 596-1982, ANSI/IEEE Std 675-1982, ANSI/IEEE Std 683-1976 (Reaff 1981), ANSI/IEEE Std 726-1982 and IEEE Std 758-1979 (Reaff 1981), plus introductory material and a glossary of CAMAC terms.

**Covers only mechanical features and connector pin assignments.

§ In preparation.

NOTE: *Availability of Documents*
ANSI   Sales Department, American National Standards Institute, 1430 Broadway, New York, NY 10018.
IEEE   IEEE Service Center, 445 Hoes Lane, Piscataway, New Jersey 08854, USA.
IEC   International Electrotechnical Commission, 1, rue de Varembé, CH-1211 Geneva 20, Switzerland.
DOE and TID Reports   National Bureau of Standards, Washington, D.C. 20234, USA, Attn: L. Costrell.
EURATOM   Office of Official Publications of the European Communities, P.O. Box 1003, Luxembourg.
ESONE   Commission of the European Communities, CGR-BCMN, B-2440 GEEL, Belgium, Attn: ESONE Secretariat, H. Meyer.

An American National Standard

IEEE Standard Modular Instrumentation and Digital Interface System (CAMAC)

1. Introduction

This standard is intended to serve as a basis for a range of modular instrumentation capable of interfacing transducers and other devices to digital controllers for data and control. It consists of mechanical standards and signal standards that are sufficient to ensure physical and operational compatibility between units regardless of source.

The standard fully specifies a data bus (Dataway) by means of which instruments and other functional modules can communicate with each other, with peripherals, with computers, and with other external controllers as shown in Fig K1A. This serves to drastically reduce both the variety and quantity of interfacing required in a single installation and provides a considerable degree of computer independence. The multiplicity of computer dependent and instrument dependent interfaces shown in Fig K1B is typically replaced by a single computer dependent interface as in Fig K1C with one side of the interface speaking the computer language and the other side the standardized Dataway language. It will be noted that the Dataway is strictly a digital interface.

A single crate as shown in Fig K1A can typically accommodate up to twenty-three separate modules plus controller. Both single-crate systems and multiple-crate systems can be assembled. Fig K1D shows a multiple-crate system in which data is transferred in parallel, utilizing a parallel highway and a parallel highway driver. Fig K1E shows a serial system in which data is transferred bit or byte serial. The parallel system is especially useful where very high data rates are encountered, whereas the serial system is advantageous for industrial control and other applications where long distances are involved such that wiring cost is an important consideration. Standard parallel and serial highway configurations based on Figs K1D and K1E are being processed.

Maximum benefit is derived from using this standard as a whole. Selected portions may find application in additional areas. Other devices and buses, such as that of IEEE 488, can be readily incorporated into the system through an interfacing module.

2. Interpretation

> Statements that specify mandatory aspects of the system are enclosed in blocks such as this. They are usually accompanied by the word *must*.

The word *should* indicates a recommended or preferred practice which is to be followed unless there are sound reasons to the contrary.

The word *may* indicates a permitted practice, leaving freedom of choice to the designer.

Appendix E defines the various CAMAC categories.

3. Basic Features

This specification is intended to serve as a basis for a range of modular instrumentation capable of linking transducers and other devices with digital controllers or computers. It consists of mechanical standards and signal standards that are sufficient to ensure compatibility between units from different sources of design and production.

The basic features of the system are as follows:

(1) It is a modular system, with functional units which can be combined to form equipment assemblies.

(2) The functional units are constructed as plug-in units and are mounted in a standard crate.

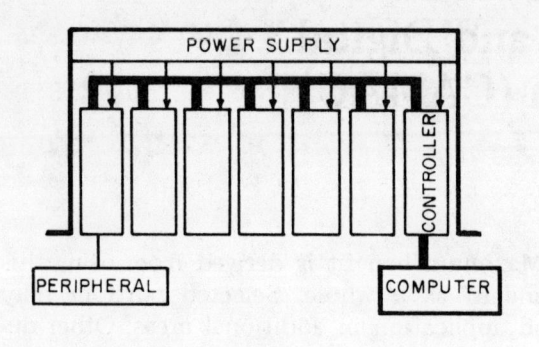

Fig K1A
Standard Arrangement. Data bus (Dataway)
is heavy black line going to all modules

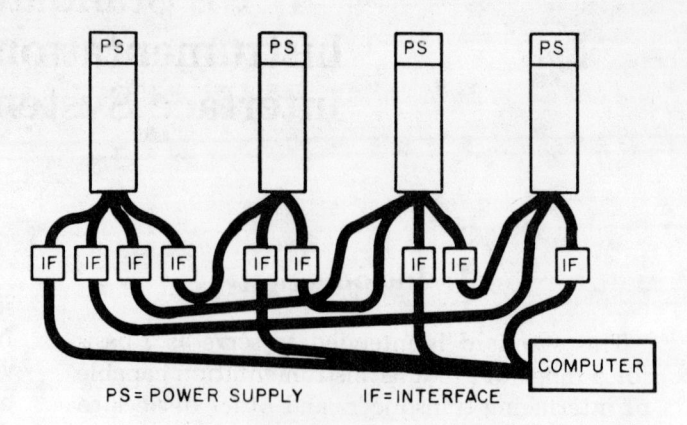

Fig K1B
Multiplicity of Interfaces Required by
Nonstandard Arrangements

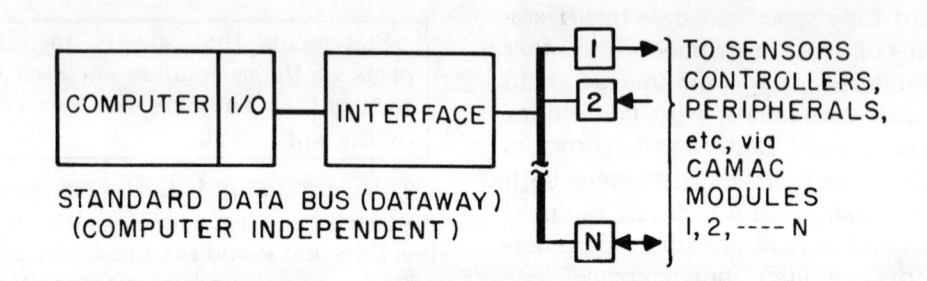

Fig K1C
Typical System With Standard Data Bus
(Dataway) and Dedicated Interface

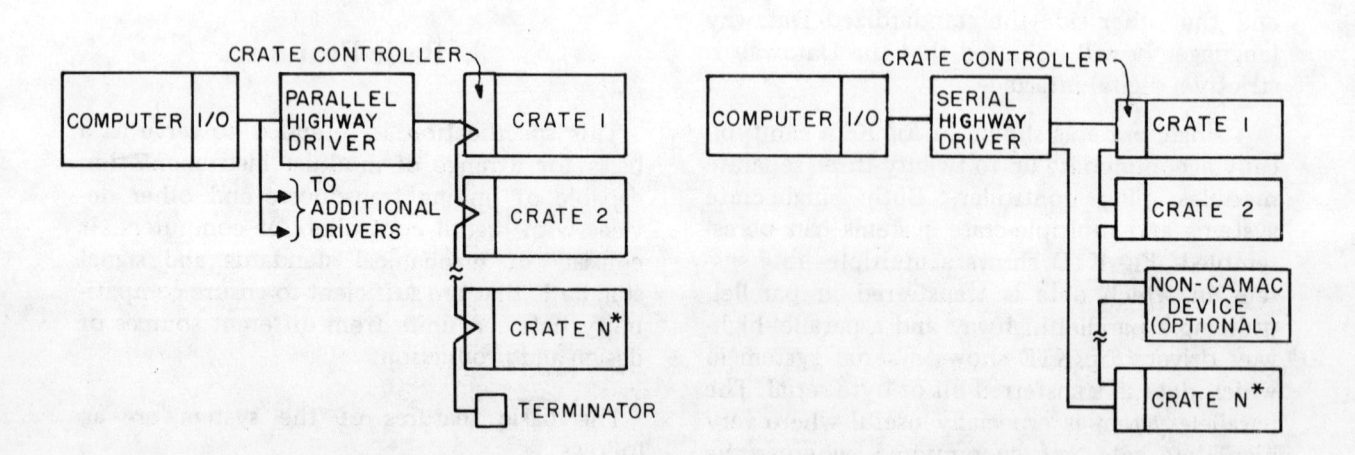

Fig K1D
Parallel Highway System
*Typically N = 7 max

Fig K1E
Serial Highway System
*Typically N = 62 max

(3) The mechanical structure is designed to exploit the high-component packing density possible with integrated circuit packages and similar devices.

(4) Each plug-in unit makes direct connection to a standard Dataway. This highway forms part of the crate and conveys digital data, control signals, and power. The standards of the Dataway are independent of the type of plug-in unit or computer used.

(5) The system has been designed so that an assembly consisting of a crate and plug-in units can be connected to an on-line digital computer. However, the use of a computer is entirely optional and no part of this specification depends upon its presence in the system.

(6) External connections to plug-in units may conform to the digital or analog signal standards of associated transducers, computers, etc, or to the recommended standards given in this standard.

(7) Assemblies of crates may be interconnected by means of parallel or serial highways.

(8) No license or other permission is needed in order to use this standard.

4. Mechanical Characteristics

CAMAC is a modular system. Equipment assemblies are formed by mounting appropriate plug-in units in a standard chassis or crate. Each plug-in unit occupies one or more mounting stations in the crate. At each station there is an eighty-six contact connector socket giving access to the CAMAC Dataway, a data highway which forms part of the crate. The Dataway consists mainly of bus lines for data, control, and power.

Drawings for the manufacture of CAMAC compatible crates and plug-in units can be derived from the definitive dimensions given in Figs 1-3 for crates, Fig 4 for plug-in units, and Fig 5 for Dataway connector plugs and sockets.

Recommended dimensions for ventilated crates, NIM adaptors, and printed wiring cards for plug-in units are given in the nonmandatory Figs 6-8, respectively.

All dimensions in these figures are in millimeters unless indicated otherwise.

4.1 The Crate. The crate mounts in a 19 inch rack and has up to twenty-five stations for plug-in units on a pitch of 17.2 mm. Each station has upper and lower guides for the runners of a plug-in unit, an eighty-six contact Dataway connector socket, and a tapped hole for the fixing screw of a plug-in unit. Modules conforming to the NIM specification (Type N Module of IEC Publ 482), also, ERDA Report TID - 20893 (see Bibliography) can be mounted in the crate on their basic pitch of 34.4 mm (see Section 4.3).

> Unless indicated otherwise, all crates must conform to Fig 1-3 and those parts of Fig 5 defining the connector socket.

Sections 4.1.1 and 4.1.2 are comment on these figures.

4.1.1 Dimensions. Fig 1 shows the front view of a basic twenty-five station crate which occupies the minimum height of $5U$ ($U = 44.45$ mm). Crates may have less than twenty-five stations, which, as indicated by Note 3 on Fig 1, need not be positioned symmetrically.

The lower cross-member has holes tapped ISO·M4 pitch 0.7 for the fixing screws of CAMAC plug-in units, and intermediate holes tapped UNC 6-32 for the lower fixing screws of NIM units. The upper cross-member may also have holes for the fixing screws of NIM units. The positions of these holes for CAMAC and NIM units, relative to the left-hand edge of the front aperture, are given in Fig 1 by the formulas for dimensions z and w, respectively.

The positions of the centers of the guides, also relative to the left-hand edge of the aperture, are given by the formula for dimension x in Fig 1. Detail A shows the entry into a guide. The dimensions of the lead-in are not specified.

Detail B gives dimensions specified for 19 inch rack-mounting equipment by the International Electrotechnical Commission in IEC Publ 297 and also specified in Electronic Industries Association Std RS - 310-B (ANSI C83.9-1972).

Fig 2 is a plan view of the lower guides in the crate. In order to remove any heat generated in the plug-in units, it is necessary to provide adequate ventilation through the bottom and top of the crate. The unobstructed area between adjacent guides, both at the top and bottom of the crate, is not permitted to be less than 15 cm^2 and should preferably be distributed over the full depth of the crate from the front cross-members to the Dataway assembly. If crates such as that shown in Fig 1 with height $5U$ ($U = 44.45$mm per IEC Publ 297), are mounted above or below other equipment, (including other similar crates), it may be necessary to use intermediate deflectors, etc, to ensure adequate ventilation. Alternatively, the crate may be extended to include additional ventilation features, as described in Section 4.1.3.

Fig 3 is a sectioned side view on the offset line d-d in Fig 1, passing through the center of an upper guide and a ventilating space between lower guides. The front faces of the upper and lower cross-members constitute the vertical datum of the crate. This datum is set back from the front face of the crate by a distance e, typically between 3 and 4 mm, so that the front panels of plug-in units do not project beyond the front of the crate. The backs of the crate-mounting flanges are typically, but not necessarily, aligned with the datum.

The front ends of the upper and lower guides may be set back from the vertical datum. The guides extend sufficiently far toward the rear of the crate to ensure that the connector plug of a plug-in unit is guided into the entry of the connector socket.

The minimum overall depth of the crate provides mechanical protection for the Dataway assembly. The side panels are shorter than the frontal height of the crate (see dimensions a in Figs 1, 3, and 6,) to permit the use of typical runners for supporting the crate in the rack. This reduction in height extends at least to within 25 mm of the rear face of the rack-mounting flanges of the crate.

The running surface of the lower guide constitutes the crate horizontal datum. The Dataway assembly is not permitted to extend upward more than 135 mm from this horizontal datum, so that there is unrestricted access to the upper part of the rear of plug-in units.

The positions of the connector sockets are defined with respect to the three datum lines of the crate. The center lines of the sockets are defined with respect to the left-hand edge of the front aperture by dimension y in Fig 1. The vertical datum of the sockets is shown relative to the vertical datum of the crate in Figs 2 and 3, and the horizontal datum of the sockets relative to the horizontal datum of the crate in Fig 3.

SUPPLEMENTARY INFORMATION: The design of the rear of the crate should be such as to protect the Dataway from damage. At the same time, in the space above the Dataway, access must be provided to the rear of the plug-in units. Fig D5 of Appendix D shows a typical interface housing unit to provide this protection and to accommodate a typical power supply. (See also Appendix D.)

4.1.2 *Dataway Connector Sockets.* The Dataway connector sockets have two rows of forty-three contacts on a pitch of 0.1 in (2.54 mm). Mandatory and recommended dimensions of the sockets are given in Fig 5, together with additional commonly used dimensions upon which the designs of many existing crates and Dataway assemblies have been based.

The vertical datum of the connector sockets is the nominal position of the leading edge of the connector plug of a plug-in unit fully inserted into the crate. The position of the vertical datum is defined in (5.5) of Fig 5 with respect to other functional features of the socket. In some commonly used sockets the plane of the mounting face coincides with the vertical datum of the connector socket, but this is not necessarily so.

The maximum forward projection of the connector socket in front of the vertical datum is shown in (5.5) of Fig 5. The shapes of the straight or curved chamfers that guide the connector plug into the socket are shown in (5.6), (5.7), (5.8) of Fig 5. Within the minimum width shown for each chamfer the angle between any tangent to the chamfer and the line of entry of the connector plug does not exceed 60°.

If the front aperture of the crate extends to the inner surface of the right-hand side panel (as in Figs 1 and 2), the adjacent connector socket cannot exceed the recommended width of 12 mm. Elsewhere, sockets up to the maximum width of 17.2 mm can be used.

The dimensions of the contacts of the connector socket are shown in (5.4) of Fig 5. The position of each edge is defined by a dimension d, D relative to the horizontal datum of the socket, and is completely independent of the positions of all other edges on both rows of contacts.

Alternatively, a connector socket with point contacts may be used, in which case the distance between each point contact and the horizontal datum of the connector socket is $(2.56 + 2.54k) \pm 0.13$.

4.1.3 *Optional Features of the Crate.* The height of the crate may be extended by an integral number of U units (U = 44.45 mm), in Fig

6, in order to provide an entry for cool air, which then flows up between the guides, and an exit for any warm air that may be rising from equipment below.

A crate may have fewer than twenty-five stations. The width of the front aperture is $17.2s^{+0.3}_{-0.0}$ mm for s stations, and formulas given in Fig 1 are used for locating the guides, connector socket, etc, at each station.

Power supply units may be mounted at the rear of a CAMAC crate. The overall depth of a crate with rear-mounted power supplies may be limited by the depth of the rack. A recommended maximum depth of 525 mm is shown in Fig 3. A power supply unit is not allowed to extend upward above the maximum height of the Dataway assembly. It should not obstruct the entry or exit of the ventilating air flows in a crate such as that shown in Fig 6. The width of a rear-mounted power supply is limited to 447 mm.

4.2 Plug-In Units. Basically a plug-in unit consists of a front panel with fixing screw, top and bottom runners that slide in the guides of the crate, and an eighty-six contact Dataway connector plug. The connector plug is typically an integral part of a printed-wiring card but may be a separate male connector mounted at the rear of the plug-in unit. A plug-in unit may occupy more than one station and, if so, may have more than one set of runners and more than one connector plug.

> Unless indicated otherwise, all plug-in units must conform to Fig 4 and those parts of Fig 5 defining the connector plug.

The following sections are comments on these figures.

4.2.1 *Dimensions*. The horizontal datum of a plug-in unit is the edge of the lower runner. The vertical datum is the rear face of the front panel. The upper and lower parts of the rear face should be in contact with the crossmembers of the crate when the plug-in unit is fully inserted. Note 4 of Fig 4 therefore requires that the upper and lower 11 mm of the rear face of the front panel be free of projections, other than the fixing screws.

Fig 4 shows the dimensions of single-width and double-width plug-in units and gives general formulas for the front-panel widths of units.

It is recommended that the fixing screw should also provide a jacking action to assist in overcoming the withdrawal forces of the connector socket. The fixing screw of a single-width plug-in unit is located on the center line of the front panel. If a multiple-width unit has only one fixing screw, and this has a jacking action, the screw should be positioned to give the most effective pull for withdrawal of the plug-in unit (hence it should be at the same station as a single connector or approximately symmetrical with respect to two or more connectors).

Above the maximum height of the Dataway assembly there can be projections at the rear of the plug-in unit, extending more than 290 mm from the vertical datum. Below this height, in order to provide clearance for the connector socket, only the connector plug is allowed to extend beyond 290 mm.

There should be adequate ventilation through the bottom and top of each plug-in unit to remove any heat generated within the unit.

4.2.2 *Dataway Connector Plug*. The dimensions of the connector plug are shown in ⑤.₁, ⑤.₂, ⑤.₃ of Fig 5.

The full eighty-six contacts are always present and extend to the extreme edge of the plug. In order to avoid the risk of damage to the contact plating of connector sockets by exposed abrasives in the substrate of the connector plug contacts must not expose substrate.

Chamfers are provided at the top and bottom of the connector socket and are therefore not needed at the top and bottom corners of the connector plug where the maximum permitted chamfer is 1×1 mm. For at least 13 mm from the edge of the plug the contacts are straight and plated.

SUPPLEMENTARY INFORMATION: A chamfer of approximately 0.1 mm × 0.1 mm is preferred on the top and bottom corners of the Dataway connector plug to further reduce the possibility of momentary mismating of the edge connector with the Dataway connector socket. Mismating can result in destruction of the circuits when plug-in units are inserted in energized crates. It will be noted that the Dataway connector socket includes an entrance ramp.

A chamfer of 0.3 mm max × 0.3 mm max is permitted on the two vertical edges of the connector (see 5.2, Section a-a of Fig 5).

The dimensions of the contacts of the connector plug are shown in ⑤.₃ of Fig 5. The position of each edge is defined by a dimension h, H relative to the horizontal datum and is completely independent of the position of all other edges on both sides of the plug. The lowest contact on each side of the plug may be

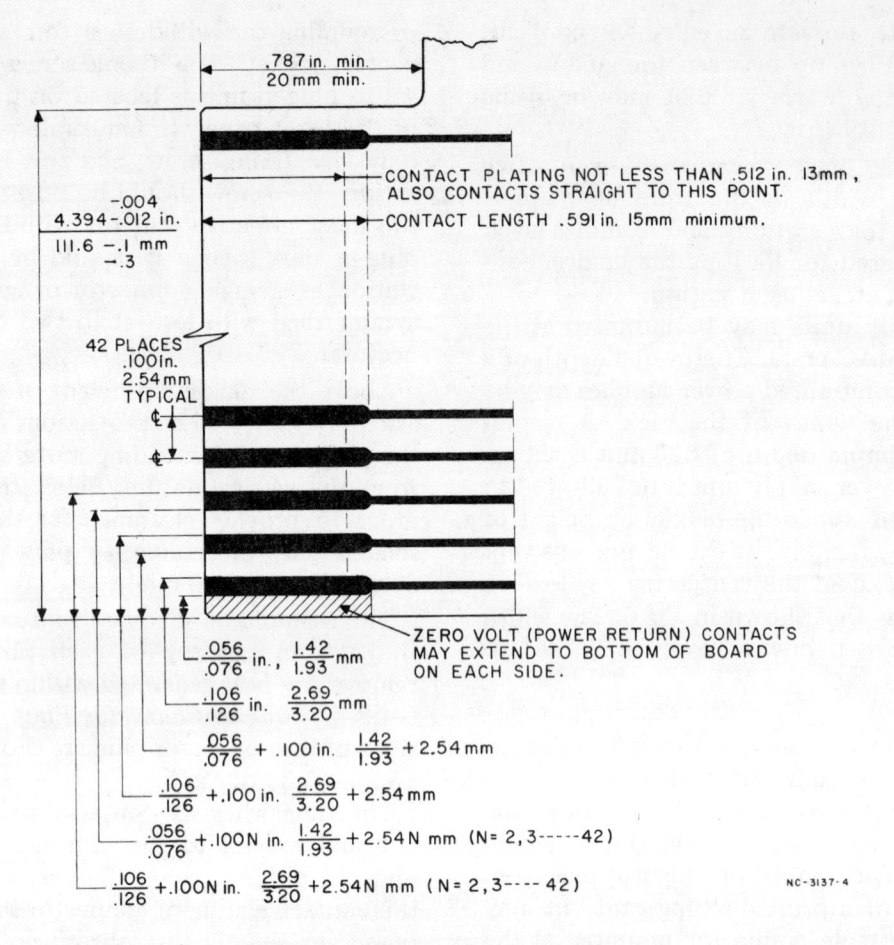

**Fig K4.2.2
Typical Dataway Connector Plug**

extended to the horizontal datum in order to reduce the impedance of the 0 V line.

SUPPLEMENTARY INFORMATION: Fig K4.2.2 shows details of a typical Dataway connector plug derived from Figs 5 and 8.

4.2.3 *Insertion of the Plug-In Unit into the Crate.* In the initial stages of insertion the plug-in unit is supported by the lower guide in the crate. The upper runner, although within the guide, has some vertical clearance. When the plug-in unit is fully inserted, the connector plug is located by the connector socket, and the front panel is supported by the securing screw. The top and bottom runners are then within the guides and approximately parallel to them, but both have some vertical clearance. The transition between these two states is described in detail below.

The dimensions of the guides and runners (Figs 1 and 4) ensure that the plug-in unit moves freely and is guided so that the leading edge of the connector plug enters the chamfers of the connector socket. The lower corner of the leading edge of the plug comes into contact with the chamfer at the bottom of the connector socket. Further insertion of the plug-in unit lifts the connector plug until its lower edge rests on the horizontal datum face of the connector socket. Even a connector plug with the maximum permitted 1 × 1 mm chamfer will have been lifted into correct alignment before any electrical contact occurs between the connector plug and socket. The position of maximum insertion without electrical contact, even with a maximum thickness plug, is defined in (5.5) of Fig 5 with respect to the vertical datum of the connector socket.

Before this point has been reached it will have been possible to engage the fixing screw in the corresponding tapped hole in the lower cross-member of the crate. This can be facilitated by having a tapered end to the screw, so that the front panel is lifted into the correct alignment.

DIGITAL INTERFACE SYSTEM (CAMAC)

Further insertion of the plug-in unit brings the contacts of the plug and socket into engagement, and the insertion force of the connector is encountered. The recommended maximum insertion and withdrawal forces are 80 N for each connector plug. Forces in excess of this can cause difficulty in inserting and withdrawing the plug-in unit and can also result in damage.

Fig 5, (5.5), defines, with respect to the vertical datum of the connector socket, the line beyond which there is reliable contact between corresponding contacts on the plug and socket, even with a plug of minimum thickness.

Finally, when the plug-in unit is fully inserted in the crate, the leading edge of the connector plug is nominally at the vertical datum of the connector socket and the lower datum face of the front panel of the plug-in unit is in contact with the lower cross-member of the crate. However, the forces due to the connector socket and fixing screw are not in line and tend to lift the connector plug off the horizontal datum of the socket, in which case there may be clearance between the upper datum face of the front panel and the upper cross-member. Fig 5, (5.5), ensures that there is adequate clearance beyond the extreme position of the connector plug, by defining a minimum distance between the vertical datum of the socket and any internal obstruction.

4.2.4 *Printed-Wiring Card.* Fig 8 gives recommended dimensions for a printed-wiring card suitable for use with typical (but not necessarily all) commercially available frameworks for plug-in units conforming to this standard.

4.2.5 *Other Connectors.* Connectors or other components such as switches may be mounted on the front panel or at the rear of the plug-in unit. Rear mounted components must be located above the maximum height limit of the Dataway assembly.

For subminiature coaxial connectors the Type 50CM in accordance with Fig 13 is strongly recommended.

There may, however, be special circumstances requiring the use of other connectors in order to suit a specific external equipment with which the plug-in unit is closely associated.

SUPPLEMENTARY INFORMATION: *Digital Signal Classes and Standard Markings for Coaxial Connectors.* Appendix B provides information for the guidance of users.

Auxiliary Connectors. Preferred auxiliary connectors for some applications, together with recommended contact assignments, are described in Appendix C. These connectors are useful for communicating from and to plug-in units. This list is not intended to be restrictive since it is recognized that connectors must be suitable for the applications.

4.3 Adaptor for NIM Units. Plug-in units conforming to the NIM specification (Type N Module of IEC Publ 482) can be inserted into the guides of a CAMAC crate. In order to supply power to a NIM unit, which is shorter than a CAMAC plug-in unit, an adaptor is required between the Dataway connector socket and the connector on the NIM unit. The essential dimensions of such an adaptor are given in Fig 7.

4.4 The Dataway. Communication between plug-in units takes place through the Dataway. This passive multiwire highway is incorporated into the crate and links the Dataway connector sockets at all stations. The Dataway consists of signal lines and power lines, as shown in Table 1.

The extreme right-hand station, as viewed from the front of the crate, has the special role of control station. The data lines in the Dataway are accessible at the remaining normal stations, but not at the control station.

Most signal lines are bus lines linking corresponding contacts of the Dataway connector sockets at all normal stations and, in some cases, the control station. There are also individual lines, each linking one contact at a normal station to one contact at the control station. At each station there are contacts for unspecified uses. Two of these contacts are linked across all normal stations to form free bus lines. The remainder are available as patch contacts, but do not have specified Dataway wiring. The Dataway construction may extend these patch contacts, and others associated with the individual lines and certain bus lines, to more readily accessible patch points to which patch connections can be attached.

Table 1
Standard Dataway Usage

Title	Designation	Contacts	Use at a Module
Command			
Station Number	N	1	Selects the module (individual line from control station)
Subaddress	$A1, 2, 4, 8$	4	Selects a section of the module
Function	$F1, 2, 4, 8, 16$	5	Defines the function to be performed in the module
Timing			
Strobe 1	$S1$	1	Controls first phase of operation (Dataway signals must not change)
Strobe 2	$S2$	1	Controls second phase (Dataway signals may change)
Data			
Write	$W1-W24$	24	Bring information to the module
Read	$R1-R24$	24	Take information from the module
Status			
Look-at-Me	L	1	Indicates request for service (individual line to control station)
Busy	B	1	Indicates that a Dataway operation is in progress
Response	Q	1	Indicates status of feature selected by command
Command Accepted	X	1	Indicates that module is able to perform action required by the command
Common Controls			*Operate on all features connected to them, no command required*
Initialize	Z	1	Sets module to a defined state (Accompanied by $S2$ and B)
Inhibit	I	1	Disables features for duration of signal
Clear	C	1	Clears registers (Accompanied by $S2$ and B)
Nonstandard Connections			
Free bus lines	$P1, P2$	2	For unspecified uses
Patch contacts	$P3-P5$	3	For unspecified interconnections. No Dataway lines
Mandatory Power Lines			*The crate is wired for mandatory and additional lines*
+24 V dc	+24	1	
+ 6 V dc	+ 6	1	
− 6 V dc	− 6	1	
−24 V dc	−24	1	
0 V	0	2	Power return
Additional Power Lines			*Lines are reserved for the following power supplies:*
+12 V dc	+12	1	
−12 V dc	−12	1	
Clean Earth	E	1	Reference for circuits requiring clean earth
Supplementary −6 V	$Y1$	1	See Section 10.
Supplementary +6 V	$Y2$	1	See Section 10.
Reserved undesignated		3	
Total		86	

DIGITAL INTERFACE SYSTEM (CAMAC)

ANSI/IEEE
Std 583-1982

Table 2
Contact Allocation at a Normal Station
(Viewed From Front of Crate)

Bus line	Free Bus line	P1	B	Busy	Bus line
Bus line	Free Bus line	P2	F16	Function	Bus line
Individual patch contact		P3	F8	Function	Bus line
Individual patch contact		P4	F4	Function	Bus line
Individual patch contact		P5	F2	Function	Bus line
Bus line	Command Accepted	X	F1	Function	Bus line
Bus line	Inhibit	I	A8	Subaddress	Bus line
Bus line	Clear	C	A4	Subaddress	Bus line
Individual line	Station Number	N	A2	Subaddress	Bus line
Individual line	Look-at-Me	L	A1	Subaddress	Bus line
Bus line	Strobe 1	S1	Z	Initialize	Bus line
Bus line	Strobe 2	S2	Q	Response	Bus line
Twenty-four Write Bus lines		W24	W23		
$W1$ = least significant bit		W22	W21		
$W24$ = most significant bit		W20	W19		
		W18	W17		
		W16	W15		
		W14	W13		
		W12	W11		
		W10	W9		
		W8	W7		
		W6	W5		
		W4	W3		
		W2	W1		
Twenty-four Read Bus lines		R24	R23		
$R1$ = least significant bit		R22	R21		
$R24$ = most significant bit		R20	R19		
		R18	R17		
		R16	R15		
		R14	R13		
		R12	R11		
		R10	R9		
		R8	R7		
		R6	R5		
		R4	R3		
		R2	R1		
Power Bus lines	−12 V dc	−12	−24	−24 V dc	
	Reserved (c)*		−6	−6 V dc	
	Reserved (a)*	—	—	Reserved (b)*	
	Supplementary −6 V**	Y1	E	Clean Earth	
	+12 V dc	+12	+24	+24 V dc	
	Supplementary +6 V**	Y2	+6	+6 V dc	
	0 V (Power Return)	0	0	0 V (Power Return)	

*Reserved (c) was previously assigned to +200 V dc,
 Reserved (a) to 117 V ac Line, and
 Reserved (b) to 117 V ac neutral, all non-mandatory voltages.
Effective 1 January 1978, these assignments were cancelled in order to avoid hazardous voltages on the connectors.
**See Section 10.

Table 3
Contact Allocation at the Control Station
(Viewed From Front of Crate)

Individual patch contact		P1	B	Busy	Bus line
Individual patch contact		P2	F16	Function	Bus line
Individual patch contact		P3	F8	Function	Bus line
Individual patch contact		P4	F4	Function	Bus line
Individual patch contact		P5	F2	Function	Bus line
Bus line	Command Accepted	X	F1	Function	Bus line
Bus line	Inhibit	I	A8	Subaddress	Bus line
Bus line	Clear	C	A4	Subaddress	Bus line
Individual patch contact		P6	A2	Subaddress	Bus line
Individual patch contact		P7	A1	Subaddress	Bus line
Bus line	Strobe 1	S1	Z	Initialize	Bus line
Bus line	Strobe 2	S2	Q	Response	Bus line

Twenty-four individual Look-at Me lines (L1 from Station 1, etc)		L24	N24	Twenty-four individual Station Number lines, (N1 to Station 1, etc)
		L23	N23	
		L22	N22	
		L21	N21	
		L20	N20	
		L19	N19	
		L18	N18	
		L17	N17	
		L16	N16	
		L15	N15	
		L14	N14	
		L13	N13	
		L12	N12	
		L11	N11	
		L10	N10	
		L9	N9	
		L8	N8	
		L7	N7	
		L6	N6	
		L5	N5	
		L4	N4	
		L3	N3	
		L2	N2	
		L1	N1	

Power Bus lines	−12 V dc	−12	−24	−24 V dc	
	Reserved (c)*		−6	−6 V dc	
	Reserved (a)*	—	—	Reserved (b)*	
	Supplementary −6 V**	Y1	E	Clean Earth	
	+12 V dc	+12	+24	+24 V dc	
	Supplementary +6 V**	Y2	+6	+6 V dc	
	0 V (Power Return)	0	0	0 V (Power Return)	

*Reserved (c) was previously assigned to +200 V dc,
  Reserved (a) to 117 V ac Line, and
  Reserved (b) to 117 V ac Neutral, all non-mandatory voltages.
  Effective 1 January 1978, these assignments were cancelled in order to avoid hazardous voltages on the connectors.
**See Section 10.

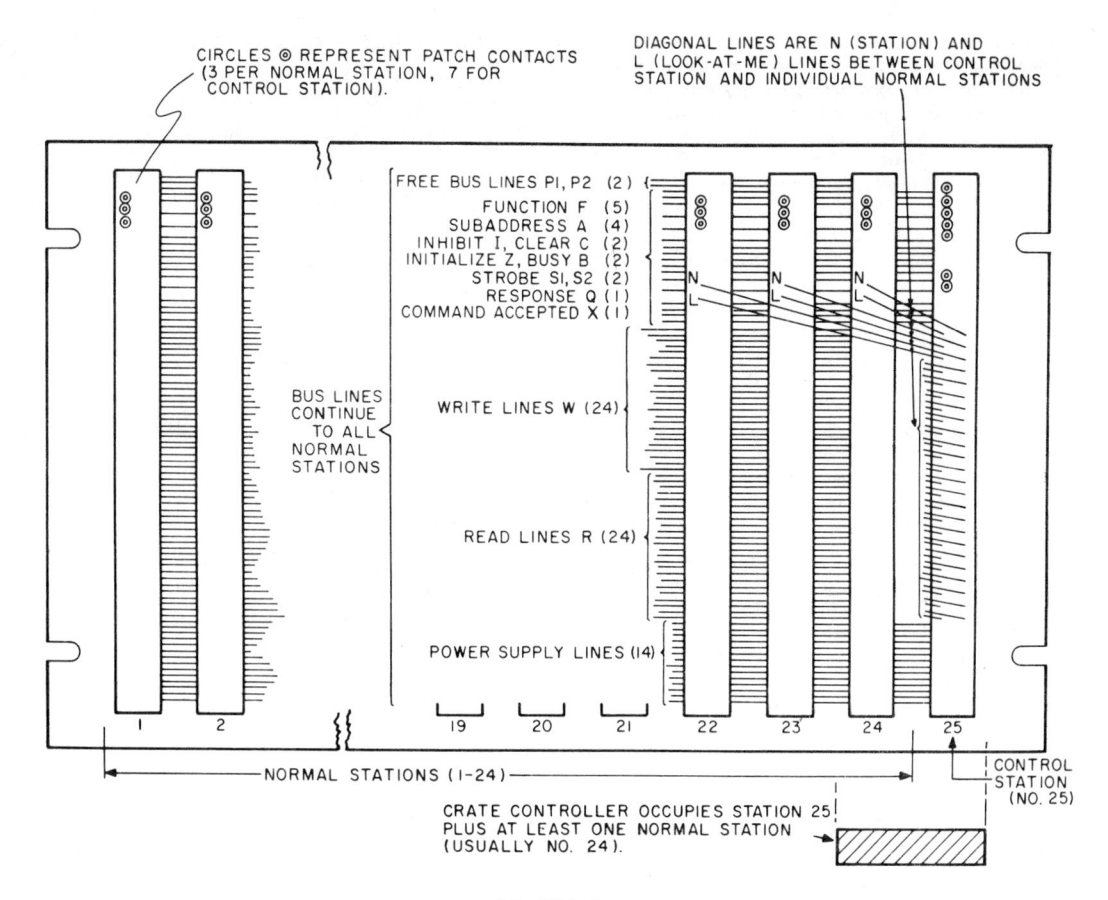

DIGITAL INTERFACE SYSTEM (CAMAC)   ANSI/IEEE
                                                                                    Std 583-1982

**Fig K4.4
Dataway Wiring, Front View of Twenty-Five
Station Crate**

The power lines link corresponding contacts of the Dataway connector sockets at all stations. The power return line (0 V) links two contacts in parallel at all stations.

> The assignment of contacts at the Dataway connector and their connections to bus lines, individual lines, and patch contacts must be as shown in Table 2 for normal stations and Table 3 for the control station. The control station must be to the right of all normal stations.
>
> The method of construction of the Dataway must be consistent with the signal standards for signal lines (see Section 7) and with the maximum current loads specified for the power lines (see Section 8).

Apart from this, the construction of the Dataway is not specified. Appropriate techniques include printed wiring on flexible or rigid substrates (with and without ground planes) and soldered or wrapped wiring. Particular attention should be given to the cross-coupling between signal lines, and to their capacitance to ground.

SUPPLEMENTARY INFORMATION:
  Dataway Wiring. Fig K4.4 shows the Dataway wiring in a crate with twenty-five stations. The twenty-four normal stations (station numbers 1–24) of the crate are wired identically. The wiring of the control station (number 25) is different from that of the normal stations. Since a crate controller must have access to the Read *R* and Write *W* bus lines, it must occupy at least one normal station in addition to the control station.

The Dataway lines (and contacts) fall into the four categories described below:

(1) *All Station Bus Lines* linking corresponding Dataway connector contacts at all stations, including the control station. There are thirty-one such bus lines as follows:

Function lines *F*	5	Busy *B*	1
Subaddress lines *A*	4	Response *Q*	1
Inhibit *I*	1	Strobe *S1, S2*	2
Clear *C*	1	Command Accepted *X*	1
Initialize *Z*	1	Power Supply lines	14

21

(2) *Normal Station Bus Lines* linking corresponding Dataway connector contacts at all normal stations, but not connected to the control station. There are fifty such bus lines as follows:

   Write lines W 24
   Read lines R 24
   Free bus lines $P1, P2$ 2

(3) *Individual Lines* between the control station and each normal station. There are two such lines for each normal station, the N line (Station Number line), by which the control station addresses specific normal stations, and the L line (Look-at-Me line), by which specific stations advise the control station that they desire attention. Thus twenty-four N lines and twenty-four L lines connect to the control station.

(4) *Patch Contacts* (for each Dataway station) that are not connected to the Dataway and are thus available for patch connections. There are three patch contacts for each normal station and seven for the control station.

Dataway Patch Contact Terminals. It is mandatory that patch contacts (such as wire-wrap terminals) be provided for $P3-P5$ of normal stations and $P1-P7$ of the control station. In addition, it is preferred practice to have such contact terminals provided also for I, C, and the 0 V (Power Return) buses and for each N and L line.

5. Use of the Dataway Lines

> Each line of the Dataway must be used in accordance with the mandatory requirements detailed in the following sections and summarized in Table 1.

A typical Dataway operation involves at least two plug-in units, one of which acts as a controller and the other as a controlled module. In this standard the terms controller and module have the following specific meanings. Controller refers to a unit occupying the control station and at least one normal station. Module refers to a unit occupying one or more normal stations. Both receive signals from some Dataway lines and generate signals on others in accordance with the definitions given in Appendix A. (In practice there can be special cases of units that combine some properties of a controller with some properties of a module.)

There are two types of Dataway operations. During *command operations* the controller generates a command consisting of signals on individual Station Number lines to specify one or more modules, on the Subaddress bus lines to specify a subsection of the module, and on the Function bus lines to specify the operation to be performed. During *unaddressed operations* there is no command, but the controller generates one of the common control signals on the Initialize or Clear bus lines, and this operates on all modules connected to the bus line. During command operations and unaddressed operations the controller generates a signal on the Busy bus line. The Busy signal is available at all stations to indicate that a Dataway operation is in progress. Two timing signals, Strobes $S1$ and $S2$, are generated in sequence on separate bus lines during command operations. Only Strobe $S2$ is mandatory during unaddressed operations, but $S1$ may also be generated.

During a Dataway command operation there may be a read data transfer from a module to the controller, or a write data transfer from the controller to a module, or neither.

In response to a read command the addressed module establishes Read data signals which are available to the controller from the time of Strobe $S1$ onwards. In response to a write command the addressed module accepts Write data signals from the controller at the time of Strobe $S1$.

The addressed module indicates by a signal on the Command Accepted bus line whether it is able to perform the action required by the command. It may also transmit one bit of status information on the Response bus line. The controller accepts the Command Accepted and Response signals at the time of Strobe $S1$.

Any module may generate a signal on its individual Look-at-Me line to indicate that it requires attention.

Three common control signals are available at all stations, without requiring addressing by a command, in order to Initialize all units (typically after switch-on), to Clear data registers, or to Inhibit features such as data taking.

The use of each Dataway line is defined in the following sections. The relationship between signals, in order to generate specific commands, is defined in Section 6, and the electrical signal standards, including timing, are defined in Section 7.

The sequence of events during command operations is described in Section 7.1.3.1 and shown in Fig 9. The sequence during unaddressed operations is described in Section 7.1.3.2 and shown in Fig 10.

SUPPLEMENTARY INFORMATION: *Block Transfers.* A block transfer may be defined as a specified sequence of CAMAC operations initiated by a single, more powerful command (a block transfer command). It is possible to design a large number of block transfer sequences (modes) which are compatible with this standard. However, a relatively small number of modes can accomplish nearly all block transfers which are usually needed. To enhance compatibility among modules, controllers, and software, it is recommended that only the defined modes be used.

5.1 Commands. The state of the signals on the individual Station Number lines (specifying a module or modules), the four Subaddress lines (specifying a subsection of the module) and the five Function lines (specifying the type of operation) constitute a command.

The command signals are maintained for the full duration of the Dataway operation. They are accompanied by a signal on the Busy bus line, which indicates to all units that a Dataway operation is in progress.

> Plug-in units must not rely on the state of signals on the Subaddress and Function lines when no command operation is in progress.

5.1.1 *Station Number N*

> Each normal station is addressed by a signal on an individual Station Number line N_i which comes from a separate contact at the control station (see Tables 2 and 3). The stations are numbered in decimal from the left-hand end as viewed from the front, beginning with Station 1 (addressed by $N1$).

There is no restriction on the number of stations that can be addressed simultaneously and the design of the modules should permit this.

5.1.2 *Subaddress A8, A4, A2, A1*

> Different sections of a module are addressed by signals on the four A bus lines. These signals are decoded in the module to select one of up to sixteen subaddresses, numbered in decimal from $A(0)$ to $A(15)$.

The subaddress may be used to select, for example, a register within the module, or a feature that is to control the Response signal Q, or a section of the module that is to be operated on by functions such as Enable, Disable, and Execute. The use made of the subaddress within a module is discussed in relation to the function codes in Section 6.

> Each subaddress code used in a module must be fully decoded in the module. Full decoding means that all four Dataway Subaddress signals are used in the decoding process.

The subaddress codes are designated $A(0)$, $A(1)$, $A(2)$, $A(3)$, etc, to distinguish them from the individual Subaddress lines $A1$, $A2$, $A4$, and $A8$. For example, the Subaddress signals $A1 = 1$, $A2 = 1$, $A4 = 0$, and $A8 = 0$ represent the code $A(3)$.

5.1.3 *Function F16, F8, F4, F2, F1*

> The function to be performed at the specified subaddress in the selected module or modules is defined by the signals on the five F bus lines. These signals are decoded in the module to select one of up to thirty-two functions, numbered in decimal from $F(0)$ to $F(31)$. The definitions of the thirty-two function codes are summarized in Table 4 and are detailed in Section 6 in relation to the command structure.

In some systems the controller partially decodes the Function signals in order to determine whether a data transfer is required to a module (writing) or from a module (reading). Multiple-station addressing allows operations with the same function code to be performed simultaneously in more than one module. These features depend on some standardization in the assignment of function codes.

The function codes are subdivided into three groups, involving read operations, write operations, and operations with no transfer of data. The standard function codes have defined actions in modules and controllers. There are also reserved codes, for any future additions to the standard codes, and nonstandard codes whose use is not defined in detail.

> Each function code used in a module must be fully decoded in the module. Full decoding means that all five Dataway Function signals are used in the decoding process.

The function codes are designated $F(0)$, $F(1)$, $F(2)$, $F(3)$, etc, to distinguish them from the individual function lines $F1$, $F2$, etc. For example, the Function signals $F1 = 1$, $F2 = 0$, $F4 = 0$, $F8 = 1$, and $F16 = 1$ represent the code $F(25)$.

5.2 Strobe Signals $S1$ and $S2$. During each command operation the controller generates two Strobe signals $S1$ and $S2$ in sequence on separate bus lines. In response to these timing signals plug-in units initiate various actions appropriate to the command that is present on the Dataway.

> Both Strobe signals must be generated during each command operation.
>
> Plug-in units must not take irreversible action based on the command or data signals until the time of $S1$. Actions concerned with the acceptance of data and status information from the R, W, Q, and X lines must be initiated at the time of $S1$ (with the possible exception given below). Other actions may also be timed by $S1$, but must not change the state of signals on the R and W lines.
>
> Any actions that can change the state of Dataway Read or Write signals must be initiated by the second strobe $S2$. For example, $S2$ must be used if it is required to clear a register whose output is connected to the Dataway.

Modules normally accept write data in response to $S1$, because the data signals are permitted to change at the time of $S2$. However, modules may accept write data in response to $S2$ under special circumstances, but this is not recommended.

During unaddressed Dataway operations the controller generates Strobe $S2$ to indicate when modules accept the common control signal. Strobe $S1$ may also be generated, but this is not mandatory, and modules cannot rely on it.

> Strobe $S2$ must be generated during each unaddressed operation.

5.3 Data. All information carried by the Read and Write lines is conveniently described as data, although it may be information concerned with status or control features in modules. Information that is transferred to or from a control register in a module is thus regarded as data.

Up to 24 bits may be transferred in parallel between the controller and the selected module. Independent lines are provided for the read and write directions of transfer.

If the bits of a data word have different numerical significance, line R_n should be used for a higher order bit than R_{n-1}, and W_n for a higher order bit than W_{n-1}.

It is recommended that controllers have 24 bit capability. For particular applications assemblies are permitted in which the controller has a word length less than 24 bits and the modules have an equal or smaller word length.

> Plug-in units must not rely on the state of signals on the Read and Write bus lines when no command operation is in progress.

5.3.1 *Write Lines W1-W24*

> The controller generates data signals on the W bus lines during each write operation. The W signals must reach a steady state before $S1$ and must be maintained until the end of the operation, unless modified by $S2$. Strobe $S1$ must be used by the modules to strobe the data unless there are very strong technical reasons for choosing $S2$ (see Section 5.2).

The W lines serve a few data sources (typically only one controller) and many data receivers.

5.3.2 *Read Lines R1-R24*

> Data signals are set up on the R bus lines by the module during a read operation. The R signals must reach a steady state before $S1$ and must be maintained for the full duration of the Dataway operation, unless the state of the data source is changed by $S2$. The controller must initiate action concerned with the acceptance of data from the R lines at the time of the Strobe $S1$ and must not take irreversible action before this.

Table 4
The Function Codes

Code $F(\ )$	Function	Use of R and W Lines	$F16$	$F8$	$F4$	$F2$	$F1$	Code $F(\ )$
0	Read Group 1 register	Functions using the R lines	0	0	0	0	0	0
1	Read Group 2 register		0	0	0	0	1	1
2	Read and Clear Group 1 register		0	0	0	1	0	2
3	Read Complement of Group 1 register		0	0	0	1	1	3
4	Nonstandard		0	0	1	0	0	4
5	Reserved		0	0	1	0	1	5
6	Nonstandard		0	0	1	1	0	6
7	Reserved		0	0	1	1	1	7
8	Test Look-at-Me	Functions not using the R or W lines	0	1	0	0	0	8
9	Clear Group 1 register		0	1	0	0	1	9
10	Clear Look-at-Me		0	1	0	1	0	10
11	Clear Group 2 register		0	1	0	1	1	11
12	Nonstandard		0	1	1	0	0	12
13	Reserved		0	1	1	0	1	13
14	Nonstandard		0	1	1	1	0	14
15	Reserved		0	1	1	1	1	15
16	Overwrite Group 1 register	Functions using the W lines	1	0	0	0	0	16
17	Overwrite Group 2 register		1	0	0	0	1	17
18	Selective Set Group 1 register		1	0	0	1	0	18
19	Selective Set Group 2 register		1	0	0	1	1	19
20	Nonstandard		1	0	1	0	0	20
21	Selective Clear Group 1 register		1	0	1	0	1	21
22	Nonstandard		1	0	1	1	0	22
23	Selective Clear Group 2 register		1	0	1	1	1	23
24	Disable	Functions not using the R or W lines	1	1	0	0	0	24
25	Execute		1	1	0	0	1	25
26	Enable		1	1	0	1	0	26
27	Test Status		1	1	0	1	1	27
28	Nonstandard		1	1	1	0	0	28
29	Reserved		1	1	1	0	1	29
30	Nonstandard		1	1	1	1	0	30
31	Reserved		1	1	1	1	1	31

Function Signals

The R lines serve a few data receivers (typically only one controller) and many data sources.

5.4 Status Information. Status information is conveyed by signals on the Look-at-Me L, Busy B, Response Q, and Command Accepted X lines.

5.4.1 *Look-at-Me L*

> The Look-at-Me lines, like the N lines, are individual connections from each normal station to separate contacts at the control station ($L1$ from station 1, etc).

Any module may generate a signal on its individual line L_i to indicate that it requires attention. Modules that occupy more than one station may indicate different demands by signals on the appropriate L lines.

The L signal generated by a module may represent demands for attention originating from more than one Look-at-Me source (LAM source) in the module. A LAM structure by which various demands can be selected and grouped to form the L signal is shown in Fig 11 and described below. All modules that generate L have the mandatory features shown in Fig 11 and may incorporate additional features for more complex demand handling.

Individual bits of the LAM status register are set by the corresponding LAM sources, and are cleared by appropriate commands and by the initialize signal. The outputs (LAM status) may be examined collectively by reading the state of all outputs [Read Group 2 at $A(12)$], or individually by the command Test Status with the appropriate subaddress $A(i)$. Each LAM status should be individually enabled and disabled (for example by a LAM mask) to form the corresponding LAM request. The LAM requests are examined collectively by reading the state of all LAM requests [Read Group 2 at $A(14)$], or individually by the command Test LAM with the appropriate subaddress $A(i)$. The internal Look-at-Me signal (L signal), derived from the OR combination of the LAM requests, is tested by the command Test LAM, with subaddress $A(k)$ distinguishing this from tests of individual LAM requests. Finally, the output of this signal as the Dataway L signal is inhibited while the module is addressed by N, possibly in conjunction with certain F and A codes or groups of codes. (See Section 5.4.1.3.)

SUPPLEMENTARY INFORMATION:
Examples of LAM Logic. Figs K5.4.1A—K5.4.1C are examples of LAM logic in modules. Fig K5.4.1A shows an example of LAM logic for a module with a single source of Look-at-Me. For modules having multiple sources of Look-at-Me, operations on the LAMs may be addressed via subaddresses (as in Fig K5.4.1B) or via bits in the data word associated with the operation (as in Fig K5.4.1C).

5.4.1.1 *Look-at-Me: Clear, Disable, and Test*

> Provision must be made for resetting each bit of the LAM status register individually, either by a Clear Look-at-Me operation $F(10)$ see Section 6.2.3), or by a Selective Clear Group 2 operation $F(23)$ see Sections 5.4.1.2 and 6.3.6). All LAM status bits must be reset collectively by the Initialize signal (see Section 5.5.1).

If the LAM request calls for some specific action (for example, reading the contents of a data register), then the corresponding bit of the LAM status register should also be cleared when the appropriate Dataway operation is performed.

> A module that has generated $L = 1$ must not clear the LAM status register until it receives an appropriate command or the Initialize signal.

Each module that generates L should have a means of enabling and disabling the LAM requests. This may be done by loading and clearing a mask register, or by Enable and Disable commands.

> All LAM requests that can be disabled by commands must also be disabled by the Initialize signal Z.
>
> A module that generates L must have a means of testing the L signal by a Test Look-at-Me operation [Function code $F(8)$ (see Section 6.2.1), with a subaddress distinguishing this from tests of individual LAM requests]. If there are several LAM sources, the corresponding LAM requests must also be capable of being examined either by Test Look-at-Me operations associated with appropriate subaddresses or by reading a LAM request pattern in a read operation.

DIGITAL INTERFACE SYSTEM (CAMAC)　　　　　　　　　　　　　　　ANSI/IEEE
　　　　　　　　　　　　　　　　　　　　　　　　　　　　　　　　　　Std 583-1982

NOTES:

(1) Features mandatory for this LAM system shown with an asterisk.

(2) Subaddresses may be chosen arbitrarily, for example, to match subaddresses associated with the LAM sources. In this example, $A(15)$ is shown for Test Look-at-Me for uniformity with more complex modules. See Figs K5.4.1A–K5.4.1C

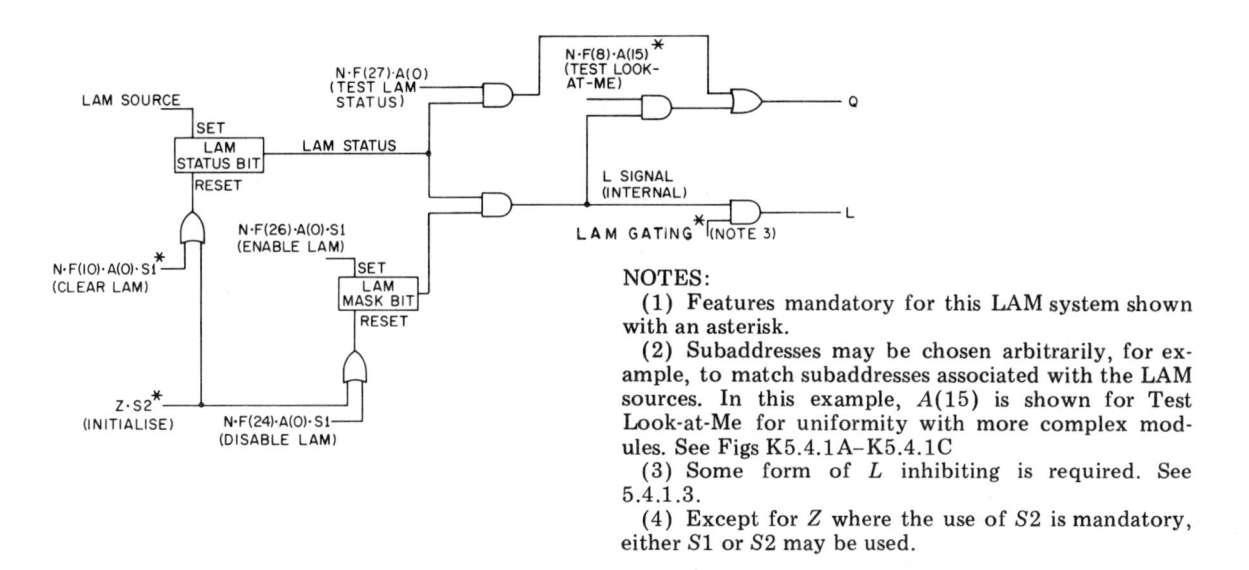

(3) Some form of L inhibiting is required. See 5.4.1.3.

(4) Except for Z where the use of $S2$ is mandatory, either $S1$ or $S2$ may be used.

Fig K5.4.1A
Example of Single-Source Look-at-Me Logic

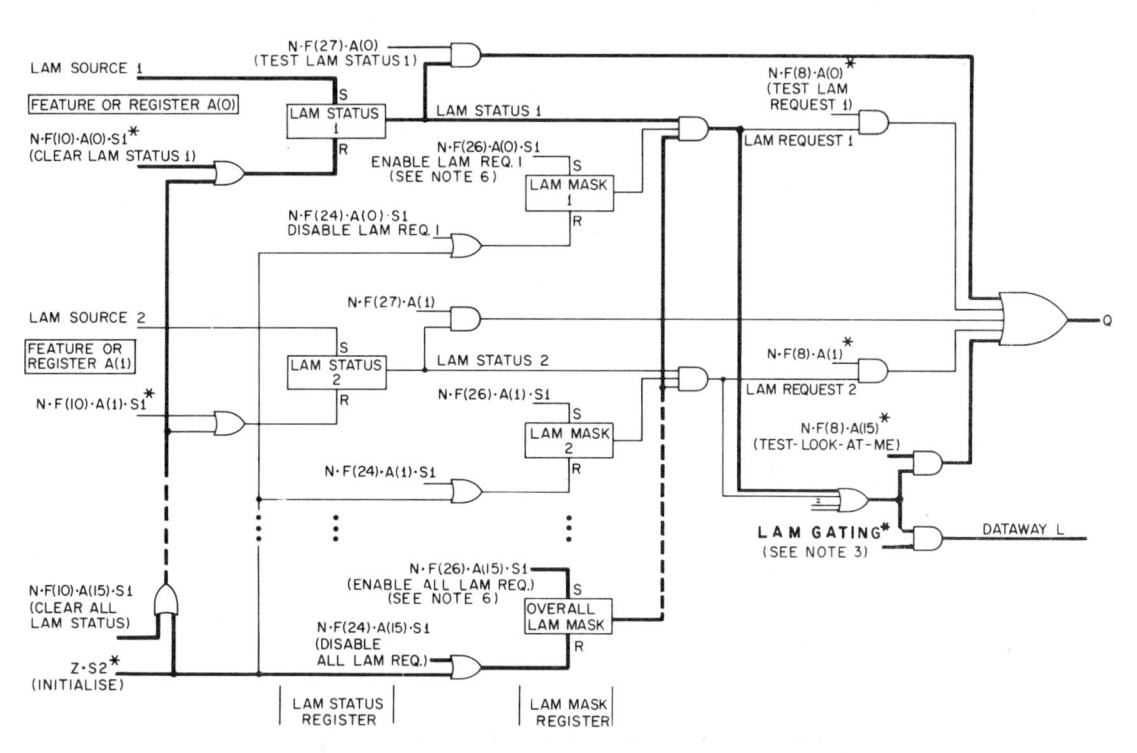

NOTES:

(1) Mandatory features for a module which performs operations on LAMs via subaddresses are shown with an asterisk.

(2) Although choice of subaddress is arbitrary, it is generally useful to use the same subaddress for the commands associated with a given LAM source. For example, the figure shows subaddress $A(0)$ used for LAM source 1; $A(1)$ for LAM source 2, etc. Subaddress $A(15)$ is suggested for operations which affect all LAM sources, for example, Enable/Disable all LAM requests and Test L signal.

(3) and (4) See Notes 3 and 4 for Fig K5.4.1A.

(5) The heavy lines facilitate comparison with the single-source logic shown in Fig K5.4.1A.

(6) For modules that include both individual LAM masks and an overall LAM mask (as above), the module information summaries must point out that an individual LAM request is enabled only if its individual mask and the overall mask have both been set.

Fig K5.4.1B
Example of Multiple-Source Look-at-Me Logic
Showing Operations on LAMs via Subaddresses

NOTES:
(1) Features which are mandatory for a module which performs operations on LAMs via databits are shown with an asterisk.
(2) Although $A(15)$ is shown, choice of subaddress for Test Look-at-Me is arbitrary. All other subaddresses shown are those recommended in Section 5.4.1.2.
(3) and (4) See Notes (3) and (4) for Fig K5.4.1A.
(5) Appropriate logic should be provided to ensure that LAM Status Bits cannot be set during Dataway command cycles for Read LAM status and Read LAM request.

**Fig K5.4.1C
Example of Multiple-Source Look-at-Me Logic
Showing Operations on LAMs via Databits**

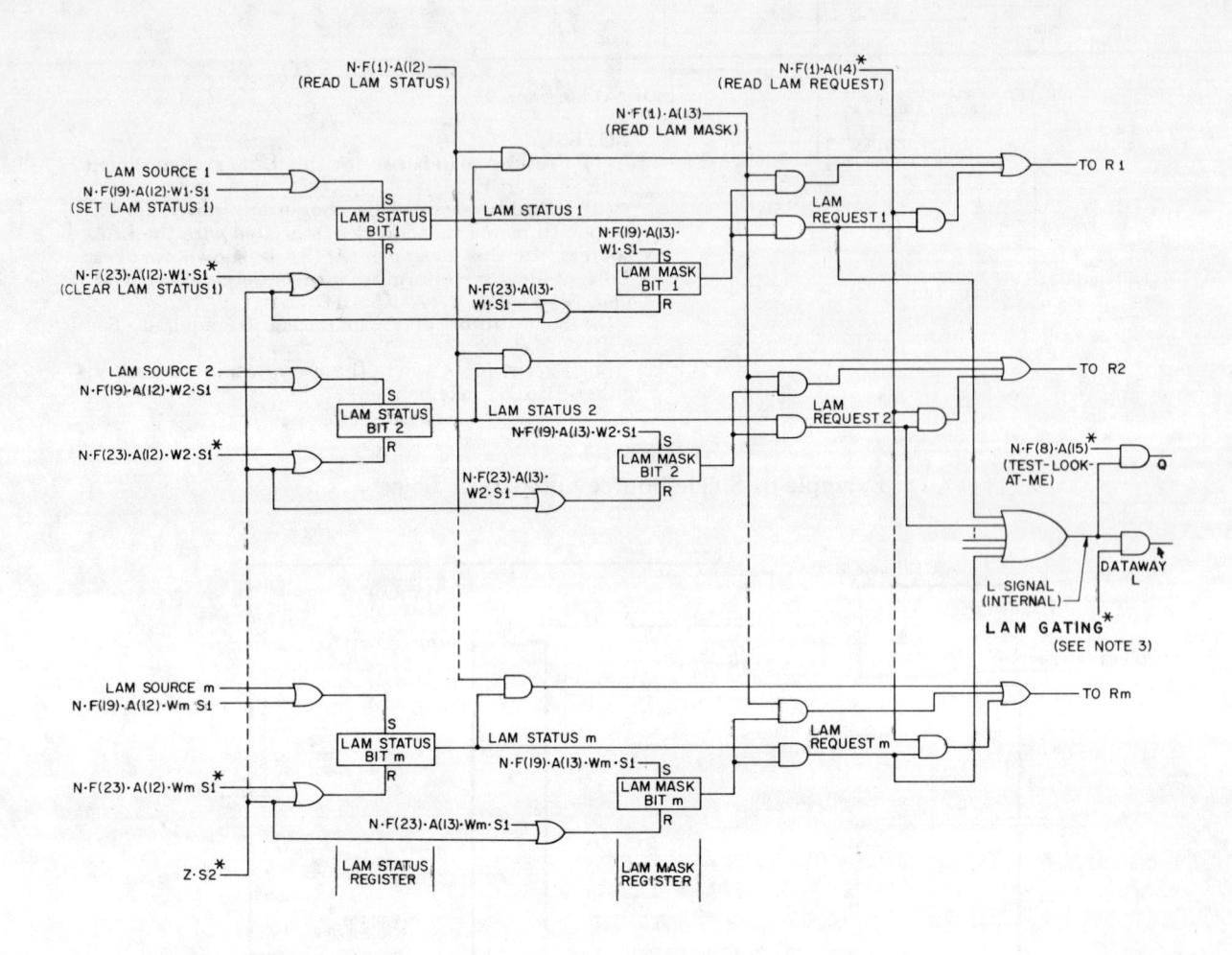

5.4.1.2 *Look-at-Me: Commands for Access.* Modules may contain registers for LAM information. These registers are not mandatory but if included they should be accessed as Group 2 registers at the following subaddresses:

LAM status register $A(12)$
LAM mask register $A(13)$
LAM request register $A(14)$

Corresponding bit positions should be associated with the same LAM source.

The state of each data bit read from the LAM status or LAM request register is the same as the state of the Q response that would be obtained by a Test Status or Test Look-at-Me operation.

The data word read from $A(12)$ should have LAM status information in the low-order bits and may also contain other status information. Each bit of the data word loaded into $A(13)$ should be in the 1 state to enable the corresponding LAM request, and in the 0 state to disable it.

The operations used to access LAM information may be divided into two classes. One class consists of Read $F(1)$, Write $F(17)$, Clear $F(11)$, Selective Set $F(19)$, and Selective Clear $F(23)$ addressed to the Group 2 LAM registers described above. This class is preferred for operations on modules with many LAM sources. The other class consists of Clear LAM $F(10)$, Enable $F(26)$, Disable $F(24)$, Test Status $F(27)$, and Test LAM $F(8)$ addressed to specific demands. This class is preferred for modules with few LAM sources. In the one class a LAM source, LAM (i), is associated with bit position (i) in data words, and in the other class with subaddress $A(i)$. For ease of programming it is recommended that all the operations related to a particular LAM source should belong to only one class.

5.4.1.3 *Look-at-Me: Gating*

> When a module that is generating $L_i = 1$ receives a command that will cause it to cease doing so, it must inhibit the L signal or the appropriate LAM request. The inhibit condition must be effective before Strobe $S1$ and must be maintained until the end of the Dataway operation.

This requirement may be met very simply by inhibiting the L signal output when a module is addressed by any command ($L_i = 0$, when $N_i = 1$). Unaddressed modules can thus initiate $L_i = 1$ at any time, but addressed modules cannot do so until the end of the current Dataway operation.

The requirements may be met more precisely by inhibiting only those LAM status signals that are canceled by the current command. The ability to initiate $L_i = 1$ during a Dataway operation is thus extended to all LAM requests that are not being canceled. This requires the recognition of $N_i = 1$ with the specific functions and subaddresses, and the generation of appropriate inhibit conditions.

The requirement may be interpreted in ways intermediate between these extremes. For example, the ability to initiate $L_i = 1$ can be extended to all LAM requests during most Dataway operations if the L signal output is inhibited when the module is addressed by $N_i = 1$ together with appropriate, easily identifiable groups of functions and subaddresses.

The mandatory statement above replaces the earlier requirement in EUR 4100e (1969) that all modules gate off their L signal output during every operation ($L_i = 0$, when $B = 1$). It allows L signals to be initiated at any time, maintained continuously, and removed in advance of Strobes $S1$ and $S2$. This counteracts delays elsewhere in systems and leads to improved performance in handling demands for attention.

Units conforming to this standard and to the earlier requirement in EUR 4100e (1969) can generally be used together in systems where the improved performance is not required.

5.4.1.4 *Look-at-Me: Use for Operation Synchronization in Block Transfers, Supplementary Information.* A module which is using a LAM for synchronizing individual operations of a block transfer sets LAM = 1 only when it is ready to effect the operation. During the associated Dataway cycle, the module sets LAM = 0 unless it is ready to effect another transfer.

Modules designed with this feature should have a means of patching this synchronizing signal onto the L line of the module, onto a patch pin, or onto a panel connector.

This synchronization method is equally applicable to Stop and Stop-on-Word operations. (See sections 5.4.3.3 and 5.4.3.4.)

5.4.2 *Busy B.* The Busy signal is used to interlock various aspects of a system that can compete for the use of the Dataway. The signal $B = 1$ indicates to all units that a Dataway operation is in progress.

> The Busy signal $B = 1$ must be generated during each Dataway command operation (when N signals are also generated) and during unaddressed operations (when Z or C are generated).

5.4.3 *Response Q.* During every command operation the addressed module may generate a signal on the Q bus line to indicate the status of any selected feature of the module.

> The controller must initiate action concerned with the acceptance of the status information from the Q line at the time of Strobe $S1$ and must not take irreversible action before this.
>
> In read and write operations (see Sections 6.1 and 6.3) addressed modules must establish the signal $Q = 0$ or $Q = 1$ before Strobe $S1$ and must maintain it until at least Strobe $S2$.

In Test Look-at-Me operations (see Section 6.2.1) addressed modules may initiate the Q signal at any time during a Dataway operation if the status of the appropriate LAM request changes. If $Q = 1$ has been initiated, it will remain static until the end of the operation since this command is not allowed to reset LAM status.

In all operations other than read, write, and Test Look-at-Me the Q signal is permitted to change at any time. There is a risk that status information will be missed if $Q = 1$ is initiated between Strobes $S1$ and $S2$ during an operation in which the module resets the status condition at the time of Strobe $S2$.

In any operation the Q signal conveys only one bit of information, which should be clearly defined for each subaddress and function code used by the module.

Examples of the use of the Q signal during read and write operations are given in Sections 5.4.3.1, 5.4.3.2, and 5.4.3.3, which define three methods of transferring blocks of data. These three methods are summarized in the following table. However, the status information transmitted by the addressed module during read and write operations is not restricted to these examples.

Response	Q Mode Address Scan	Q Mode Repeat	Q Mode Stop
$Q = 1$	register present	register ready	within block
$Q = 0$	register absent	register not ready	end of block

[See ANSI/IEEE Std 683-1976 (R 1981)]

5.4.3.1 *Use of Q for Block Transfers: Address Scan Mode*

> If a module contains registers that are intended to be accessed sequentially in Address Scan mode, they must be located at consecutive subaddresses starting at $A(0)$. During read and write operations the module must generate $Q = 1$ at all subaddresses at which these registers are present and $Q = 0$ at the first unoccupied subaddress, if any. A module with n such registers must therefore generate $Q = 1$ at $A(0)$ to $A(n-1)$. If $n<16$, the module must generate $Q = 0$ at $A(n)$.

The Address Scan mode of block transfer is used for data transfers to or from an array of modules that do not necessarily occupy consecutive stations or all subaddresses. The state of Q during each operation is used by the controller to determine the station number and subaddress for the next operation. When $Q = 1$, the subaddress is incremented, with carry-over into the station number. When $Q = 0$, the subaddress is set to $A(0)$, and the station number is incremented. This allows unoccupied stations within an array.

The block transfer may be terminated by the controller on reaching a specified word count (recommended) or address.

5.4.3.2 *Use of Q for Block Transfers: Repeat Mode*

> If a module contains a register that is intended to be accessed in Repeat mode it must generate $Q = 1$ during a read or write operation if the register is ready to participate in a data transfer. It must generate $Q = 0$ during a read or write operation if the register is not ready to participate in a data transfer.

DIGITAL INTERFACE SYSTEM (CAMAC)

ANSI/IEEE
Std 583-1982

Table K5.4.3.5
Single Module, Single Address Block Transfers

Part I — Method for Performing Stop mode CAMAC Block Transfers (This is the recommended mode)

Q	Position of operation relative to a block of n words	Transfer Direction	Module	Interface	Computer
Q = 1	1 to n	Read	Data word transmitted	Pass data word Keep channel open	Store data word in computer memory
		Write	Data word accepted	Data word already passed. Keep channel open	Data word already delivered
Q = 0	$n+1$	Read	No significant data word transmitted	Do not pass data word. Close channel	No action required
		Write	Data word not accepted	Close channel. Data word already passed	Data word to be recovered for retransmission

Part II — Method for performing Stop-on-Word mode CAMAC Block Transfers.
  This is a special-application mode only. (Cannot be used with many earlier interfaces and modules.)

Q	Position of operation relative to a block of n words	Transfer Direction	Module	Interface	Computer
Q = 1	1 to $(n-1)$	Read	Data word transmitted	Pass data word. Keep channel open	Store data word in computer memory
		Write	Data word accepted	Data word already passed. Keep channel open	Data word already delivered
Q = 0	n	Read	Word* transmitted	Pass word.* Close channel	Store word* in computer memory
		Write	Word* accepted	Close channel. Word* already passed	Word* already delivered

*The word may be a data-word, status-word or a dummy word.

The Repeat mode of block transfer is used when data transfers to or from one register have to be related to the state of readiness of the register or of associated external equipment. The response $Q = 0$ indicates that the same operation should be repeated until the register becomes ready ($Q = 1$). The operation may be repeated continuously if the module samples the state of readiness in a way that tolerates the B and N signals being maintained continuously. There is a risk that a system will lock up if an operation is repeated indefinitely while waiting for $Q = 1$.

5.4.3.3 Use of Q for Block Transfers: Stop Mode

> If a module contains a register that is intended to be accessed in Stop mode, it must generate $Q = 1$ during each read or write operation while the block of data is being transferred and must generate $Q = 0$ during any further operations after the end-of-block condition has been encountered.

The Stop mode of block transfer is used when data transfers to or from one register have to be terminated by an end-of-block indication from the module. A block of data transfers accompanied by $Q = 1$ is followed by at least one further operation accompanied by $Q = 0$ to indicate end-of-block.

5.4.3.4 Use of Q for Block Transfers: Stop-on-Word mode, Supplementary Information.
If a module contains a register intended to be accessed in Stop-on-Word mode, it generates $Q = 1$ on each read or write operation while the block of data is being transferred, except that it generates $Q = 0$ on the last data transfer within the block.

When data transfers to or from a module are to be terminated by an end-of-block indication from the module, the Stop mode (Section 5.4.3.3) is recommended. Stop-on-Word mode may be used for exceptional circumstances where Stop mode is inappropriate.

5.4.3.5 Use of Q for Block Transfers, Supplementary Information.
Table K5.4.3.5 summarizes the characteristics of modules and controllers to be used in Stop and Stop-on-Word modes. It also lists the expected action of the computer or other system controller that initiates the blocks.

5.4.4 Command Accepted X

> Whenever a module is addressed during a command operation, it must generate $X = 1$ on the Command Accepted bus line if it recognizes the command as one that it is equipped to perform, either within the module or in association with external equipment. The signal on the X line must reach a steady state before $S1$ and must be maintained until $S2$. Controllers that accept the X signal must do so at the time of $S1$.

The signal $X = 0$ should indicate a serious malfunction, for example, a module that is not present, not powered, lacks external connections, or is not equipped to perform the required action. The action taken by the controller in response to $X = 0$ may, for example, be to call for intervention by the operator or operating system.

Modules designed earlier when the X line was reserved for future allocation can be adapted for use in systems that depend on $X = 1$ by generating $X = N_i$.

SUPPLEMENTARY INFORMATION:
Command Accepted X Disabling. CAMAC computer interfaces, (including crate controllers) that include provisions for monitoring the Command Accepted X response should also include a mode of operation in which an $X = 0$ response does not result in an automatic alarm. This mode is necessary to permit "normal" operation of a system that includes early plug-in units that do not have provision for generating or transmitting the Command Accepted signals. Such plug-in units always "respond" with $X = 0$. When performing an Address Scan block transfer, the combination $Q = 0$, $X = 0$ should not result in an automatic system alarm.

Interpretation of Command Accepted X Response for Q-Controlled Block Transfers

(1) *Address Scan (Section 5.4.3.1).* During the operation by a controller of an Address Scan algorithm, there are two conditions where $X = 0$ does not indicate a malfunction. In both these conditions, $Q = 0$ responses are also generated. This means that, during an Address Scan, controllers should interpret $X = 0$ as an indication of serious malfunction only if $Q = 1$. These two conditions are described as follows:

As stated in Section 5.4.3.1, the addressing of each register in a sequence intended to be used in the Address Scan mode of block transfer must result in a response of $Q = 1$. The addressing of the first register location (if any) beyond the sequence must result in a $Q = 0$ response. This is the mechanism used to locate the end of the sequence. If the first location beyond the sequence is vacant, addressing it also results in an $X = 0$ response. However, if this register location is used for a non-Address Scan register, then, as it is addressed, an $X = 1$ response is proper, even though the Q response must be 0. ($Q = 1$, $X = 0$ is always a fault condition).

The operation of an algorithm used by the controller during Address Scan may result in addressing an empty station. The response to such a command is always $X = 0$.

(2) *Repeat (Section 5.4.3.2)*. During Repeat mode, the state of the Q response conveys information on the readiness of the addressed register to participate in a data transfer. ($Q = 1$ means ready; $Q = 0$ means not ready). Regardless of the logic state of the Q response, the Command Accepted response should be $X = 1$.

(3) *Stop (Section 5.4.3.3)*. During Stop mode the state of the Q response conveys information as to whether the transfer of the current data word to or from the addressed register is within or without a data block. ($Q = 1$ means within; $Q = 0$ means without.) Regardless of the logic state of the Q response, the Command Accepted response should be $X = 1$.

5.5 Common Controls Z, C, I.

During unaddressed Dataway operations either Initialize Z or Clear C is generated by the controller and received by each unit connected to the appropriate bus line.

The common control signal Inhibit I is not associated with Dataway operations. It may be generated at any time and is received by each unit connected to the I bus line.

> The Initialize Z and Clear C signals must be accompanied by Busy B and Strobe $S2$ signals, in a timing sequence as described in Section 7.1.3.2 and shown in Fig 10. The sequence is permitted to include Strobe $S1$, but units must not rely on the generation of $S1$ with Z or C.

5.5.1 Initialize Z.
The Initialize signal is intended to be used during system startup.

> Initialize must have absolute priority over other signals. In response to $Z = 1$ all data and control registers must be set to a defined initial state, all LAM status registers must be reset, and, if possible, all LAM requests must be disabled.
>
> Units that generate Z must also initiate a sequence including Busy, Strobe $S2$, and Inhibit (see Section 5.5.2).
>
> Units that accept Z must gate it with Strobe $S2$ as a protection against spurious signals on the Z line.

5.5.2 Inhibit I

> The signal $I = 1$ must inhibit any feature to which it is connected in a module.

The designer is free to choose which activities (for example, data taking) within a module are inhibited by this signal. The signal may be gated or routed within the module, for example, as a result of a previous command or by patch wiring.

> When any unit generates the Initialize signal $Z = 1$, it must also generate $I = 1$. The Inhibit signal accompanying Z must be established by time t_1 (see Fig 10) and must be maintained for at least the duration of the Z signal. All units that generate I and can maintain $I = 1$ must respond to $Z \cdot S2$ by generating and maintaining $I = 1$ until specifically reset.

5.5.3 Clear C

> The signal $C = 1$ must clear all registers and bistables to which it is connected.
>
> Units that generate C must also initiate a sequence including Busy and Strobe $S2$.
>
> Units that accept C must gate it with Strobe $S2$ as a protection against spurious signals on the C line.

The designer is free to choose which registers and bistables are cleared in response to $C \cdot S2$. The signal may be gated or routed within the module, for example, as a result of a previous command or by patch wiring.

5.6 Nonstandard Connections P1-P7.
Five contacts $P1$-$P5$ on the Dataway connector at each normal station, and seven contacts $P1$-$P7$ at the control station, are available for unspecified uses.

5.6.1 Free Bus Lines P1, P2

> The contacts $P1$ and $P2$ at all normal stations must be linked by two Free bus lines.
>
> Each plug-in unit is permitted to generate signals on either or both of these lines, or to accept signals from them. Within the plug-in unit there must be means by which any access to these lines can be disconnected or disabled.
>
> Signals on the Free bus lines must either conform to Section 7.1.4 and Table 7 (with distributed pull-ups and freedom to vary the number of inputs and outputs) or to Section 7.1.2 and Table 6 (with one pull up on each line, not necessarily in the controller, and with current standards as for Read or Write lines).

No standard uses are defined for the Free bus lines. Conflicts between various uses can be resolved by making appropriate disconnections within units (for example, by wired or plug-in links).

In early versions the contacts P1 and P2 at normal stations were defined as individual patch points. Therefore some older units may have used these contacts in ways inconsistent with the bused connections required by this standard.

5.6.2 *Patch Contacts P3-P7.* Contacts P3-P5 at each normal station and P1-P7 at the control station are not wired to Dataway lines. They are available for patch connections to other of these contacts, to optional patch points on certain Dataway lines, to the 0 V line, or to external equipment.

> Patch connections must not be essential for the operation of the main features of general-purpose units.
>
> The signals on patch connections must conform to Section 7.1.4 and Tables 5 and 7.

The patch contacts on the Dataway connector socket may either be directly accessible for making patch connections or may be wired to separate patch points. An earthed patch point, or access to the 0 V line, should be provided at each station. Patch connections may also be made to the I, C, N, and L lines at each station, but the permitted loadings on these lines (see Table 6) restrict the number of such additional connections.

5.7 Power Lines

> The Dataway must include lines for all the mandatory and additional power supplies shown in Table 1.

The Dataway need not include lines for the supplementary +6 V and −6 V supplies (Y2 and Y1), but its construction should be such that bus lines and wiring in accordance with Section 10 can be installed on Y1 and Y2 if required.

Details of the voltage tolerances and permitted loadings of the power lines are given in Section 8. See also Supplementary Information in Section 8 regarding supplementary 6 V power lines Y1 and Y2.

6. Dataway Commands

A command consists of signals on the Station Number lines, the Subaddress lines, and the Function lines. During each command operation the crate controller generates the appropriate command, accompanied by the Busy signal $B = 1$ and by the two Strobes $S1$ and $S2$. Data may be transferred on the Read or Write lines in response to the command. Modules generate the Command Accepted signal $X = 1$ when they recognize the command (see Section 5.4.4). Modules may also transfer one bit of status information on the Response line (see Section 5.4.3).

The following sections define the mandatory actions by modules and controllers in response to each command, including mandatory transfers of data and status information via the Dataway. The term register is used here, and in the summary of the Function codes in Table 4, to indicate an addressable data source or receiver, without necessarily implying that it has a data storage property.

> During a Dataway command operation, modules and controllers must perform the actions specified for the particular command. They may also perform additional internal actions, but these must not involve transferring to or from the Dataway any data or status information other than that specified for the command. Additional internal actions must not convert one standard command into another standard command.

The function codes $F(0)-F(3)$, $F(9)$, $F(11)$, $F(16)-F(19)$, $F(21)$, and $F(23)$ allow the registers in a module to be divided into two distinct sets, known as Group 1 and Group 2, so that it is possible to operate on two sets of sixteen registers. Within each group the appropriate register is selected by the subaddress. Information concerning status or system organization, or requiring restricted access, should be held in Group 2 registers (for example, see Section 5.4.1.2).

If a module allows a descriptor (module characteristic) to be read, the command used should be Read Group 2 Register, $F(1) \cdot A(15)$.

6.1 Read Commands: Function Codes $F(0)$-$F(7)$.

Read commands are identified by the combination $F16 = 0$ and $F8 = 0$ in the function code. All read commands involve the transfer of data and status information from a module to the controller via the Read, Q, and X lines (see mandatory statements in Sections 5.3.2, 5.4.3, and 5.4.4).

Recommendations for the use of the Q signal in Read operations are given in Section 5.4.3.

6.1.1 Read Group 1 Register, Code F(0)

> This command transfers to the controller the contents of a register in the first group in the module. The contents of the register are not changed.

The required register within the group is selected by the subaddress.

6.1.2 Read Group 2 Register, Code F(1)

> This command transfers to the controller the contents of a register in the second group in the module. The contents of the register are not changed.

The required register within the group is selected by the subaddress.

6.1.3 Read and Clear Group 1 Register, Code F(2)

> This command transfers to the controller the contents of a register in the first group in the module. The contents of the register are cleared at time $S2$.

The required register within the group is selected by the subaddress.

6.1.4 Read Complement of Group 1 Register, Code F(3)

> This command transfers to the controller the ones complement of the contents of a register in the first group in the module. The contents of the register are not changed.

The required register within the group is selected by the subaddress.

The command is provided mainly as a means of error detection. A read transfer with $F(0)$ or $F(2)$ can be checked by preceding it with a transfer from the same register with $F(3)$. The two data words received by the controller should be complementary. A write transfer with $F(16)$ can be checked by following it with a read transfer from the module with $F(3)$. The data words sent and received by the controller should be complementary.

6.1.5 Other Read Commands, Codes F(4)-F(7).

These commands transfer the contents of a register in the module to the controller. Codes $F(4)$ and $F(6)$ are available as nonstandard functions. Codes $F(5)$ and $F(7)$ are reserved for extensions of the standard functions.

6.2 Control Commands: Function Codes $F(8)$-$F(15)$.

This first group of control commands is identified by $F8 = 1$ and $F16 = 0$ in the function code. Information is not transferred on either the R or W lines. However, status information may be conveyed on the Q line in response to any of these commands. The signal on the Q line is permitted to change at any time. It is strobed into the controller at time $S1$ and may, except in operations with code $F(8)$, be reset by strobe $S2$. There is a risk that information can be lost due to Q signals appearing between $S1$ and $S2$.

6.2.1 Test Look-at-Me, Code F(8)

> This command transfers to the controller a signal on the Response line Q, representing the state of the L signal or a LAM request in the module (see Sections 5.4.1 and 5.4.1.1). The response must be $Q = 0$ if the feature is in the 0 state or is prevented, by masking or gating, from contributing to a 1 state L signal. The LAM status must not be reset by this command.

The feature to be tested (the L signal or a particular LAM request) is selected by the subaddress.

6.2.2 Clear Group 1 Register, Code F(9)

> This command clears the contents of a register in the first group in the module.

The required register within the group is selected by the subaddress.

6.2.3 Clear Look-at-Me, Code F(10)

> This command resets a LAM status in the module (see Section 5.4.1).

The required LAM status is selected by the subaddress. The Q signal may indicate the status of any selected feature in the module.

6.2.4 Clear Group 2 Register, Code F(11)

> This command clears the contents of a register in the second group in the module.

The required register within the group is selected by the subaddress.

6.2.5 Other Control Commands, Codes F(12)-F(15).
These commands do not transfer data on the R or W bus lines. Codes F(12) and F(14) are available for use as nonstandard functions. Codes F(13) and F(15) are reserved for extensions to the standard functions.

6.3 Write Commands: Function Codes F(16)-F(23).
Write commands are identified by the combination F16 = 1 and F8 = 0 in the function code. All write commands involve the transfer of data from the controller to a module via the Write bus lines, and status information from a module to the controller via the Q and X lines (see mandatory statements in Sections 5.3.1, 5.4.3, and 5.4.4).

Recommendations for the use of the Q signal in write operations are given in Section 5.4.3.

6.3.1 Overwrite Group 1 Register, Code F(16)

> This command forces each bit of a register in the first group in the module to the same state as the corresponding data bit transmitted by the controller.

The required register within the group is selected by the subaddress.

The effect of this command is to write data bit W_i into bit M_i of the Group 1 register. Thus:

$$M_i := W_i$$

6.3.2 Overwrite Group 2 Register, Code F(17)

> This command forces each bit of a register in the second group in the module to the same state as the corresponding data bit transmitted by the controller.

The required register within the group is selected by the subaddress.

The effect of this command is to write data bit W_i into bit M_i of the Group 2 register. Thus:

$$M_i := W_i$$

6.3.3 Selective Set Group 1 Register, Code F(18)

> This command operates on selected bit positions of a register in the first group in the module. The bit positions are selected by 1 bits in a data word transmitted by the controller, and their contents are set to the 1 state. The contents of unselected bit positions are unchanged.

The required register within the group is selected by the subaddress.

The effect of this command is to form the inclusive OR function of the data bit W_i and the bit M_i in the Group 1 register. Thus:

$$M_i := W_i + M_i$$

This can also be regarded as overwriting the 1 bits from the data word.

6.3.4 Selective Set Group 2 Register, Code F(19)

> This command operates on selected bit positions of a register in the second group in the module. The bit positions are selected by 1 bits in a data word transmitted by the controller, and their contents are set to the 1 state. The contents of unselected bit positions are unchanged.

The required register within the group is selected by the subaddress.

The effect of this command is to form the inclusive OR function of the data bit W_i and the bit M_i in the Group 2 register. Thus:

$$M_i := W_i + M_i$$

This can also be regarded as overwriting the 1 bits from the data word.

6.3.5 Selective Clear Group 1 Register, Code F(21)

> This command operates on selected bit positions of a register in the first group in the module. The bit positions are selected by 1 bits in a data word transmitted by the controller, and their contents are cleared to the 0 state. The contents of unselected bit positions are unchanged.

The required register within the group is selected by the subaddress.

The effect of this command is to form the function of the data bit W_i and the bit M_i in the Group 1 register:

$$M_i := \overline{W}_i \cdot M_i$$

6.3.6 *Selective Clear Group 2 Register, Code F(23)*

> This command operates on selected bit positions of a register in the second group in the module. The bit positions are selected by 1 bits in a data word transmitted by the controller, and their contents are cleared to the 0 state. The contents of unselected bit positions are unchanged.

The required register within the group is selected by the subaddress.

The effect of this command is to form the function of the data bit W_i and the bit M_i in the Group 2 register:

$$M_i := \overline{W}_i \cdot M_i$$

6.3.7 *Other Write Commands, Codes F(20) and F(22).* These codes are available for use as nonstandard functions that operate on some or all bits of a register in the module in accordance with the data transmitted by the controller.

6.4 Control Commands: Function Codes F(24)-F(31). This second group of control commands is identified by $F8 = 1$ and $F16 = 1$ in the function code. Information is not transferred on either the R or W bus lines. However, status information may be conveyed on the Q line in response to any of these commands. The signal on the Q line is permitted to change at any time. It is strobed into the controller at time $S1$ and may, except in operations with code $F(27)$, be reset by Strobe $S2$. There is a risk that information can be lost due to Q signals appearing between $S1$ and $S2$.

6.4.1 *Disable, Code F(24)*

> This command disables a feature of the module or masks off a signal. The action is initiated by Strobe S1 or S2.

The feature that is disabled, for example a LAM request or data input, is selected by the subaddress.

The Disable command is preferably used to disable a feature that is enabled by another command, such as Enable $F(26)$.
(Compare with Execute, Section 6.4.2, which does not form a pair with Enable.)

6.4.2 *Execute, Code F(25)*

> This command initiates or terminates an action when Enable or Disable is not appropriate. The initiation or termination occurs at the time of Strobe $S1$ or $S2$. The Execute command must not be used to set a feature of the module that requires a Disable command $F(24)$ to reset it, nor to reset a feature that requires an Enable command $F(26)$ to set it.

The action that is to be executed, or the feature of the module to which it is to be applied, is selected by the subaddress.

Execute may be used, for example, to initiate the generation of a pulse. The operation Increment Preselected Registers, which was defined for $F(25)$ in earlier specifications, is one of the possible uses of the Execute command.

6.4.3 *Enable, Code F(26)*

> This command activates or enables a feature of the module or unmasks a signal. The action is initiated by Strobe $S1$ or $S2$.

The feature that is to be enabled, for example, a LAM request or data input, is selected by the subaddress.

The Enable command is preferably used to enable a feature that is disabled by another command, such as Disable $F(24)$.
(Compare with Execute, Section 6.4.2, which does not form a pair with Disable.)

6.4.4 *Test Status, Code F(27)*

> This command produces on the Q line a response corresponding to the status of a feature of the module. The feature, which is selected by the subaddress, may be a LAM status but must not be a LAM request or L signal (use Test Look-at-Me, Section 6.2.1). The feature must not be reset by the Test Status command.

6.4.5 *Other Control Commands, Codes F(28) to F(31).* These commands do not transfer information on the *R* or *W* bus lines. Codes *F*(28) and *F*(30) are available for use as nonstandard functions. Codes *F*(29) and *F*(31) are reserved for extensions to the standard functions.

6.5 External Representation of the Command. A command is represented on the Dataway by the 5 b function code, the 4 b subaddress code, and signals on the appropriate *N* lines. The standard does not define the form in which the command should be transmitted externally (for example, between a computer and a crate). It will generally be convenient to use the same function and subaddress codes. An external 5 b code for *N* could be more convenient than the 24 b form used internally. For example, the binary code 00001 could correspond to Station Number 1 and therefore generate a signal on line *N*1. The other station numbers would then correspond to the binary codes in sequence.

Fewer than thirty-two codes are needed for addressing individual stations. Spare codes are therefore available, for example, to select multiaddressing modes. For example, one code may address all modules simultaneously, and another may allow the *N* lines to be controlled by a Station Number register. Other codes may be used to address registers, etc, within the controller.

7. Signal Standards

The standards specified in this section apply to signals into and out of plug-in units through:

(1) The Dataway (including timing standards for the main signals associated with Dataway operations)

(2) Nonstandard connections *P*1–*P*7 via the Dataway connector

(3) Other connectors on the front panel or at the rear of the unit above the Dataway (with separate standards for terminated and unterminated digital signals and for analog signals)

The signal standards do not restrict the freedom of designers to use other signals or conventions within units.

7.1 Digital Signals on the Dataway. The potentials for the binary digital signals on the Dataway lines have been defined to correspond with those for compatible current sinking logic devices [for example, the TTL (transistor-transistor logic) and DTL (diode-transistor logic) series]. The signal convention has, however, been chosen to be negative logic. The high state (more positive potential) corresponds to logic 0 and the low state (near ground potential) corresponds to logic 1. Intrinsic OR outputs are thus available from standard product ranges.

It is an essential feature of the Dataway that many units may have signal outputs connected to the Read, Command Accepted, and Response lines. Outputs onto these lines therefore require intrinsic OR gates. The same principle is extended to other lines (command, Write, etc) in order to allow more than one controller-like unit in a crate.

> Signal outputs from all plug-in units onto all Dataway lines must be delivered through intrinsic OR gates. Each line must be provided with an individual pull-up current source to restore the line to the 0 state in the absence of an applied 1 signal.
>
> The rise and fall times at signal outputs to Dataway lines must not be less than 10 ns, in order that cross-coupling of signals on the Dataway is not excessive.

7.1.1 *Voltage Standards for Dataway Signals*

> All Dataway signals must conform to the voltage levels shown in Table 5.

Table 5
Voltage Levels of Dataway Signals

	0 State	1 State
Accepted at input	+2.0 to +5.5 V	0 to +0.8 V
Generated at output	+3.5 to +5.5 V	0 to +0.5 V

7.1.2 *Current Standards for Dataway Signals*

> All Dataway signals must conform to the standards for input and output currents shown in Table 6.

DIGITAL INTERFACE SYSTEM (CAMAC)

ANSI/IEEE
Std 583-1982

Table 6
Standards for Signal Currents Through Dataway
Connectors and for Pull-Up Current Sources

DESIGNATION OF DATAWAY SIGNAL LINE	N	L	Q, R, X	W, A, F, B, Z, C, I	S1, S2
Line in '1' state at +0·5V Minimum current sinking capability (current drawn from line) of each unit generating the signal.	6·4mA	16mA		Controllers 1·6 (25−s)mA 36·8mA typical Other Units 9·6+1·6(25−s)mA 48·0mA typical	58+1·6(25−s)mA 96·4mA typical
Line in '1' state at +0·5V Maximum current fed into line by each unit receiving the signal.	3·2mA each unit, 6·4mA total (Note 1).	Unit with pull-up current source: 11·2mA Units without pull-up current source 1·6mA each: 4·8mA total (Note 1)		1·6s mA	
Line in '0' state at +3·5V Minimum pull-up capability (current fed into line) of the unit with pull-up current source.		100(25−s)µA	2·3mA typical for controllers 2·4mA typical for other units		9·9mA
Line in '0' state at +3·5V Maximum current drawn from line by each unit without pull-up current source.	200µA		100s µA		
Location of pull-up current source	Unit generating the signal.	One unit receiving the signal.		Controller	
Pull-up current Ip, **from positive potential** Line in '1' state at +0·5V		6mA ≤ Ip ≤ 9·6mA		38mA ≤ Ip ≤ 58mA	
Pull-up current Ip, **from positive potential** Line in '0' state at +3·5V		2·5mA ≤ Ip		10mA ≤ Ip	

Where appropriate, the current passing through the Dataway connector of a plug-in unit is defined as a function of the width of the unit (s stations). Values are given, as examples, for typical controllers (s = 2, control station and one normal station) the other units (s = 1).
NOTE 1: Although only the controller and one module are connected directly to each N and L line, additional units may be connected via patch points or auxiliary connectors.

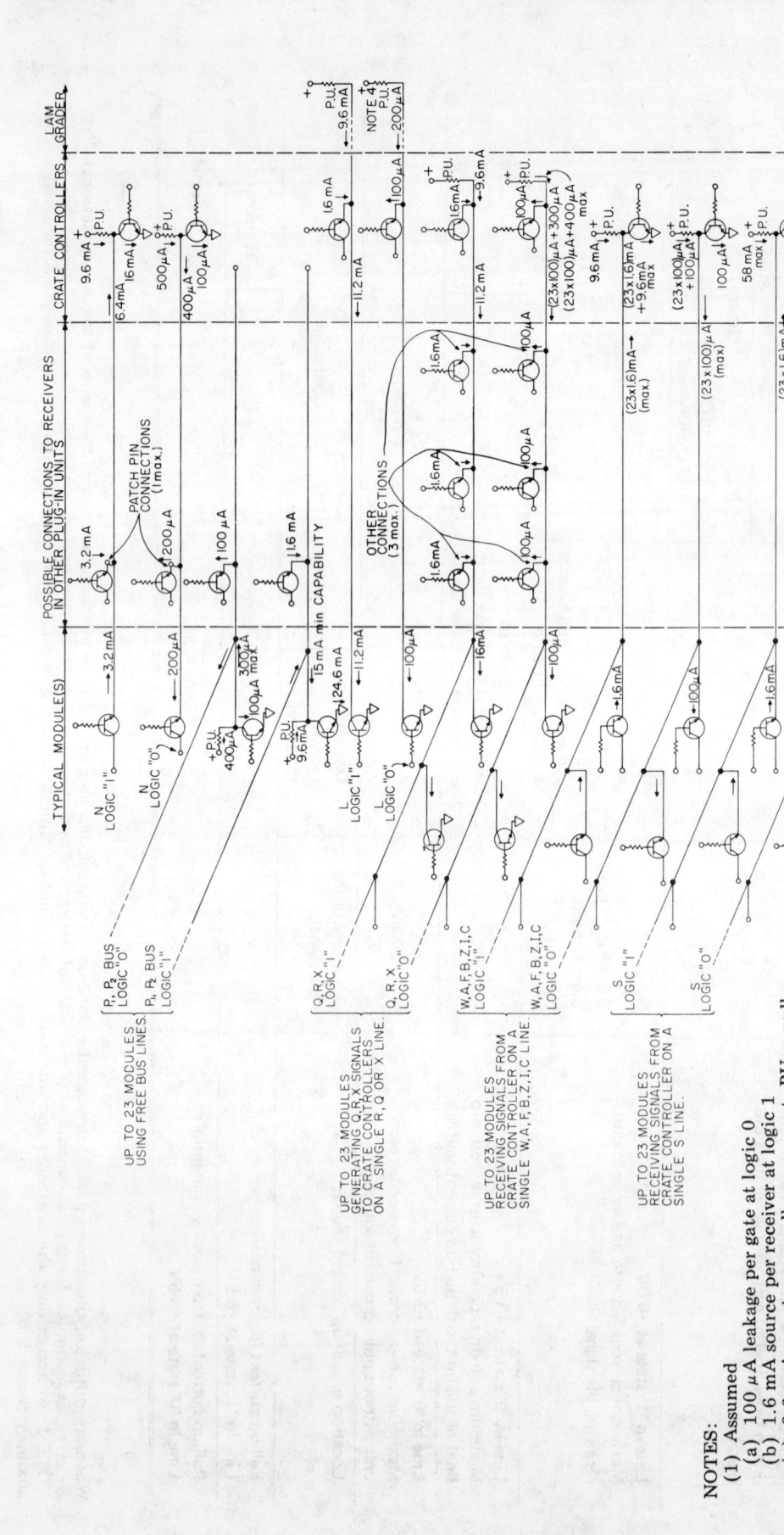

**Fig K7.1.2
Current Distribution, CAMAC Dataway**

NOTES:
(1) Assumed
   (a) 100 μA leakage per gate at logic 0
   (b) 1.6 mA source per receiver at logic 1
   (c) 9.6 mA maximum pull up current, PU = pull up resistor
   (d) crate controller two units wide, other modules one unit wide
(2) Maximum possible connections not shown
(3) ▽ indicates logic common via power return bus
(4) Pull-up capability must be 2.3 mA minimum
(5) P_1 and P_2 may also use W/R current distributions

Pull-up current sources for all standard Dataway bus lines are located in the controller (occupying the control station and at least one other station) so as to ensure that there is one and only one current source per line. The pull-up current sources for the N lines are located in the unit generating the signals and for the L lines in a unit receiving the signals so that the individual lines may be joined or grouped within these units if desired.

The Strobe signals $S1$ and $S2$, which time all actions in modules, have larger pull-up currents than other signals in order to give improved transition times and immunity against cross-coupling from other Dataway lines.

There is no restriction on the number of modules that can be addressed simultaneously (see 5.1.1). The signal output from modules to the Dataway Q, R, and X lines should therefore be capable of operating in intrinsic OR mode (see 7.1). Thus, it should be possible for one module to drive a line to the logic '1' state while all other modules are generating logic '0' outputs to the same line.

SUPPLEMENTARY INFORMATION:

Current Distribution, CAMAC Dataway. Table 6 gives the standard for signal currents through Dataway connectors, and for pull-up current sources, while Table 7 gives the current standards for patch contacts. Fig K7.1.2 is included here to present these current standards pictorially, although the maximum possible connections are not shown.

7.1.3 *Timing of Dataway Signals.* The sequence of events during a Dataway command operation is shown in Fig 9 by means of simplified signal waveforms. Section 7.1.3.1 is an explanation of Fig 9.

The sequence of events during a Dataway unaddressed operation is shown in Fig 10 and an explanation is given in Section 7.1.3.2.

In both figures the shaded areas indicate the permitted variation in the timing of each signal. The vertical edge of each shaded area corresponds to an ideal signal without delay. The sloping edge corresponds to a signal that reaches the appropriate threshold (0.8 V or 2.0 V) after the maximum permitted delay.

> The performance of all plug-in units and Dataway assemblies must be consistent with the timing requirements shown in Figs 9 and 10.

7.1.3.1 *Timing of Dataway Command Operations.* The sequence of events during a command operation is shown in Fig 9.

During the operation command and data signals may take up either the 1 state or the 0 state. For convenience Fig 9 shows only signals that take up the 1 state, but similar timing requirements apply to those that take up the 0 state.

The Busy signal and the various command signals need not occur in exact synchronism, provided each is individually within the shaded areas of the diagram. Similar variation is permitted between the signals on the various data and status lines.

The W, R, Q, and X signals are shown as being maintained until the end of the operation, but a broken line indicates the earliest time at which they are permitted to change as a result of actions initiated by Strobe $S2$. During some operations the Q signal may change at any time.

The L signal is shown for the particular case of a module that inhibits its L signal output in response to a command that does not clear the LAM source (see Section 5.4.1.3). The signal $L_i = 1$ is therefore removed but reappears at the end of the operation.

Time markers t_0-t_{12} in Fig 9 indicate key points at which signal transitions are initiated or reach one of the threshold levels (0.8 V or 2.0 V).

At t_0 the transition of the Busy signal to the 1 state is initiated. Command signals on the N, A, and F lines also take up the 1 or 0 states as appropriate to the command.

At t_1 the Busy signal has reached the 0.8 V threshold, and all the command signals have reached the appropriate thresholds.

During the period t_1-t_2 the addressed module responds to the command, and by t_2 the appropriate X, Q and data signals are initiated. By t_3 at the latest these signals have all reached the appropriate thresholds. Any L signals that are inhibited during the operation have reached the 2.0 V threshold by t_3.

The transition of the $S1$ signal to the 1 state is initiated at t_3 and has reached the 0.8 V threshold by t_4.

At t_5 the transition of the $S1$ signal to the 0 state is initiated and reaches the 2.0 V threshold by t_6.

The transition of the $S2$ signal to the 1 state is initiated at t_6 and has reached the 0.8 V

Table 7
Current Standards for Patch Contacts

State of Line	Current To and From Patch Connections	
	Outputs	Inputs
1 State at +0.5 V	Units must be capable of drawing more than 15 mA from connection when generating 1	Unit must not feed more than 2 mA into connection
	Unit must not feed more than 300 µA into connection when generating 0	
0 State at +3.5 V	Pull-up capability (current fed into connection):	100 µA minimum 300 µA maximum

threshold by t_7. Modules may respond to $S2$ by changing the state of the R, Q, and X signals.

The transition of the $S2$ signal to the 0 state is initiated at t_8 and has reached the 2.0 V threshold by t_9, which is the end of the Dataway operation.

At t_9 the transition of the B signal to the 0 state is initiated, and the command signals may also change from their established states.

At t_{10} the B signal and Command signals have reached the 2.0 V threshold. During the period t_{10}-t_{11} the module responds to the removal of the command. By t_{11} the transitions of the W, R, Q, and X signals to the 0 state are initiated, and the inhibit is removed from the L signal. By t_{12} the L signal has reached the 0.8 V threshold, and all other signals have reached the 2.0 V threshold.

> Controllers must initiate the transitions of the command and Strobe signals at intervals not less than the minimum times shown in Fig 9. Modules must respond to the Command within the time t_1-t_2 and to the strobes in the times t_4-t_5 and t_7-t_8. The electrical characteristics of the Dataway and connections from it into plug-in units must allow transitions between the two threshold levels to take place within the times t_0-t_1, t_2-t_3, etc.
>
> The next Dataway operation must not start before t_9.

In the extreme case when the next operation starts at t_9, the time markers t_0, t_1, t_2 of the new operation coincide with t_9, t_{10}, t_{11} of the previous operation. The command and data signals of one operation may thus be removed while those of the next operation are being established. The Busy signal may be maintained continuously during a sequence of consecutive Dataway operations. Under suitable conditions any command or data signals which have the same state during successive operations may also be maintained. In the extreme case of successive operations with the same command and data there could be a complete absence of signal transitions between t_0 and t_3.

7.1.3.2 *Timing of Unaddressed Operations.* The sequence of events during an unaddressed Clear or Initialize operation is shown in Fig 10.

At t_0 the transition of the Busy signal to the 1 state is initiated. In a Clear operation the transition of the C signal is also initiated at this time. In an Initialize operation the transitions of the Z and I signals are initiated.

By t_1 the B signal and, as appropriate, either Z and I or C have reached the 0.8 V threshold.

The interval t_1-t_6 allows integration of the Z or C signals within the module if required.

At t_6 the transition of the $S2$ signal to the 1 state is initiated. The $S2$ signal is established and removed as described previously. (The $S1$ signal may be generated with timing relative to t_6 as shown in Fig 9.)

The $S2$ signal reaches the 2.0 V threshold by t_9. The transitions of the B signal and C or Z to the 0 state are initiated at t_9 and reach the 2.0 V threshold by t_{10}. The Inhibit signal may be removed at t_9 or, if possible, it is maintained in the 1 state as indicated by the broken line.

**Table 8
Unterminated Signals**

Outputs*	
V_{out}	
Logic 1	Unit must generate 0 V to +0.5 V
Logic 0	Unit must generate +2.4 V to +5.5 V
I_{out}	
Logic 1 at +0.5 V	Unit must draw > 16 mA from connection
Logic 0 at +2.4 V	Unit must feed > 6 mA into connection
Inputs	
V_{in}	
Logic 1	Unit must accept 0 V to +0.8 V
Logic 0	Unit must accept +2.0 V to +5.5 V
I_{in}	
Logic 1 at +0.5 V	Unit must feed < 2.0 mA into connection
Logic 0 at +2.4 V	Unit must feed current into connection or draw < 100 µA from connection

*Not necessarily intrinsic OR

7.1.4 *Digital Signals on Nonstandard Connections*

> Signals on the Free bus lines (contacts *P1* and *P2* at normal stations) must be generated from intrinsic OR outputs and conform to the voltage standards of Table 5. They must conform to the current standards of either Table 7 or Table 6, for Read or Write lines as appropriate (see also Section 5.6.1).
>
> Signals on patch connections using contacts *P3-P5* at normal stations or *P1-P7* at the control station must be generated from intrinsic OR outputs and must conform to the voltage standards of Table 5 and the current standards of Table 7. Disconnected inputs must take up the 0 state.

In Table 7 each input and output has an individual pull-up current source to compensate for leakage current in the 0 state. This allows flexibility in the number of inputs and outputs that can be patched together.

7.2 Other Digital Signals. The standards defined as follows should normally be used for all terminated and unterminated digital signals via connectors on the front panel and at the back of plug-in units above the Dataway. There may, however, be special circumstances requiring the use of other signals, for example, to suit a specific equipment with which the plug-in unit is closely associated. (See also Appendix B.)

7.2.1 *Unterminated Signals.* Unterminated signals should conform to the standard set out in Table 8, unless there are special reasons for using other standards.

> Individual outputs must be able to withstand, without damage, a short-circuit to ground. Outputs through multiway connectors need not withstand a short circuit on all pins simultaneously.
>
> Disconnected inputs must take up the 0 state.

**Table 9
Terminated Signals**

	Logic 0	Logic 1
Outputs must deliver into 50 Ω	−2 to +2 mA	−14 to −18 mA
	Preferred −1 to +1 mA	
Inputs must accept	−4 to +20 mA	−12 to −36 mA

7.2.2 *Terminated Signals.* The characteristic impedance for terminated signals is 50 Ω. Signals terminated in 50 Ω should conform to the standard set out in Table 9, unless there are special reasons for using other standards.

Negative signs indicate currents flowing into an output circuit.

7.3 Analog Signals. Recommended standards for amplitude analog signals are to be processed later.

8. Power Line Standards

The Dataway includes bus lines for mandatory, additional, and reserved power supplies.

Designers of plug-in units may assume that the mandatory lines (+24 V, +6 V, −6 V, −24 V, and 0 V power return) are powered in every installation. (See also Section 10.)

The additional bus lines are provided for special requirements, for example, compatibility with the NIM System. There are heavy current lines for +12 V and −12 V dc, and a low-current line for an independent and isolated Clean Earth (ground) return E. The 12 V lines are not necessarily powered unless specifically required for use.

> The voltages available to plug-in units at the contacts of each Dataway connector must be within the tolerances specified in Table 10. Individual plug-in units, and assemblies of plug-in units within a crate, must not exceed the current loadings specified in Table 10.
>
> The Dataway power lines, and any wiring from them to the point at which power supplies enter the crate, must be capable of carrying the maximum current loadings permitted in the crate. The resistance between any point on the Dataway 0 V power return bus line and the point at which power supplies enter the crate must not exceed 2 mΩ.

The voltage tolerances specified in Table 10 do not define the performance of a suitable power supply unit directly. They take into account factors within the crate, for example, voltage drops due to the internal wiring of the crate and to the Dataway bus line under worst case distributions of current loading.

The maximum current loads for a plug-in unit will often be restricted by power dissipation (see Notes 2 and 3 in Table 10), or by the current-carrying capacity of the Dataway connector (Note 1) and the Dataway power lines. In a crate without forced ventilation the total power dissipation is restricted to 200 W, corresponding to 8 W per station. Under special circumstances this may be increased to 25 W per station, for example, by using forced ventilation or by taking care that the total dissipation is less than 200 W.

Table 10
Power Line Standards

Nominal Voltage on Power Line in Crate	Voltage Tolerance at Dataway Connectors	Maximum Current Loads In the Plug-in per unit width) [See Notes (1) and (3)] (A)	In the Crate [See Note (2)] (A)	Notes
Mandatory				
+24 V dc	±1.0 percent	1 A	6 A	
+6 V dc	±2.5 percent	2 A*	25 A*	
−6 V dc	±2.5 percent	2 A*	25 A*	
−24 V dc	±1.0 percent	1	6	
0 V				
Additional (as required)				
+12 V dc	±1.0 percent			As specified in TID-20893 (Latest revision)
−12 V dc	±1.0 percent			

Comments:
(1) The current carried by each contact of the Dataway connector must not exceed 3 A.
(2) The total power dissipation in a crate without forced ventilation must not exceed 200 W.
(3) The power dissipation in each station must not exceed 8 W in general or 25 W under special circumstances.
*See Section 10.

DIGITAL INTERFACE SYSTEM (CAMAC)

SUPPLEMENTARY INFORMATION:

Power Line Standards. The use of +12 V and −12 V in plug-in units should be avoided since these are nonmandatory voltages and will usually not be available on the crate power buses.

Marking of Voltage and Current Requirements on Plug-in Units. The voltage and current requirements should be clearly and permanently marked on all plug-in units, preferably on the front panels. (See also Section 10.)

Preferred Power Supply Connector and Contact Assignments. A preferred power supply connector is designated for connecting power supplies to the Dataway power lines. The connector is listed in Appendix C, together with contact assignments. Typical power bus and power return bus feed and sense wiring, consistent with these contact assignments, is shown in Figs D1 and D6 of Appendix D.

Typical Power Supply and Ventilation Unit. Appendix D describes a commonly used power supply and ventilating unit for CAMAC applications.

9. Forced Air Ventilation
(Supplementary Information)

Temperature-rise measurements and experience with operating systems have shown that forced air ventilation is generally necessary.

10. Use of Supplementary 6 V Power (Bus Lines Y1, Y2)

In special cases, plug-in units may be used that draw more than 3 A at +6 V or −6 V.

> Such plug-in units are considered special, but they must comply with the current rating for Dataway connector contacts as given in Note 1 of Table 10. The total power dissipation in the station or stations occupied by the plug-in unit must be consistent with Note 3 of Table 10.
>
> They must draw their excess (supplementary) 6 V power, not to exceed 3 A per contact, through the Y1 and Y2 contacts. The total return current in each 0 V contact, including any current from the 24 V and 12 V lines, must not exceed 3 A per contact.

In order to meet this requirement the current distribution between the two 0 V contacts may need to be controlled. An example of a suitable circuit is shown in Fig 12.

Plug-in units requiring supplementary 6 V power should only be installed in special crates in which the Y1 and Y2 contacts are bussed and wired to provide the necessary power. (See also 5.7.)

> Plug-in units requiring supplementary 6 V power must have the supplementary (Y1, Y2) voltage and current requirements clearly and permanently marked on the front panel.
>
> Crates equipped with Y1 and Y2 bus lines and wired for supplementary 6 V power must be clearly marked to indicate this.

The mandatory front panel markings showing supplementary power requirements should be reinforced by a conspicuous warning on the plug-in unit (for example, on the side panel), that it can only be used in crates equipped with Y1 and Y2 supplementary power.

ANSI/IEEE
Std 583-1982

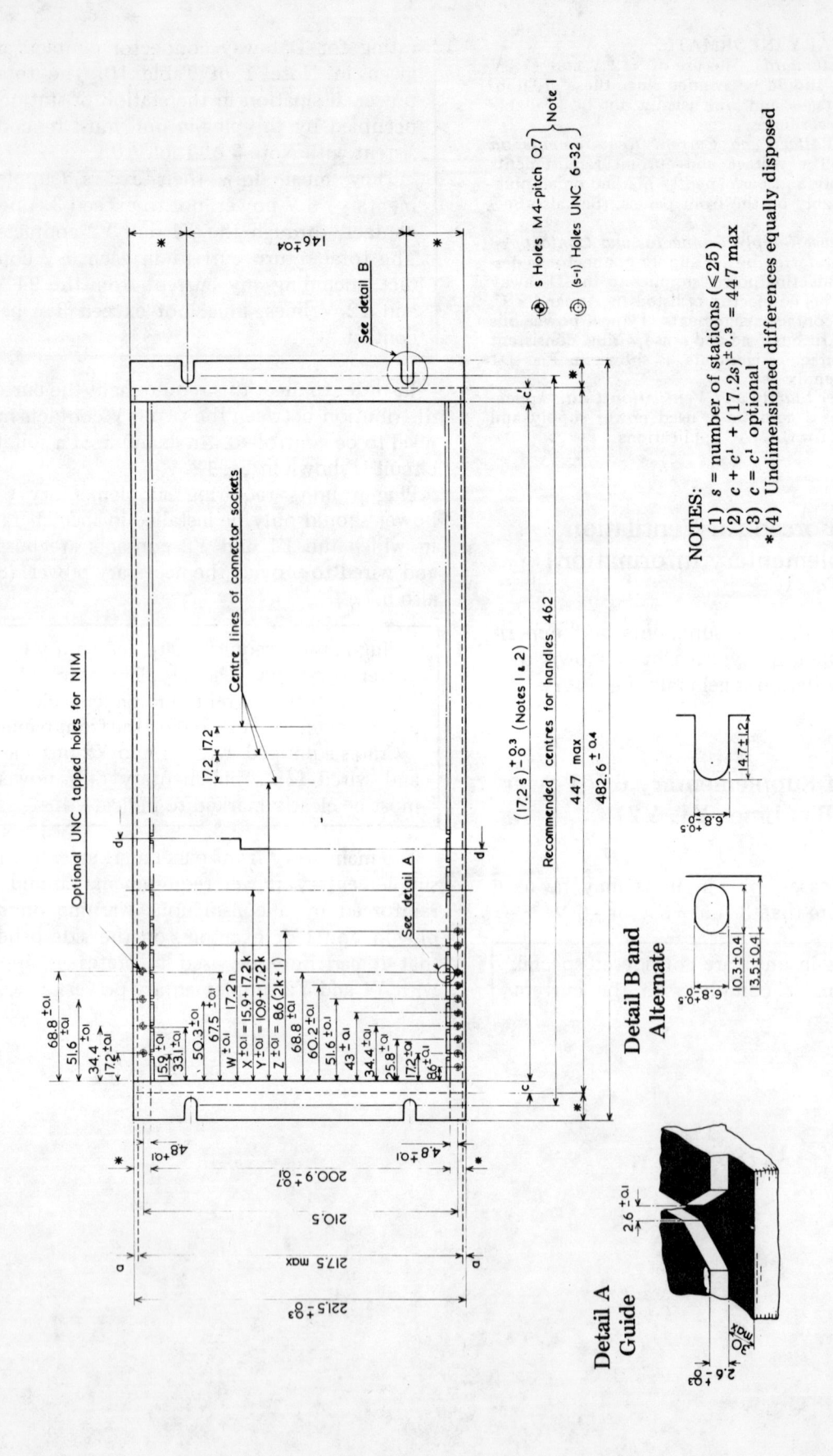

IEEE STANDARD MODULAR INSTRUMENTATION AND

Fig 1
Unventilated Crate: Front View
$k = 0, 1, 2 \cdots (s-1); n = 1, 2, 3 \cdots (s-1)$

NOTES:
(1) s = number of stations (≤ 25)
(2) $c + c^1 + (17.2s)^{+0.3}_{-0.3} = 447$ max
(3) $c = c^1$ optional
*(4) Undimensioned differences equally disposed

DIGITAL INTERFACE SYSTEM (CAMAC)

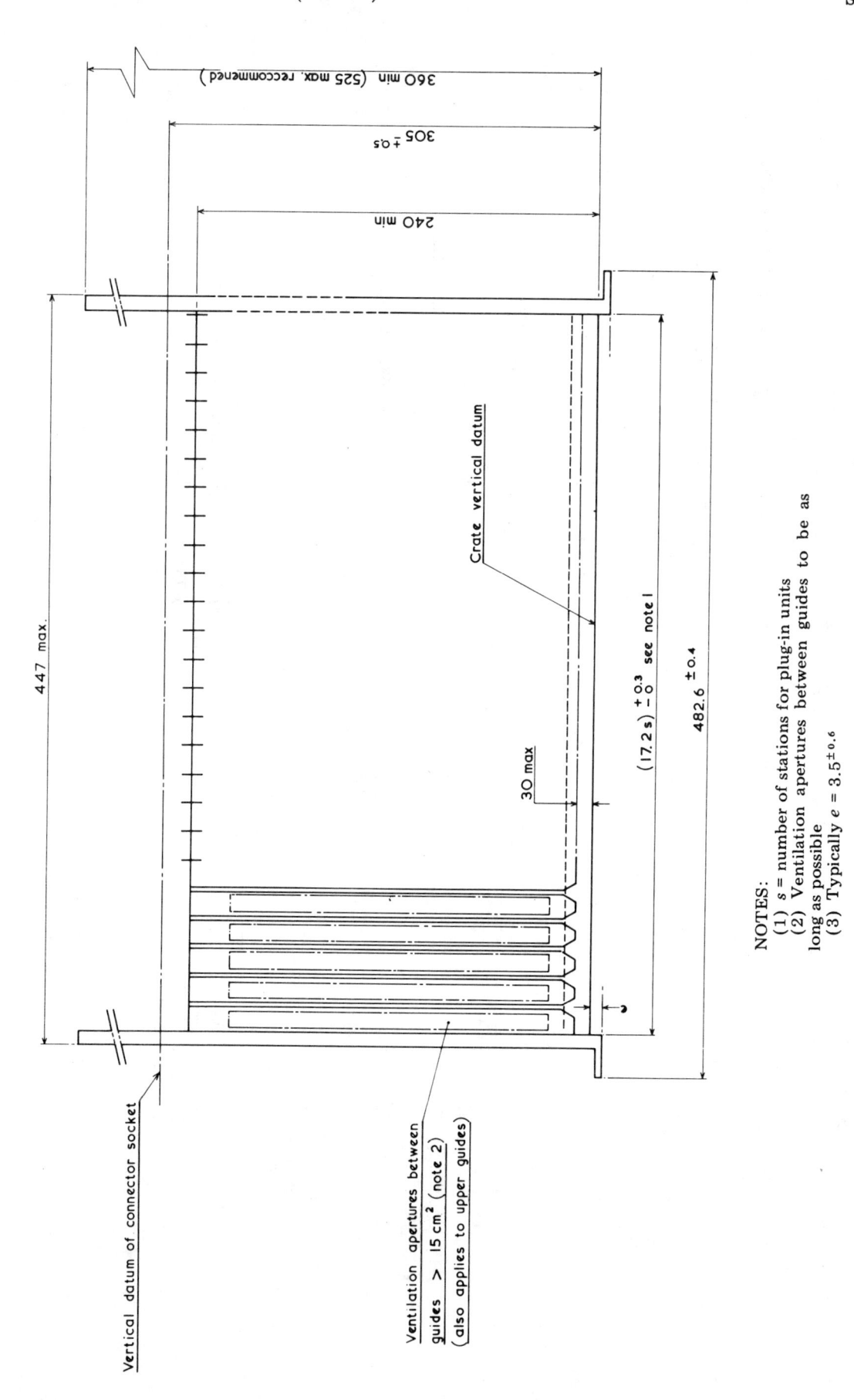

**Fig 2
Plan View of Lower Guides in Crate**

NOTES:
(1) s = number of stations for plug-in units
(2) Ventilation apertures between guides to be as long as possible
(3) Typically $e = 3.5^{\pm 0.6}$

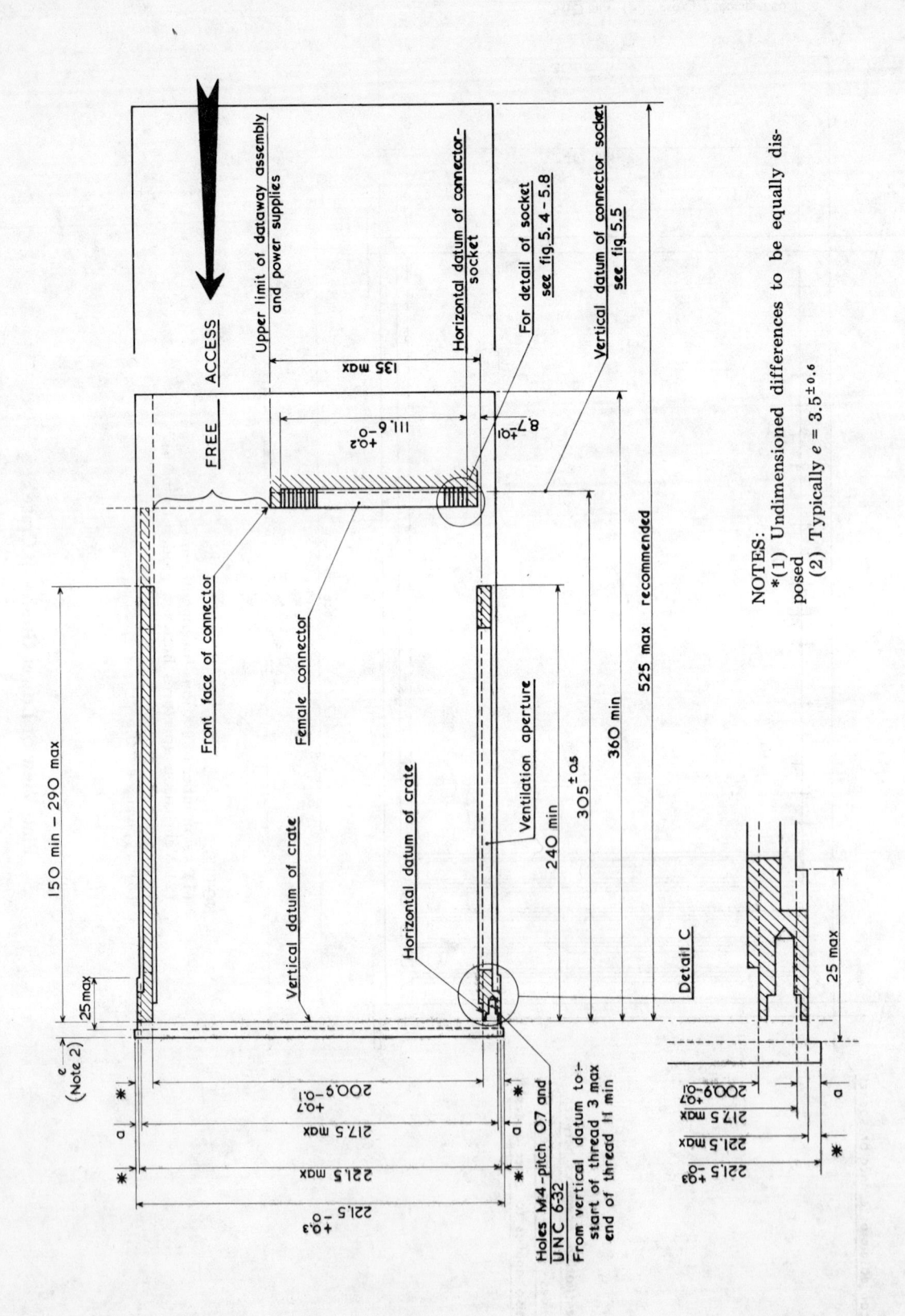

Fig 3
Crate Side View: Section *d-d* (Fig 1)

NOTES:
*(1) Undimensioned differences to be equally disposed
(2) Typically $e = 3.5^{\pm 0.6}$

DIGITAL INTERFACE SYSTEM (CAMAC)

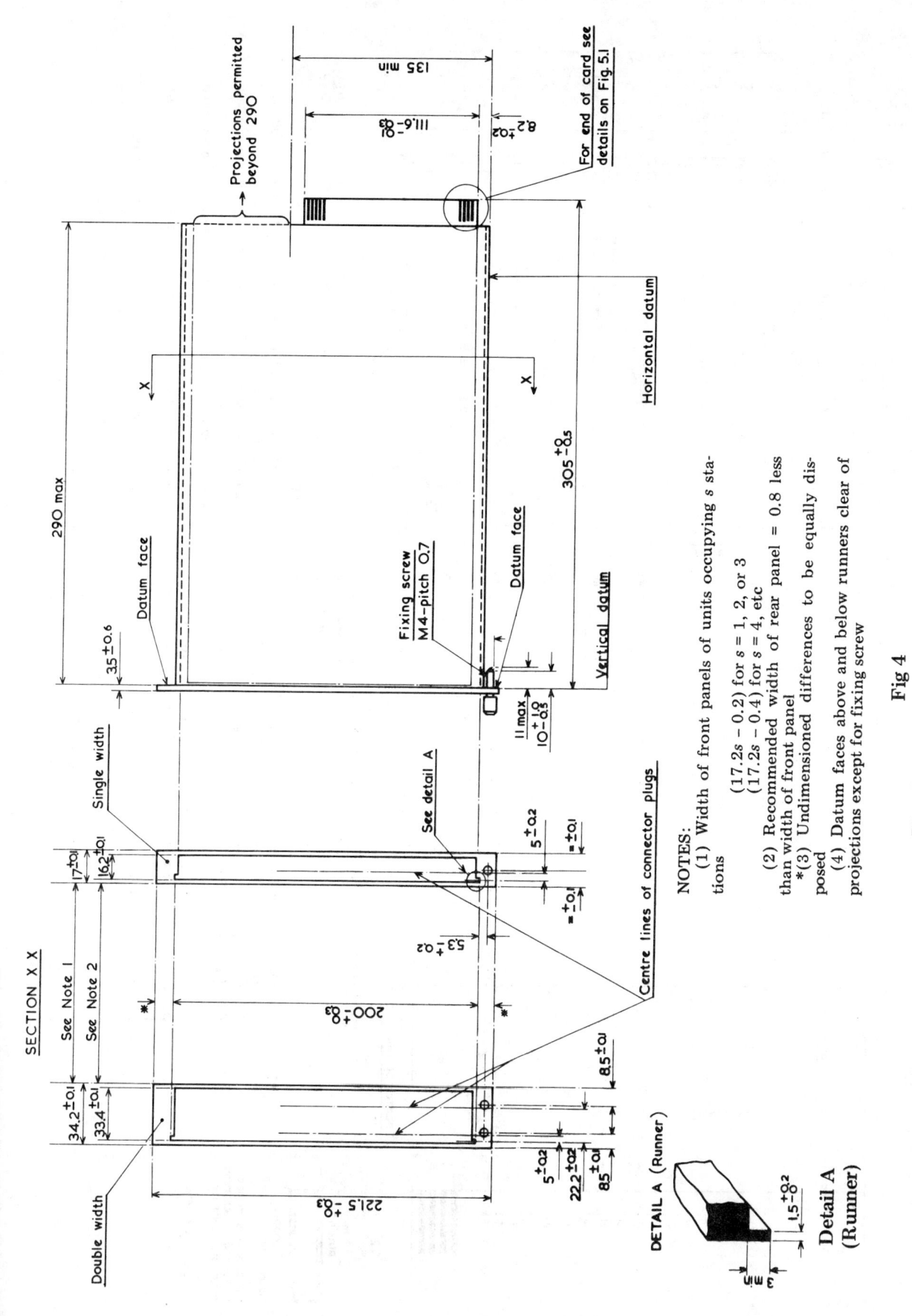

ANSI/IEEE
Std 583-1982

**Fig 4
Plug-In Unit: Side and Rear Views**

NOTES:

(1) Width of front panels of units occupying s stations

$$(17.2s - 0.2) \text{ for } s = 1, 2, \text{ or } 3$$
$$(17.2s - 0.4) \text{ for } s = 4, \text{ etc}$$

(2) Recommended width of rear panel = 0.8 less than width of front panel

*(3) Undimensioned differences to be equally disposed

(4) Datum faces above and below runners clear of projections except for fixing screw

ANSI/IEEE
Std 583-1982

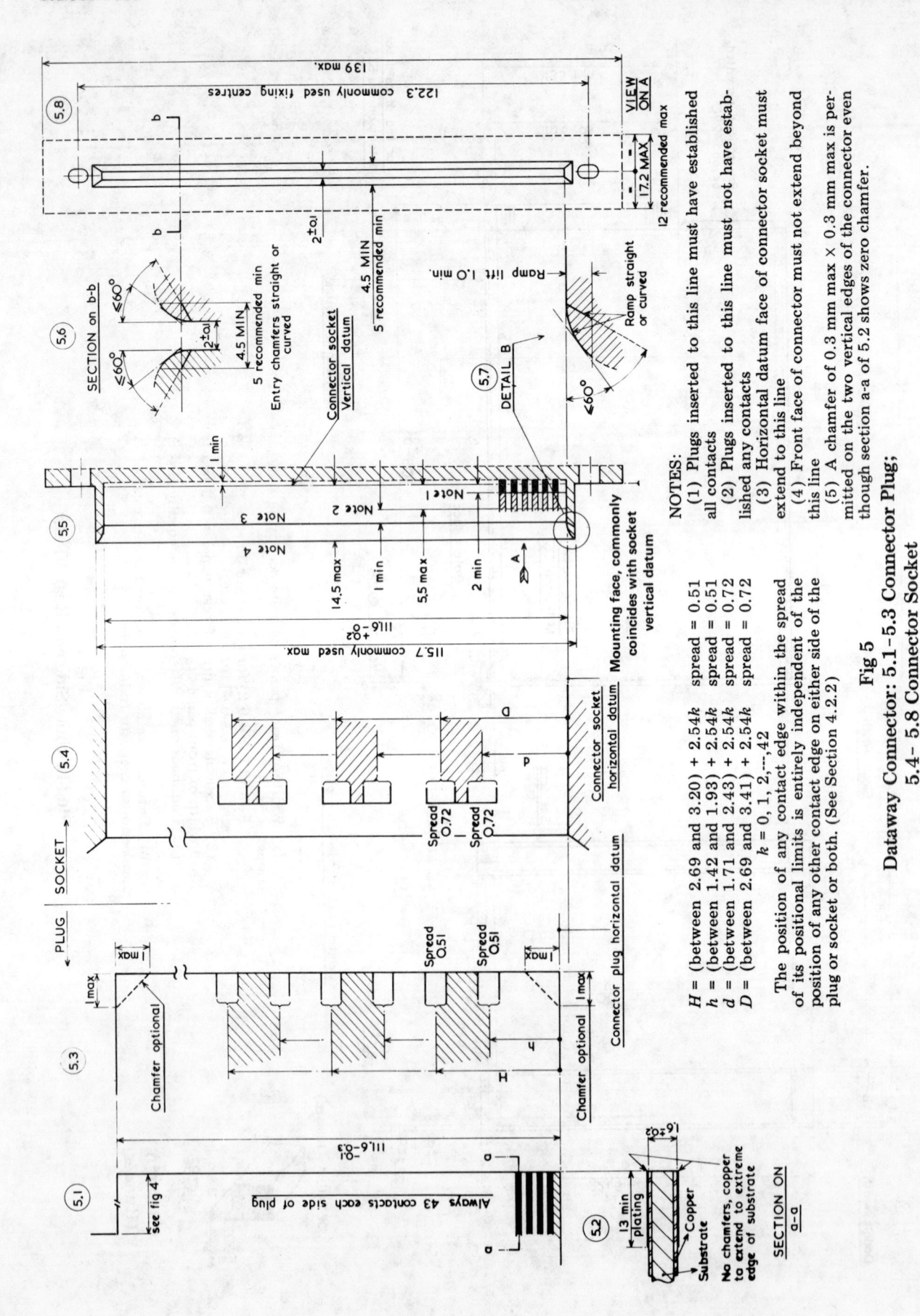

IEEE STANDARD MODULAR INSTRUMENTATION AND

NOTES:
(1) Plugs inserted to this line must have established all contacts
(2) Plugs inserted to this line must not have established any contacts
(3) Horizontal datum face of connector socket must extend to this line
(4) Front face of connector must not extend beyond this line
(5) A chamfer of 0.3 mm max × 0.3 mm max is permitted on the two vertical edges of the connector even though section a-a of 5.2 shows zero chamfer.

H = (between 2.69 and 3.20) + 2.54k spread = 0.51
h = (between 1.42 and 1.93) + 2.54k spread = 0.51
d = (between 1.71 and 2.43) + 2.54k spread = 0.72
D = (between 2.69 and 3.41) + 2.54k spread = 0.72

k = 0, 1, 2, ..., 42

The position of any contact edge within the spread of its positional limits is entirely independent of the position of any other contact edge on either side of the plug or socket or both. (See Section 4.2.2)

Fig 5
Dataway Connector: 5.1–5.3 Connector Plug;
5.4–5.8 Connector Socket

50

NOTES:
(1) For all details not shown see Fig 1
(2) $X = 44.45L$, where $L = 1, 2, 3$, etc
*(3) Undimensioned differences to be equally disposed

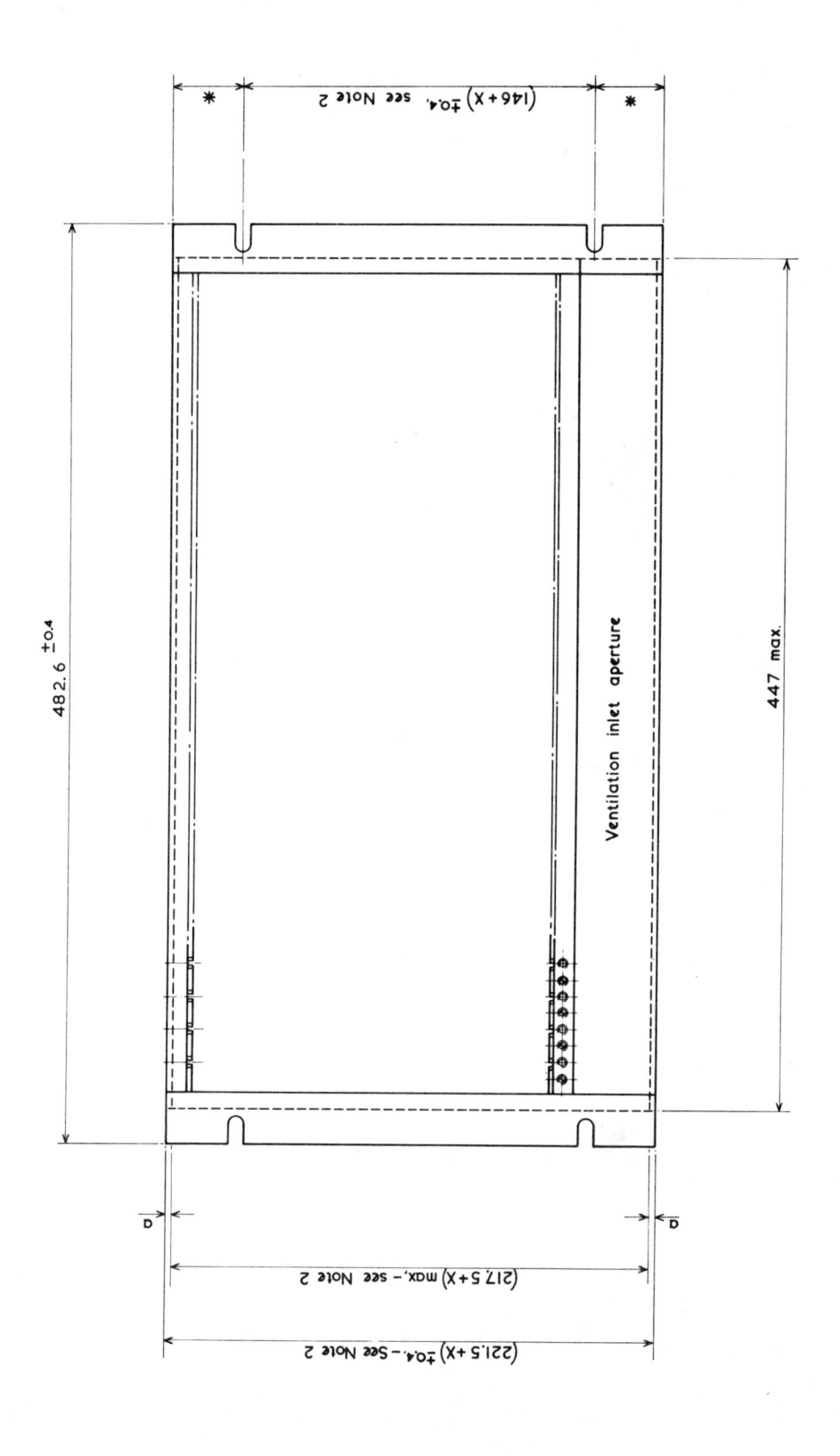

Fig 6
Ventilated Crate: Front View

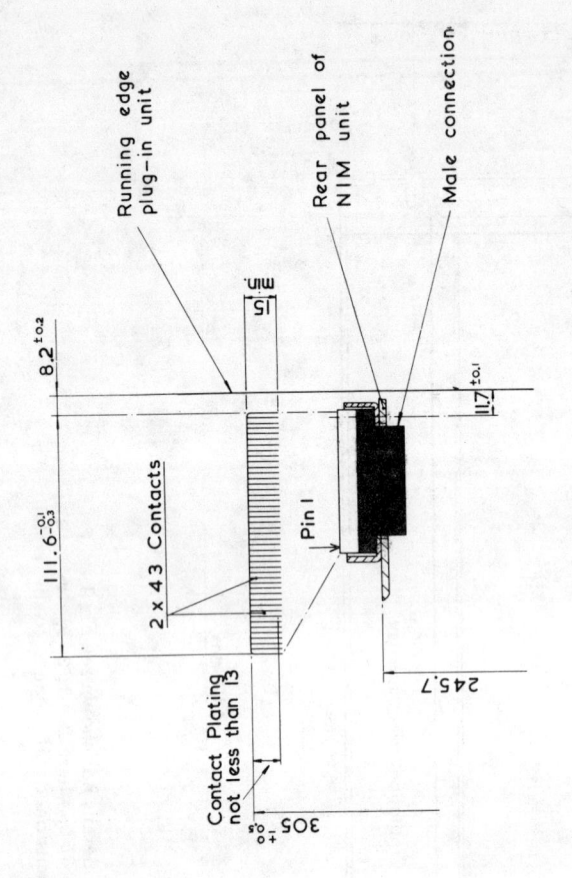

NOTE: For Contact details, see Fig 5

Fig 7
Adaptor for NIM Units

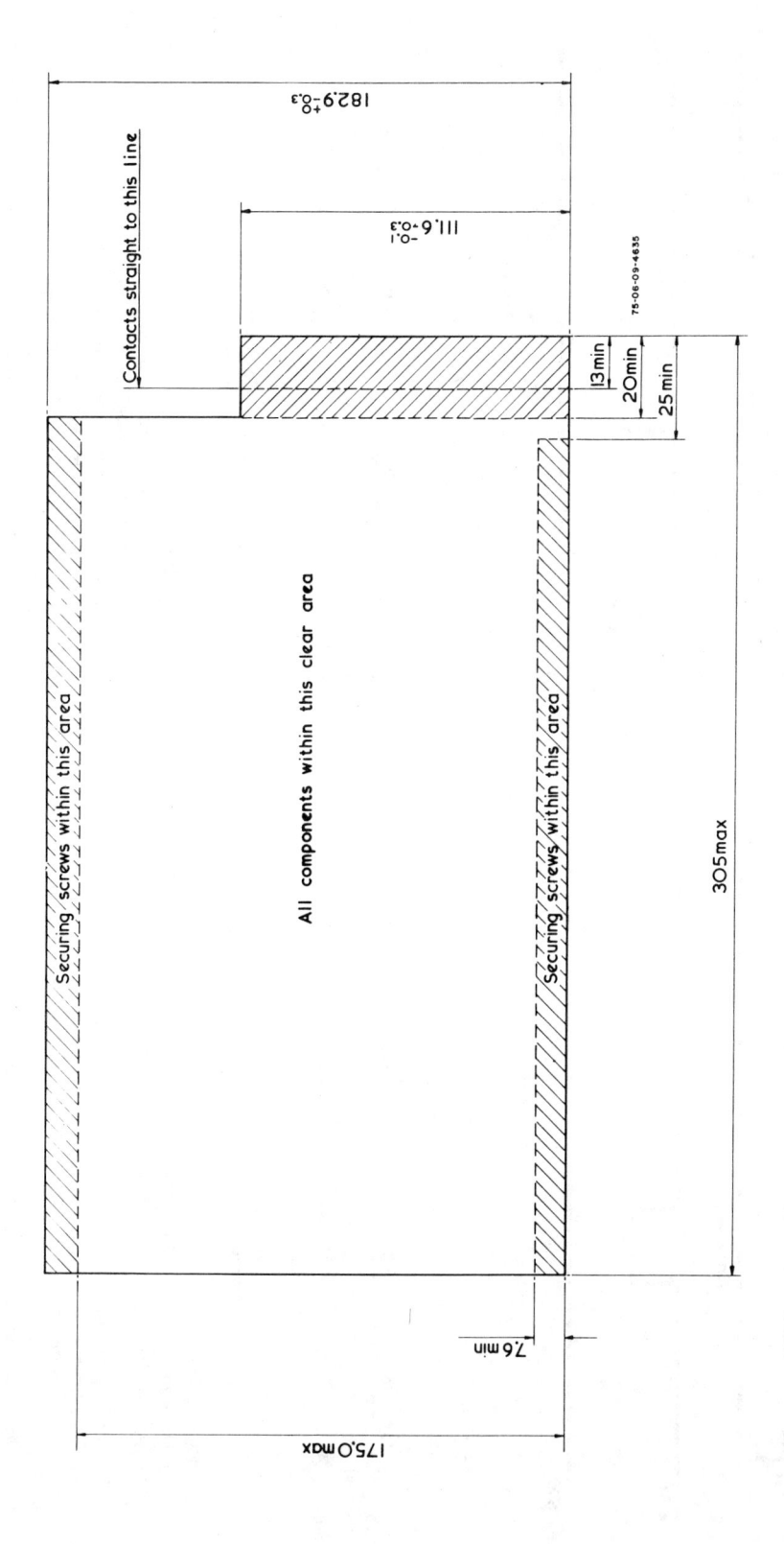

Fig 8
Typical Printed Wiring Card

ANSI/IEEE
Std 583-1982

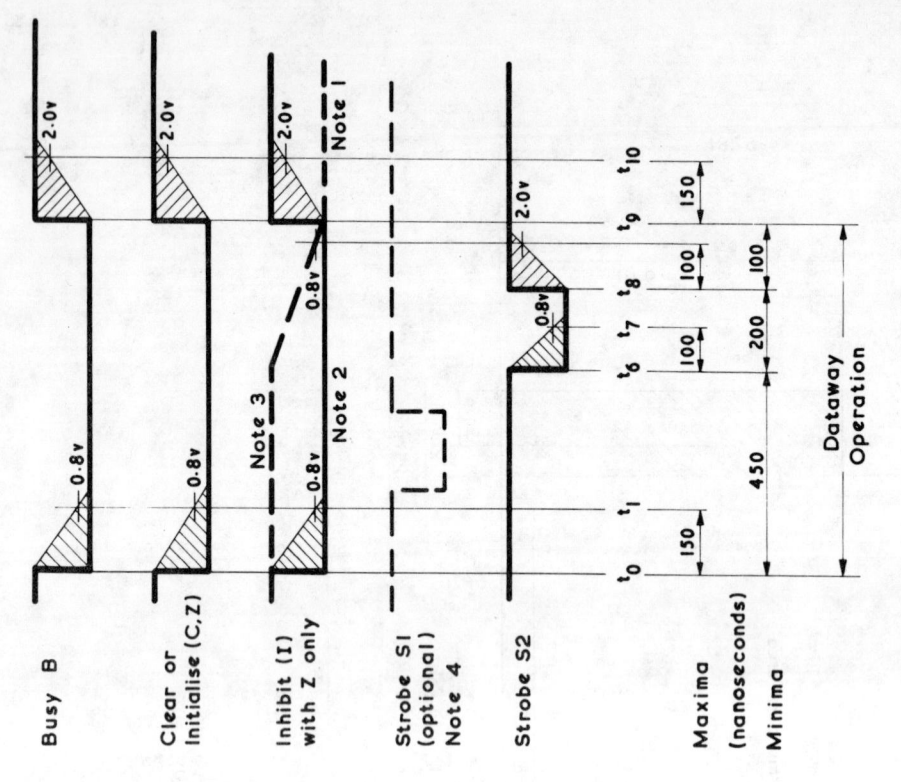

IEEE STANDARD MODULAR INSTRUMENTATION AND

Fig 10
Timing of a Dataway Unaddressed Operation

NOTES:
(1) I preferably maintained
(2) I accompanying Z
(3) I generated in response to $Z \cdot S2$
(4) Other times as in Fig 9
(5) For all signals the minimum rise or fall time is 10 ns. See 7.1.
(6) Signal transition at t_0 or t_9 may be absent if the signals on command or data lines are the same for the immediately preceding or following operations. See 7.1.3.1.

Fig 9
Timing of a Dataway Command Operation

NOTES:
(1) Data and status may change in response to S2
(2) During some operations Q may change at any time
(3) LAM status may be reset during operation
(4) L signal may be maintained during operation
(5) For all signals the minimum rise or fall time is 10 ns. See 7.1.
(6) Signal transition at t_0 or t_9 may be absent if the signals on command or data lines are the same for the immediately preceding or following operations. See 7.1.3.1.

DIGITAL INTERFACE SYSTEM (CAMAC)

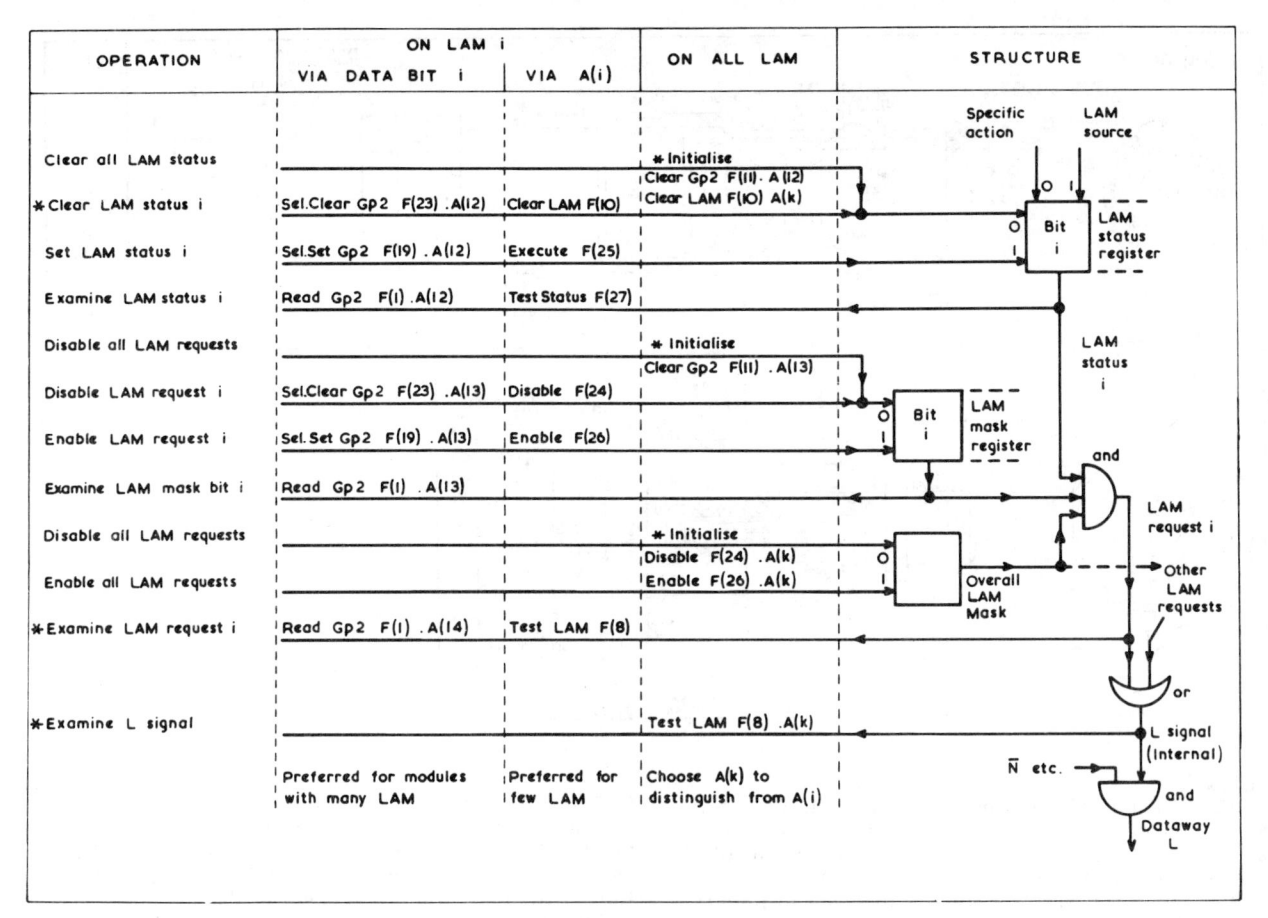

ANSI/IEEE
Std 583-1982

OPERATION	ON LAM i — VIA DATA BIT i	ON LAM i — VIA A(i)	ON ALL LAM	STRUCTURE
Clear all LAM status			*Initialise / Clear Gp2 F(11).A(12)	
*Clear LAM status i	Sel.Clear Gp2 F(23).A(12)	Clear LAM F(10)	Clear LAM F(10) A(k)	
Set LAM status i	Sel.Set Gp2 F(19).A(12)	Execute F(25)		
Examine LAM status i	Read Gp2 F(1).A(12)	Test Status F(27)		
Disable all LAM requests			*Initialise / Clear Gp2 F(11).A(13)	
Disable LAM request i	Sel.Clear Gp2 F(23).A(13)	Disable F(24)		
Enable LAM request i	Sel.Set Gp2 F(19).A(13)	Enable F(26)		
Examine LAM mask bit i	Read Gp2 F(1).A(13)			
Disable all LAM requests			*Initialise / Disable F(24).A(k)	
Enable all LAM requests			Enable F(26).A(k)	
*Examine LAM request i	Read Gp2 F(1).A(14)	Test LAM F(8)		
*Examine L signal			Test LAM F(8).A(k)	
	Preferred for modules with many LAM	Preferred for few LAM	Choose A(k) to distinguish from A(i)	

Fig 11
Some LAM Structure Options
*Mandatory (See Section 5.4.1.1)
NOTE: See also Figs K5.4.1A, B, C.

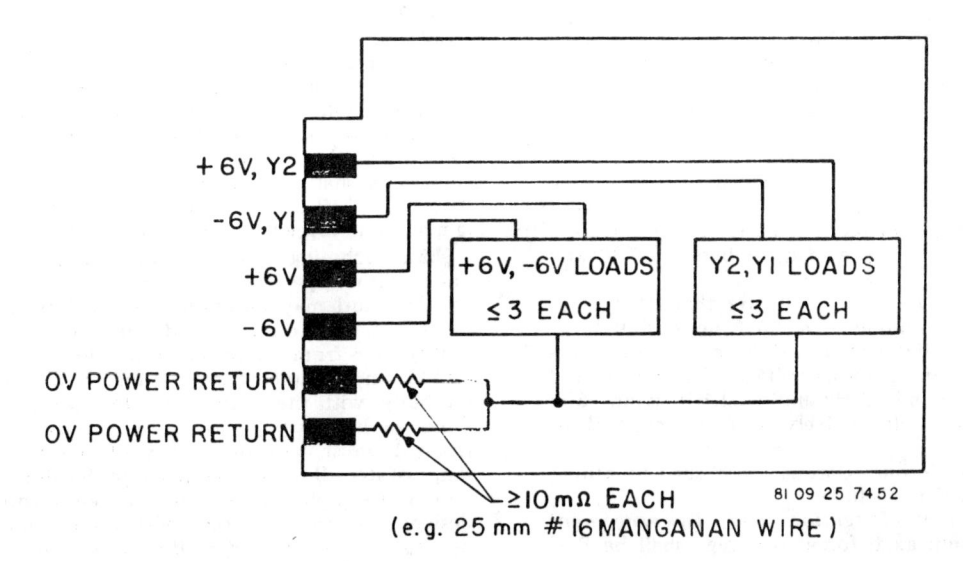

Fig 12
Typical Current Sharing with
Supplementary 6 V Power

55

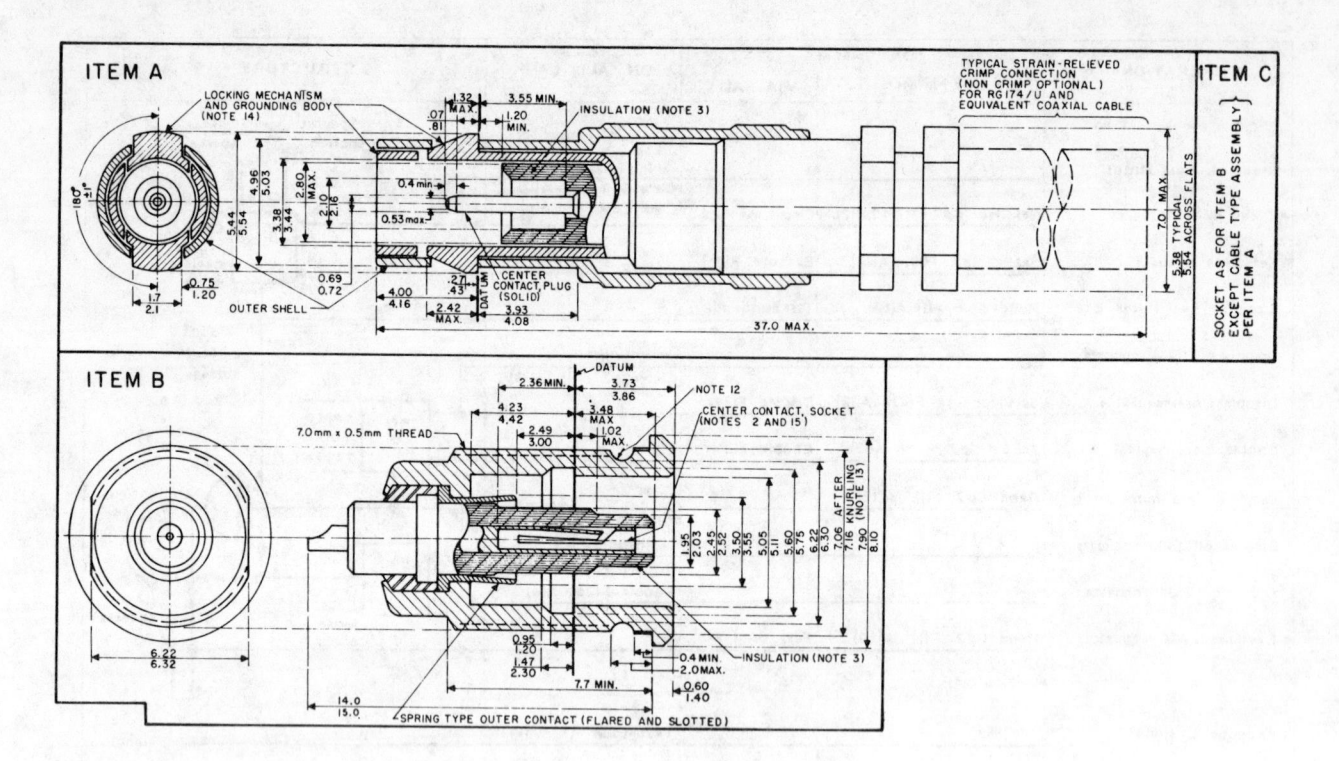

NOTES:

(1) *General.* These connectors are to be used for signal and dc voltage transmission with coaxial cables such as RG-174/U and RG-188/U. The maximum voltage reflection coefficient (VSWR−1)/(VSWR+1) of a mated connector pair consisting of item A and item C shall not exceed 0.09 for a pulse rise time (10 to 90 percent) of 150 ps in a 50 Ω system.

Normal disengagement of a mated connector shall be accomplished by applying axial pull to the sleeve of the plug. All linear dimensions are given in millimeters.

(2) *Connector Mating.* In the mated condition the combined resistance of the center contacts and the outer contacts shall not exceed 8 mΩ. Design, dimensions, and tolerances of center contact shall be such that the center contact socket will mate properly over the full range of center contact plug tolerances.

(3) *Insulation.* The insulation shall be Teflon PTFE.

(4) *Finish.* Shell: Bright Finish. Contacts: 2.5 µm minimum hard gold over nickel. The hard gold is to be gold with either nickel or cobalt as the hardening agent and capable of producing a deposit having a Knoop hardness range, at a 25 g load, of between 130 and 250.

(5) *Insulation Resistance.* Greater than 10^{12} Ω at 50 percent relative humidity at 25°C under 500 V dc.

(6) *Dielectric Withstanding Voltage.* These connectors are to have an operating voltage of 700 V dc and 500 V rms at 50 and 60 Hz and must withstand 1500 V dc for 1 min or, alternatively, 1100 V rms at 50 or 60 Hz.

(7) *Altitude/Corona.* Corona levels shall be 500 V minimum at 5000 m altitude.

(8) *Engagement Retention and Disengagement Forces.* Maximum axial force to engage shall be 9 N (2.0 lb). Minimum cable retention force shall be 90 N (20 lb). Minimum locking lug retention force shall be 90 N (20 lb). Maximum axial disengaging force applied to outer plug sleeve shall be 12 N (2.7 lb). After extraction of the plug from the socket, the diameter over the locking lugs of the plug shall return to the dimension specified in this drawing.

(9) *Durability.* A connector assembly shall be capable of being mated and unmated at least 500 times with no evidence of physical damage which could effect the mechanical or electrical performance of the connector.

(10) *Temperature Range.* The connector shall be capable of operating within the specification herein over the temperature range of −55°C to +150°C.

(11) *Corrosion.* Connectors shall be exposed to a 5 percent salt solution at 35°C for 48 h. After exposure, the connectors shall be washed, shaken, and lightly brushed and then permitted to dry for 24 h at 40°C. Connectors shall then show no sign of corrosion or pitting and shall meet the specifications herein regarding maximum force to engage and disengage.

(12) Machining Runout Relief may Assume Any Shape Within Specified Dimensions.

(13) Knurl may be cylindrical as shown, or it may taper conically to an outer diameter of 6.95/7.05 mm as it recedes from the panel mating surface.

(14) An example of locking mechanism and grounding body, with their associated dimensions, are shown. Other designs may be used, but the overall assembly (item A) must mate properly with item B over the full range of item B tolerances given on this drawing.

(15) A sample center contact socket is shown. Other designs that are consistent with note 2 and with the dimensions shown for item B may be used.

Fig 13
Coaxial Connector, Type 50CM

DIGITAL INTERFACE SYSTEM (CAMAC)

ANSI/IEEE
Std 583-1982

Appendixes

[The following Appendixes are not a part of ANSI/IEEE Std 583-1982, IEEE Standard Modular Instrumentation and Digital Interface System (CAMAC).]

Appendix A
Definitions of Module and Controller

In this standard the terms module and controller refer to plug-in units whose use, if any, of each Dataway line is consistent with the following table. A controller occupies the control station and at least one normal station. A module occupies one or more normal stations. A plug-in unit may combine some features of a module with some of a controller.

Line	Use by a Module	Use by a Controller
A	Receives	Generates
B	Receives	Generates
C	Receives	Generates
F	Receives	Generates
L	Generates	Receives
N	Receives	Generates
Q	Generates	Receives
R	Generates	Receives
S	Receives	Generates
W	Receives	Generates
X	Generates	Receives
Z	Receives	Generates

Appendix B
Digital Signal Classes and Standard Markings for Coaxial Connectors
(Supplementary Information)

For the information and guidance of users, each coaxial connector on the front or rear panel of a CAMAC plug-in unit should have a clear indication of its signal class.

Two Digital Signal Classes are Defined.

CAMAC Unterminated. Receivers and drivers for a class of unterminated signals compatible with TTL or DTL integrated circuits are defined in 7.2.1 (Table 8). (Note that the signal levels defined in Section D-6 of the NIM Specification may not be compatible with CAMAC unterminated signals because of logic level or current level differences.)

CAMAC Terminated (NIM Fast). Receivers and drivers for a class of terminated signals compatible with the NIM fast logic levels are defined in 7.2.2 (Table 9). Section 7.2.2 specifies that the signal is terminated in 50 Ω. The termination may be incorporated in the receiver. If this is the case, the input is referred to as being internally terminated. Alternatively, it may be desired to permit chaining of a commonly used signal to several plug-in units from a single driving circuit. In this case, the termination is external and the input is said to be externally terminated. The plug-in unit accepts the signal via a tap whose impedance is high compared to 50 Ω. The signal is terminated in 50 Ω at the end of the chain.

Standard Identifying Marks.

The signal class associated with each coaxial connector mounted on a front or rear panel of a CAMAC plug-in unit should be clearly indicated. It is recommended that the indications be either a letter code or a color code as shown in the following table. For example, the color code could consist of a colored ring around the connector or a washer of permanently colored material mounted under the coaxial connector. The washer must not interfere with any intended grounding of the connector body to the panel.

Pairs of connectors for externally terminated signals should be mounted close together. The panel should have an engraved, painted, or otherwise suitably affixed line between the connectors that form a pair.

Recommended Identifications for Coaxial Connectors

Signal Class	Letter Code	Color Code
CAMAC unterminated	TTL	light blue
CAMAC terminated (internally terminated)	50	black
CAMAC terminated (externally terminated)	HiZ	gray

Inputs and outputs should be clearly differentiated.

DIGITAL INTERFACE SYSTEM (CAMAC)

ANSI/IEEE
Std 583-1982

Appendix C
Preferred Connectors and Contact Assignments

Preferred Power Supply Connector

A preferred power supply connector is designated for connecting power supplies to the Dataway power lines. The connector is listed in Table C1 together with contact assignments. Typical power bus and power return bus feed and sense wiring consistent with these contact assignments is shown in Figures D1 and D6 of Appendix D to IEEE Std 583.

Table C1
Preferred Power Supply Connectors and Contact Assignments*

Original Manufacturer	AMP		Winchester	
Connector Type*	AMP-INCERT Series M		MRAC	
Number of Contacts	50		50	
Catalog Numbers				
Fixed Member PG-27 (crate connector)	201358-3		MRAC-50P-752	
Free Member PG-26 (power supply connector)	200277-4		MRAC-50S-752	
Suitable Contacts				
AWG Wire Gauge Accommodated	Pin†, §	Socket‡, §	Pin†	Socket‡
One #16 or #18	66098 or 202507	66100 or 202508	100-1016P-112 or 100-7116P	100-51016S-112 or 100-7116S
One #30	66425 or 201555	—	100-1026P-112	—
Two #18	66359 or 202725	—	100-7113P	—
Polarization Guides	200833(GP)**	200835(GS)**	111-20855(GP)**	111-20856-1(GS)**

Table C1 continued on following page.

* Other connectors fully mateable with those listed here and with at least 13 A per contact rating may be used.
† Pins to be used in Fixed Member (crate connector).
‡ Sockets to be used in Free Member (power supply connector).
§ Contacts listed here in the 66 000 series are Type III+ (formed) and those in 20 000 series are Type II (machined). Type II have lower resistance than Type III+ but are more expensive.
**GP designates ground pin. GS designates ground socket. (GP and GS are not part of manufacturers' numbers.) Two guide pins or sockets are required at each end of the connector blocks as indicated in the following Contact Assignments.

Table C1 (continued)
Contact Assignments (for Preferred Power Supply Connectors)

A	Reserved (a)*	a	−12	AA	+12 R
B	Reserved (b)*	b	−12 S	BB	+12 S
C	Reserved (c)*	c	−12 R	CC	+12
D	Reserved (d)*	d	−12 RS	DD	+24 RS
E	reserved	e	−6 S		
		f	−6 S		
F	reserved	h	−6	EE	+24 R
H	reserved	j	−6 R	FF	+24 S
J	chassis Ground, Status and	k	−6	HH	+24
	Temperature Warning Returns	m	−6 R		
		n	−6		
K	Status	p	−6 R		
L	Temperature Warning	r	+6 RS		
M	Y2 R	s	+6 S		
N	Y2 RS	t	+6 R		
P	Y2 (Supplementary +6 V)	u	+6		
R	Y2 S	v	+6 R		
S	Y1 S	w	+6		
T	Y1 (Supplementary −6 V)	x	+6 R		
U	Y1 RS	y	+6		
V	Y1 R	z	+12 RS		
W	−24				
X	−24 S				
Y	−24 R				
Z	−24 RS				

Guide Pin/Socket Assignments (Two on Each End Of Block)

Fixed Member	Block End A,B	(GS)
(crate connector)	Block End FF, HH	(GP)
Free Member	Block End A,B	(GP)
(power supply connector)	Block End FF, HH	(GS)

NOTE: R = return, S = sense, and RS = return sense.
*Reserved (c) was previously assigned to +200 V dc
 Reserved (a) to 117 V ac line, and
 Reserved (b) to 117 V ac Neutral.
Effective January 1, 1978, these assignments were cancelled in order to avoid hazardous voltages on the connectors.

C2. Auxiliary Connectors

Preferred auxiliary connectors as listed in Table C2 are designated for communicating from and to plug-in units. The fixed connector is preferably mounted with pin 1 at the bottom.

When used for balanced or unbalanced signal transmission lines, it is recommended that the signal pairs be assigned as shown in Table C3. *Return* is to be used as the complement of *signal* for a balanced pair, or as the ground reference for an unbalanced pair. The recommended location of the common ground, and also of Vcc, if used, is in the table. The contact assignments for the 52 pin and 88 pin connectors have been made to facilitate the use of ribbon cables.

Cable assemblies consisting of multiple twisted pairs of conductors can be made with individual twisted pairs assigned to pins as shown in the table. The same cable assemblies can also be used where individual signals of a group are each carried on a single conductor, with a common ground return for the group. Cable assemblies made in conformance with these recommendations should be marked to differentiate them from other, outwardly similar, cables so as to indicate to the user that the cable contains only twisted pairs, with pairs connected to all pins as shown in Table C3.

Table C2
Preferred Auxiliary Connectors†

Original Manufacturer	ITT Cannon Electric	Hughes Aircraft Company	
Connector Type	2D subminiature rectangular connector	WSS subminiature rectangular connector	
Number of Contacts	52	88	132
Polarizing Code			BN
Usable on	Front or rear panels, all widths	Front panels, all widths	Front panels down to double width
Catalog Numbers			
Fixed member (Chassis connector)	2DB52P Pin Housing	WSS 0088 S00 BN00 Socket Housing	WSS 0132 S00 BN00 Socket Housing
Locking assembly	D20418-2		
Free member (Cable connector)	2DB52S Socket Housing	WSS 0088 Pxx BNyyy* Pin Housing	WSS 0132 Pxx BNyyy* Pin Housing
Cable hood	DB24659	WAC 0088 H005 for example	WAC 0132 H005 for example
Locking assembly	D20419		

*Pxx yyy denotes type of jack screw.
†Equivalent connectors that are fully mateable with those listed here may be used.

⊗ = SIGNAL
⊘ = RETURN

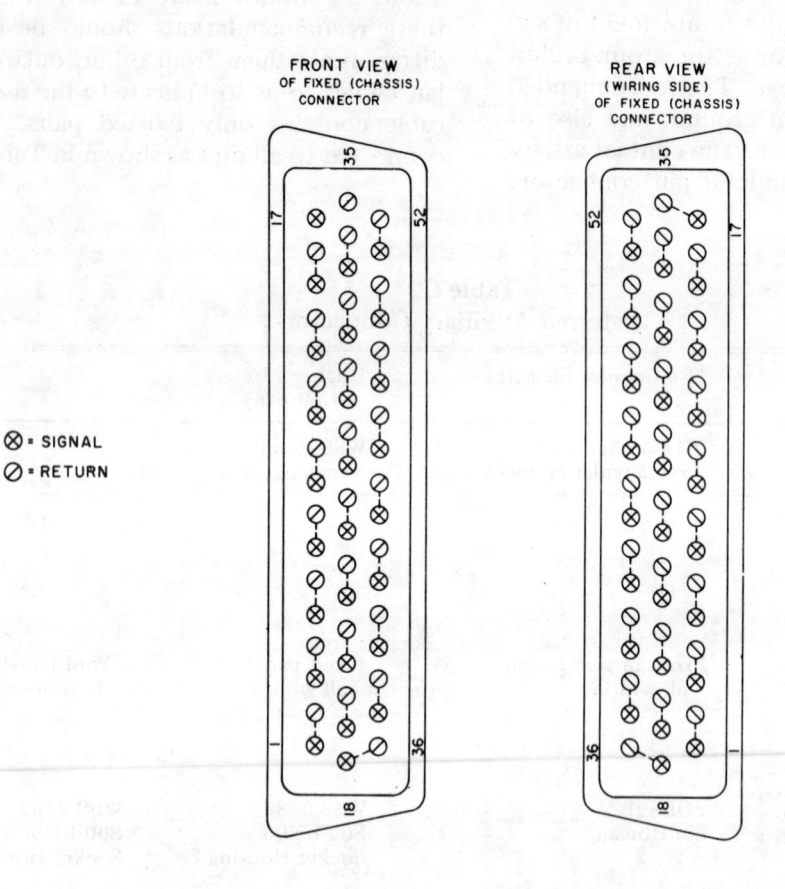

Fig C1
Contact Arrangement for Preferred 52 Pin
Auxiliary Connector

DIGITAL INTERFACE SYSTEM (CAMAC)

ANSI/IEEE
Std 583-1982

Fig C3
Contact Arrangement for Preferred 132 Pin Auxiliary Connector

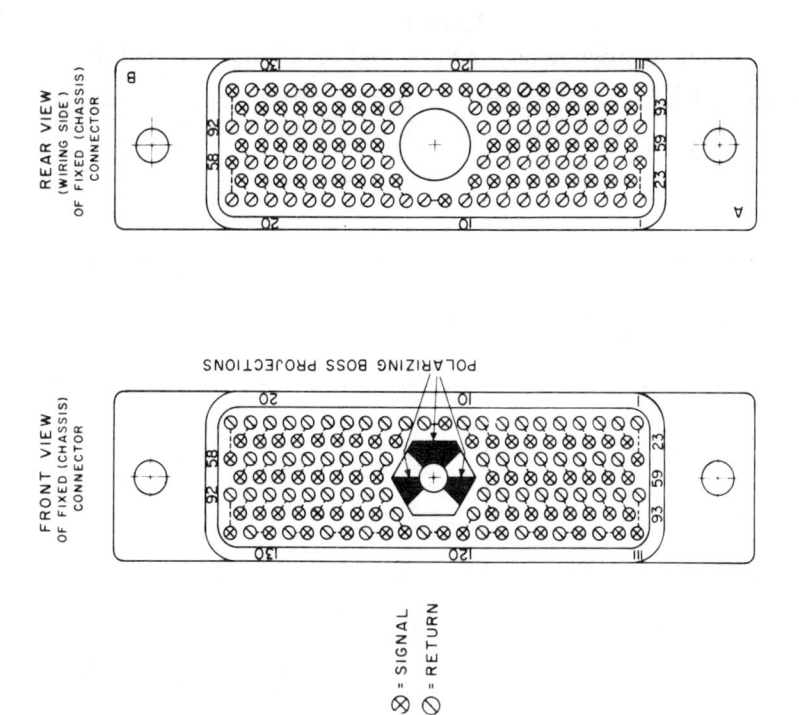

Fig C2
Contact Arrangement for Preferred 88 Pin Auxiliary Connector

Table C3
Contact Arrangement for Preferred Auxiliary Connector

PAIR NO.	52 Pin Cannon Signal Contact	52 Pin Cannon Return Contact	88 Pin Hughes Signal Contact	88 Pin Hughes Return Contact	132 Pin Hughes Signal Contact	132 Pin Hughes Return Contact
1	1	2	2	1	41	1
2	3	4	4	3	23	2
3	5	6	6	5	24	3
4	7	8	8	7	25	4
5	9	10	10	9	26	5
6	11	12	12	11	27	6
7	13	14	14	13	28	7
8	15	16	16	15	29	8
9	17	35	18	17	30	9
10	18	36	20	19	31	10
11	19	20	22	21	11	12
12	21	22	24	23	32	13
13	23	24	26	25	33	14
14	25	26	28	27	34	15
15	27	28	30	29	35	16
16	29	30	32	31	36	17
17	31	32	34	33	37	18
18	33	34	36	35	38	19
19	37	38	38	37	39	20
20	39	40	40	39	40	21
21	41	42	42	41	58	22
22	43	44	44	43	59	42
23	45	46	46	45	60	43
24	47	48	48	47	61	44
25	49	50	50	49	62	45
26	51	52	52	51	63	46
27			54	53	64	47
28			56	55	65	48
29			58	57	66	49
30			60	59	67	50
31			62	61	68	51
32			64	63	69	52
33			66	65	70	53
34			68	67	71	54
35			70	69	72	55
36			72	71	73	56
37			74	73	74	57
38			76	75	93	76
39			78	77	94	77
40			80	79	95	78
41			82	81	96	79
42			84	83	97	80
43			86	85	98	81
44			88	87	99	82
45					100	83
46					103	84
47					104	85
48					105	86
49					106	87
50					107	88
51					108	89
52					109	90
53					110	91
54					111	75
55					112	113
56					114	115
57					116	117
58					118	119
59					120	101
60					121	122
61					123	102
62					124	125
63					126	127
64					128	129
65					130	131
66					132	92
Ground (when used)	50	& 52			111	& 75
Vcc (when used)	49	& 51				

Note: RETURN is to be used as the complement of SIGNAL for a balanced pair, or as the ground reference for an unbalanced pair.

Appendix D
Typical Crate Mounted Power Supply and Ventilation Unit With Crate/Power Supply Interface Housing

D1. General

The power supply described herein is suitable for use with the Standard Modular Instrumentation and Digital Interface System (CAMAC).

This description is written in the form of a specification for the convenience of those who wish to use it for that purpose.

Due to the high operational reliability required, only the highest quality components should be employed. All semiconductor components shall be silicon and shall be encapsulated in metal or ceramic, hermetically sealed, cases. Components shall not be used beyond their design ratings. The supply shall be designed with a life expectancy of at least 5 yr. See Fig D1 for block diagram. Wiring to the right of PG-26 is not considered part of the power supply.

D2. Input

For 120 V nominal voltage, the input voltage range shall be 103 to 129 volts. For other nominal voltages the input voltage range shall be the nominal + 10 percent to - 12 percent.

Line frequency range shall be the nominal (60 Hz in the U.S.) ±3 Hz.

D3. Output

The supply is to provide four dc outputs with at least the following current ratings:

Voltage Volts	Current Amperes
+ 6.00	0-25
- 6.00	0-25
+24.00	0-6
-24.00	0-6

The four outputs shall be simultaneously available, but the currents may be limited to a minimum total output power of 294 W. The ±6.0 V supplies shall operate on a current sharing basis, such that the total combined current outputs may be limited to 25 A. Likewise, the ±24 V supplies shall be current shared and may be limited to a total combined current output of 6 A. Rated output current shall also be available to loads connected between the positive outputs and the negative outputs.

If the output power demanded should exceed a safe operating value, the supply shall protect itself.

Remote sensing shall be utilized on all outputs of this power supply. Remote sense points can be expected to be within 305 mm of the crate connector, PG-27. All wiring shall be in accordance with Fig D1.

D4. Regulation and Stability

(1) During a 24 h period the ±6.0 V outputs shall vary by not more than ±0.5 percent due to changes of input voltage and output current within the specified ranges.

(2) During a 24 h period the ±24 V outputs shall vary by not more than ±0.2 percent due to changes of input voltage and output current within the specified ranges.

(3) The long-term stability shall be such that, after a 24 h warmup, over a six month period for constant load, line, and ambient temperature conditions, the ±6.0 V output shall drift not more than ±0.5 percent; the ±24 V outputs shall drift not more than ±0.3 percent. (See Figs D3 and D4)

D5. Noise and Ripple

Noise and ripple, as measured on an oscilloscope of dc to 50 MHz bandwidth, shall not exceed 15 mV peak-to-peak.

D6. Temperature and Temperature Coefficient

The ambient temperature range is from 0°C to 50°C without derating. Ambient temperature as used throughout this specification shall be taken at a location that is not affected by the temperature of the power supply.

The output voltage coefficients for changes in ambient temperatures between 0°C and 50°C shall not exceed 0.02 percent per °C.

D7. Voltage Adjustment

The output voltage shall be adjustable over a nominal range of at least ±2 percent by means of screwdriver adjustments accessible through the rear or top of the supply. The maximum error in resetting each output voltage shall be ±0.5 percent.

D8. Recovery Time and Turn-on Turn-off Transients

The outputs shall recover to within ±0.2 percent of their steady state values within 1 ms for any change within the specified input voltage and for a 50 percent rated load current change. The peak output excursions during 1 ms shall not exceed ±5 percent of rated voltage for such line or load changes and shall be proportionately less for smaller changes.

Response to input voltage changes or to ±5 percent bus line voltage changes shall be nonoscillatory.

From turn-on the power supply outputs shall stabilize to within ±1% of their final values within 1 min for constant line, load, and ambient temperature. The outputs shall turn on in an asymptotic manner, that is, without overshoot. The transient during turn-off shall not result in any voltage exceeding 107% of nominal value when tested at any current within rating, with a resistive load.

D9. Magnetic Field Effects

A magnetic field of 50 G in any direction shall not cause performance characteristic variations of more than ±0.5 percent.

D10. Power Transformers

The power transformers shall be constructed with an electrostatic shield which is connected to the core.

D11. Terminals

All wiring shall be as shown in Fig D1.

(1) When designed for use with 117 V ac mains, a three-wire power cord of approximately 1.5 m in length shall be included. It shall have a NEMA Cap, 5-15P. The power cord may be permanently attached to the power supply, or alternatively, may terminate in a NEMA Connector Body 5-15R, mating with a NEMA Inlet 5-15P on the power supply.

(2) The dc output power shall be supplied via a connector (PG-26) as designated in Fig D1, or mating equivalent. Wire size, socket types, and pin assignments are specified in Fig D1.

D12. Protection

(1) The input of the supply shall be protected with a fuse of adequate rating in each side of the line. The fuses shall be readily accessible.

(2) The output of the supply shall be short-circuit protected by means of an electronic circuit. The current limiting threshold shall be set at least 0.2 A above the specified maximum output currents. The output voltage shall be resumed after the short has been removed. A continuous short circuit shall not damage the supply or blow a fuse.

(3) The output shall be protected by limiting circuits so that under no conditions will the ±24 V outputs exceed 34 V or the ±6.0 V outputs exceed 7.5 V. Operation of the overvoltage protection shall not damage the power supply.

(4) In no case shall a failure of any supply cause an increase in voltage of any other supply by more than 20 percent.

(5) The power supply shall not damage itself, and the conditions of $D12(3)$ shall apply if the power supply is turned on with any or all pins of PG-26 disconnected.

(6) Thermal protection circuits shall be provided to disable the supply when the temperature exceeds a safe operating value.

The maximum safe operating temperature, as measured at the thermal switch, shall be specified on the schematic circuit diagram.

D13. Crate Ventilation

This power supply shall include fans and mechanical assembly to provide forced air ventilation of a CAMAC crate. Air flow of at least 12 ft^3/min shall be directed into each of four equal crate sections extending from front to back. The air flow impedance of densely packed CAMAC modules in all twenty-five stations shall be considered in determining the minimum air flow rate.

Air shall be drawn from directly in front of the rack in which the assembly is mounted. Air filters, allowing a visual inspection from the front, shall be included. The air shall be channeled in such a way that it does not experience an appreciable temperature rise due to the heat of the power supply. The unit shall include a POWER ON-OFF switch. The switching shall be such that the fans must operate whenever the power supply is operating.

D14. Mounting

The supply shall be constructed for rack mounting immediately below a CAMAC crate in such a fashion that the ventilation requirements of this specification are achieved.

(1) Fig D2 specifies several outline dimensions and component locations to which the unit must adhere.

(2) Interface housing units (see Fig D5) mechanically adapt CAMAC crates from various sources to this power supply. They also house and protect PG-27, power busing, and the dataway connectors.

An interface housing unit is not a part of this specification. The power supply shall, however, be provided with four #10-32 captive screws in the positions shown in Fig D2 as a means of securing to an interface housing unit.

(3) The panel height of the supply is not specified. Panel height is at a premium in rack space. Trade offs between panel height and power supply costs should be optimized.

D15. Monitoring

(1) Front panel metering shall be provided to monitor the four dc voltages and their current loads. The metering shall be accurate to ±2.5 percent full scale.

The meter scales shall be calibrated with full scales of 120-135 per cent of nominal voltage and rated current values and shall have labeled markers at nominal voltage values and at rated current values.

(2) A front panel neon lamp (or suitable solid state indicator) wired as shown in Fig D1 shall be provided to indicate the ac power on condition.

(3) A front panel thermal warning light, wired as shown in Fig D1, shall be provided. It shall light whenever the temperature within the supply exceeds a value 20°C below the maximum safe operating temperature. The thermal warning lamp may be a neon lamp, as shown, or a solid state indicator may be used.

D16. Mechanical Construction

(1) Insulating materials such as printed wiring boards shall be flame retardant.

(2) All components shall be accessible for testing and replacement.

(3) All integrated circuits shall be mounted in high-quality integrated circuit sockets.

(4) Markings: Major components such as solid state devices, transformers (including leads), large capacitors, controls, and terminals shall be marked in the most readable position in the unit with respect to their identification on the schematic diagram.

D17. Circuit Diagram

Two copies of the schematic circuit diagram, which include component values, shall be provided with each supply. All semiconductor components shall be designated by Electronic Industries Association numbers or in nomenclature commonly used by semiconductor manufacturers or shall be directly replaceable by the same. Where special types are used, the schematic diagram or instruction book shall recommend a semiconductor manufacturer's equivalent that will provide satisfactory performance.

D18. Finish

All front panel metal surfaces shall be finished with a baked-on enamel or with an equally hard, chip-resistant, material. All surfaces not seen from the front may be finished similarly, or may be finished with nickel plate, iridite or other material that will assure good electrical contact and that, where necessary, is passivated against atmospheric corrosion or against electrolysis when in contact with copper or with other common finishes.

Numerals 1-25, representing station numbers in a CAMAC crate, and to identify modules inserted into a crate which may be mounted immediately above the supply, shall be printed on the front panel near the top edge. They shall be in consecutive order from left to right as viewed from the front with the numeral 13 at the front panel centerline, and shall be positioned at 17.2 mm intervals. The numerals shall be at least 4 mm in height.

D19. Test Conditions

Crate wiring between PG-27 and the Dataway power bus shall be simulated by 305 mm of lead. Sense leads and test-load leads shall be attached at this distance from PG-27, and measurements to determine adherence to these specifications shall be made at this point. Users are alerted to the fact that, in practice, performance will depend upon the actual positions of sense points and the reactive nature of loads.

D20. Optional Feature—Status Bit

A Status Bit to indicate whether the power unit is functioning normally may be provided. This optional feature, when provided, shall be standardized as follows:

(1) The Status Bit source shall be a relay which provides contact closure when in the alarm condition; shorting the Status Bit line to the power-unit chassis. Under normal operating conditions, the Status Bit line shall be an open circuit in the power unit.

Contact rating shall be minimally 50 V, 500 mA.

(2) The Status Bit alarm condition shall indicate that any one of the following conditions exist:

   (a) Any one of the voltages supplied by the unit is outside of specified voltage range.

   (b) Any one or combination of supplies is being loaded beyond specified current range.

   (c) The thermal warning switch is in the alarm condition.

The Status Bit may indicate additional alarm conditions at the option of the manufacturer, but (a), (b), and (c) must minimally be included.

(3) In the power unit, the Status Bit shall be wired to contact K of PG-26 and to a front panel 50CM coaxial connector.

D21. Figures

Fig D1 interconnection block diagram.

Fig D2 outline dimensions and illustrated unit.

Fig D3 Time and voltage characteristics ±6 V.

Fig D4 Time and voltage characteristics, ±24V

Fig D5 Interface housing unit

Fig D6 Typical Power buses and power return bus, feed and sense wiring (see note below).

NOTE: The information on Figs D5 and D6 indicate preferred practice for fabrication and assembly of CAMAC crate wiring and the interface housing unit. It is presented here because of the intimate relationship between these and the power supply and ventilation unit.

DIGITAL INTERFACE SYSTEM (CAMAC)

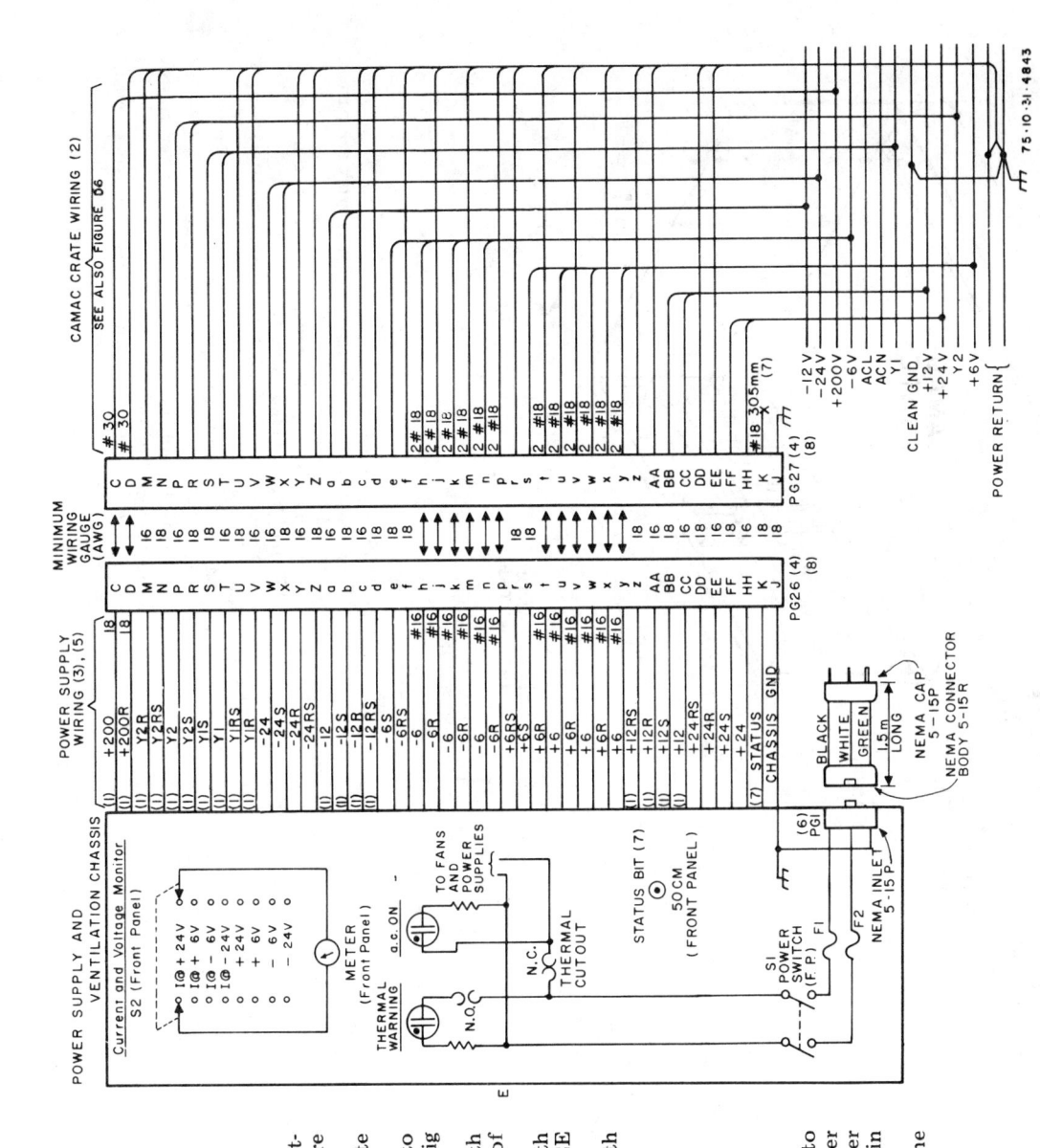

Fig D1
Interconnection Block Diagram

NOTES:

(1) Optional voltages. Pins reserved for optional voltages not supplied by this unit. If wired, minimum wire gauges shown pertain.

(2) All pins wired as shown. PG27 and CAMAC crate wiring are not provided with this power supply.

(3) Sufficient length, positioning, and flexibility to mate with crate connector mounted as shown in Fig D2 (minimum length, 305 mm).

(4) PG-26—Fixed Member (crate connector) with socket contacts in accordance with Appendix C of IEEE Std 583.

PG-27—Free Member (power supply connector) with pin contacts in accordance with Appendix C of IEEE Std 583.

Other connectors fully mateable with these and with at least 13 A per contact rating may be used.

(5) 24 indicates 24 V line
24 *R* indicates 24 V return line
24 *S* indicates 24 V sense line
24 *RS* indicates 24 V return sense line
and so forth for other voltages

(6) PG-1 optional

(7) Optional feature, Section D20.

(8) Polarization of connectors PG-26 and PG-27 is to be provided by the use of two guide pins in the corner holes of one end and two guide sockets in the corner holes of the other end of each connector block, in accordance with Appendix C of IEEE Std 583.

(9) A solid state indicator may be used in lieu of the neon lamp shown for a thermal warning lamp.

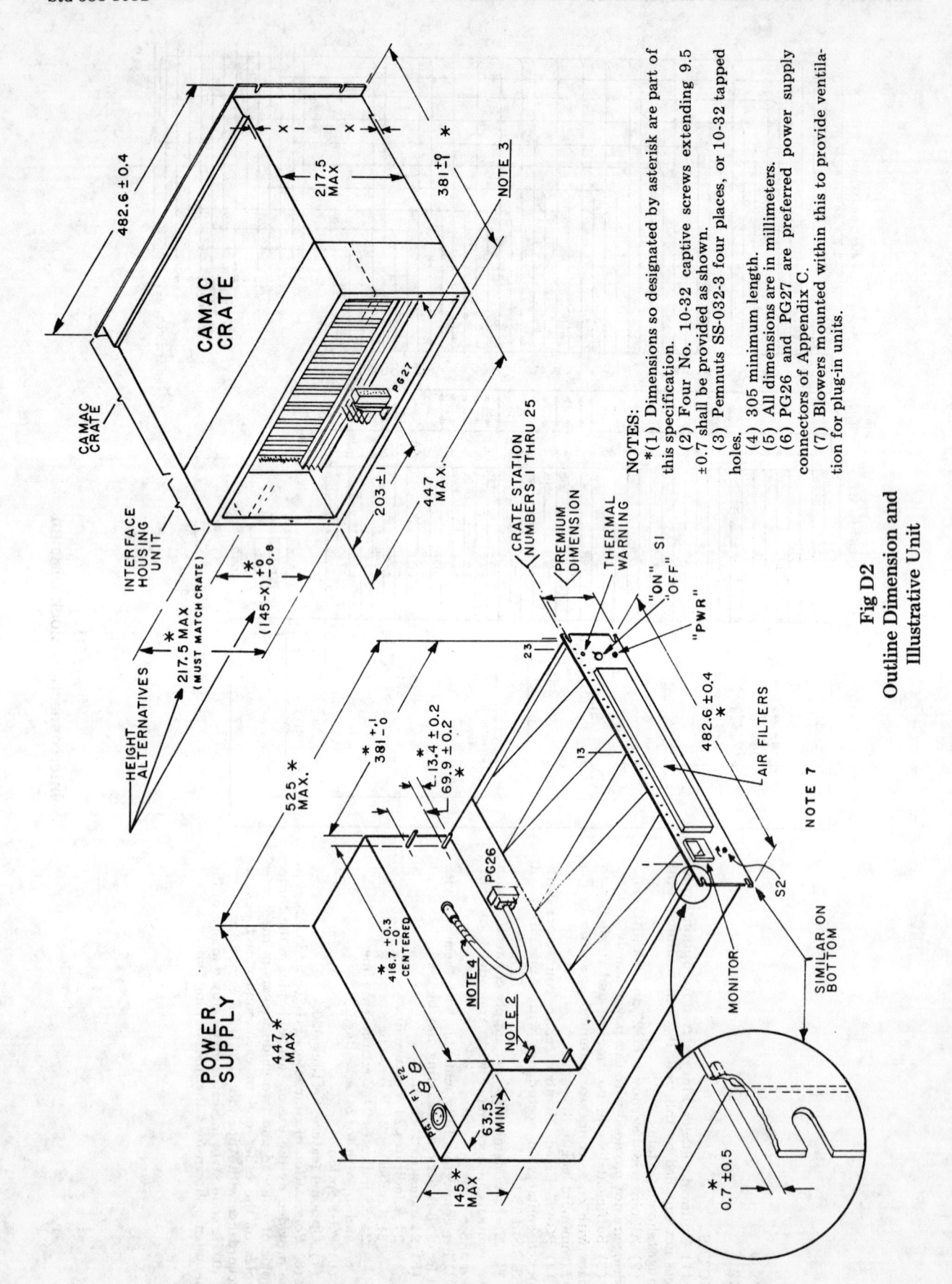

Fig D2
Outline Dimension and Illustrative Unit

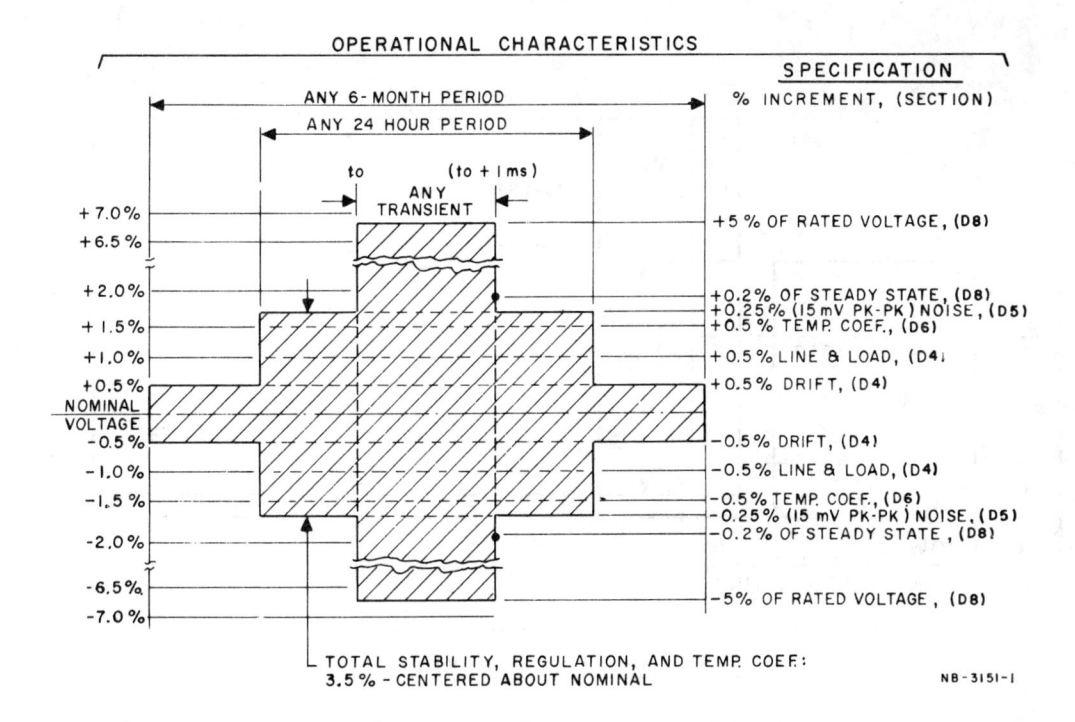

Fig D3
Time and Voltage Characteristics, ±6 V

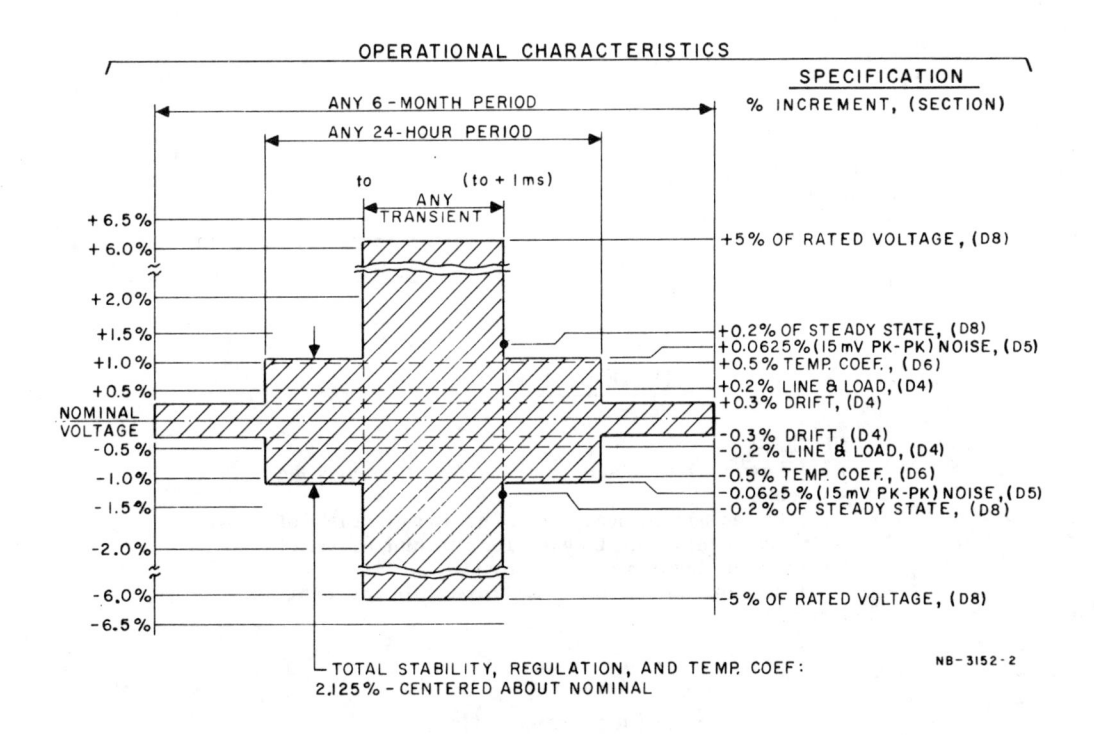

Fig D4
Time and Voltage Characteristics, ±24 V

ANSI/IEEE
Std 583-1982

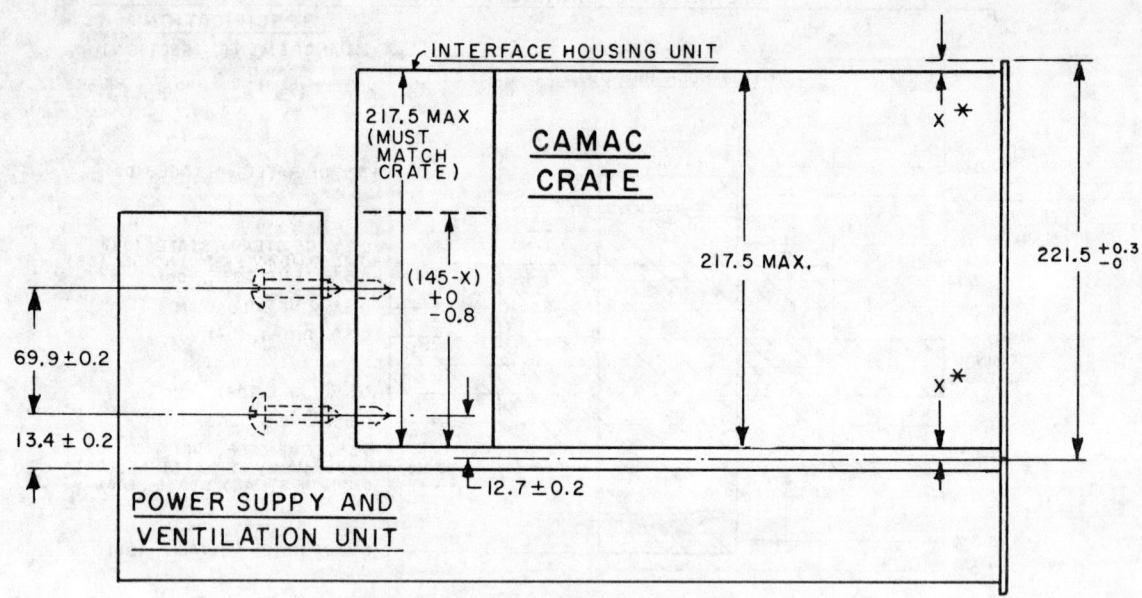

IEEE STANDARD MODULAR INSTRUMENTATION AND

ASSEMBLY – SIDE VIEW

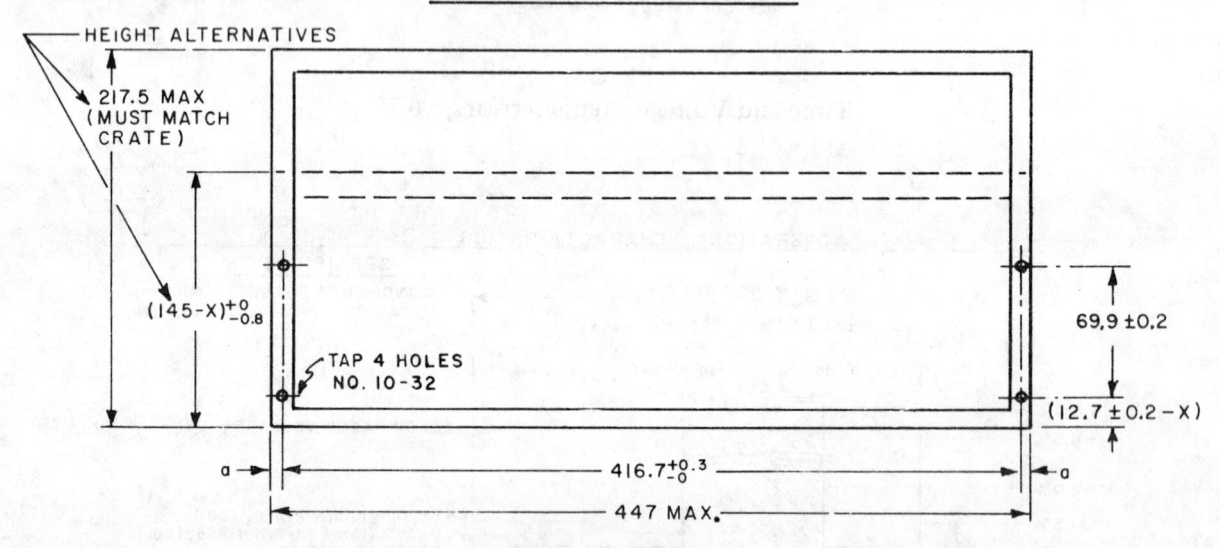

REAR VIEW

NOTES:
(1) Power section and ventilation section may be separable.
(2) Power section need not extend upward at rear of crate as shown; if not, brackets for attaching to rear of interface housing unit must be provided.
*Dimension varies with crates from different manufacturers.

NB-3154-1

**Fig D5
Interface Housing Unit**

DIGITAL INTERFACE SYSTEM (CAMAC)

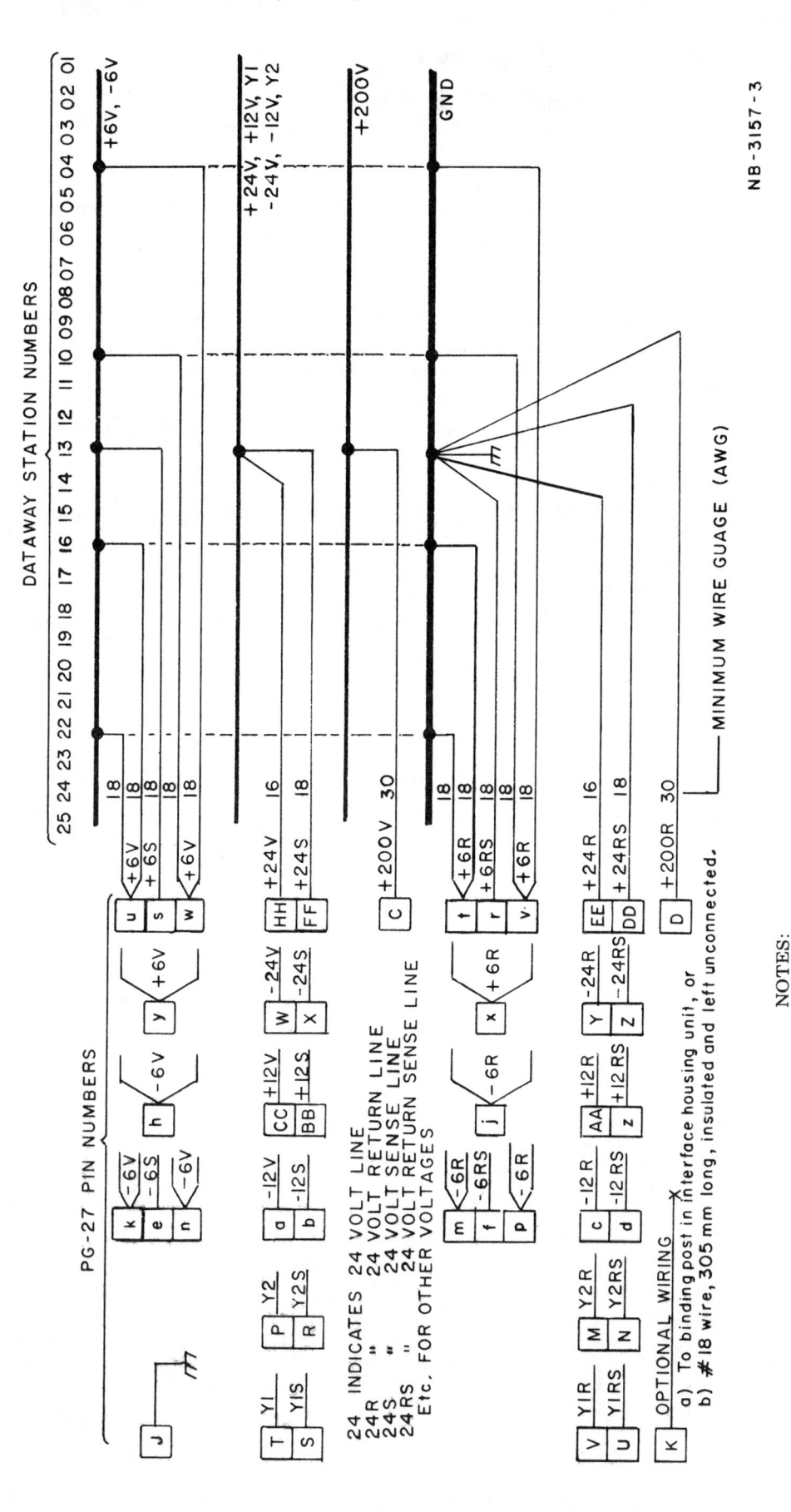

ANSI/IEEE
Std 583-1982

NB-3157-3

Fig D6
Typical Power Buses and Power Return Bus, Feed, and Sense Wiring.

NOTES:
(1) Suitable Connectors are Described in Appendix C of IEEE Std 583.
(2) See note at the bottom of Section D21.

Appendix E
Camac Categories

CAMAC plug-in unit is a functional unit that conforms to the mandatory requirements for a plug-in unit as specified in this standard.

CAMAC module is a CAMAC plug-in unit that when mounted in one or more normal stations of a CAMAC crate is compatible with this standard.

CAMAC crate controller is a functional unit that when mounted in the control station and one or more normal stations of a CAMAC crate (or CAMAC compatible crate) communicates with the Dataway in accordance with this standard.

CAMAC Dataway is an interconnection between CAMAC plug-in units which conforms to the mandatory requirements for a CAMAC Dataway as specified in this standard.

CAMAC crate is a mounting unit for CAMAC plug-in units that includes a CAMAC Dataway and conforms to the mandatory requirements for a CAMAC crate as specified in this standard.

CAMAC compatible crate is a mounting unit for CAMAC plug-in units that does not conform to the full requirements for a CAMAC crate but in which CAMAC modules can be mounted and operated in accordance with the Dataway requirements of this standard.

CAMAC crate assembly is an assembly of a CAMAC crate controller and one or more CAMAC modules mounted in a CAMAC crate (or CAMAC compatible crate), and operable in conformity with the Dataway requirements of this standard.

CAMAC system is a system including at least one CAMAC crate assembly.

A highway for a CAMAC system is an interconnection between CAMAC crate assemblies or between one or more CAMAC crate assemblies and an external controller.

Appendix F
Bibliography

[1] COSTRELL, L. Standardized Instrumentation System for Computer Automated Measurement and Control. *IEEE Transactions on Industry Applications*, vol 1A-11, May/Jun 1975, pp 319-323
LYON, W.T., and ZOBRIST, D.W. System Design Considerations When Using Computer-Independent Hardware. *IEEE Transactions on Industry Applications*, vol 1A-11, May/Jun 1975, pp 324-327.

[2] WILLARD, F.G. Interfacing Standardization in the large Control System. *IEEE Transactions on Industry Applications*, vol 1A-11, Jul/Aug 1975, pp 362-364.
JOERGER, F.A., and KLAISNER, L.A. Functional Instrumentation Modules, *IEEE Transactions on Industry Applications*, vol 1A-11, Nov/Dec 1975, pp 000-000
RADWAY, T.D. A Standardized Approach to Interfacing Stepping Motors to Computers. *IEEE Conference Record of the 1974 Ninth Annual Meeting of the IEEE Industry Applications Society*, Oct 7-10, Pittsburgh, PA, pp 257-259.
KLAISNER, L.A. A Serial Data Highway for Remote Digital Control. IEEE Conference Record of the 1974 Ninth *Annual Meeting of the IEEE Industry Applications Society*, Oct 7-10, Pittsburgh, PA, pp 453-456.
FASSBENDER, P., and BEARDEN, F. Demonstrating Process Control Standards An Exercise in success. *IEEE Conference Record of the 1974 Ninth Annual Meeting of the IEEE Industry Applications Society*, Oct 7-10, Pittsburgh, PA, pp 457-462.

[3] COSTRELL, L. The Nucleus. *IEEE Transactions on Nuclear Science*, vol NS-20, Apr 1973, p1.
COSTRELL, L. CAMAC Instrumentation System Introduction and General Description. *IEEE Transactions on Nuclear Science*, vol NS-20, Apr 1973, pp 3-8.
KIRSTEN, F.A. Operational Characteristics of the CAMAC Dataway. *IEEE Transactions on Nuclear Science*, vol NS-20, Apr 1973, pp 8-20.
HORELICK, D. and LARSEN, R. S. CAMAC: A Modular Standard. *IEEE Spectrum*, Apr 1976, pp 50–55.
KIRSTEN, F.A. A Short Description of the CAMAC Branch Highway. *IEEE Transactions on Nuclear Science*, vol NS-20, Apr 1973, pp 21-27.
LARSEN, R.S. CAMAC Dataway and Branch Highway Signal Standards. *IEEE Transactions on Nuclear Science*, vol NS-20, Apr 1973, pp 28-34.
DHAWAN, S. CAMAC Crate Controller Type A-1. *IEEE Transactions on Nuclear Science*, vol NS-20, Apr 1973, pp 35-41.
KIRSTEN, F.A. Some Characteristics of Interfaces Between CAMAC and Small Computers. *IEEE Transactions on Nuclear Science*, vol NS-20, Apr 1973, pp 42-49.
THOMAS, R.F., Jr. Some Aspects of CAMAC Software. *IEEE Transactions on Nuclear Science*, vol NS-20, Apr 1973, pp 50-68.
MACK, D.A. Summary of CAMAC: Status and Outlook, *IEEE Transactions on Nuclear Science*, vol NS-20, Apr 1973, pp 69-71.

[4] Brill, A.B; Parker, J.F; Erickson, J.J; Price, R.R.; and Patton, J.A. CAMAC Applications in Nuclear Medicine of Vanderbilt: Present Status and Future Plans. IEEE Transactions on Nuclear Science, vol NS-21, Feb 1974, pp 892-897.

[5] EURATOM Report EUR 4100e, 1972, CAMAC - A Modular Instrumentation System for Data Handling, Description, and Specification. (Earlier edition was 1969.)

[6] U.S. Energy Research and Development Administration Report TID-25875, July 1972, CAMAC - A Modular Instrumentation System for Data Handling, Description and Specification.

[7] U.S. Energy Research and Development Administration Report TID-25877, Dec 1972, Supplementary Information on CAMAC Instrumentation system.

[8] U.S. Energy Research and Development Administration Report TID-26618, Oct 1976, CAMAC Tutorial Articles.

[9] U.S. Energy Research and Development Administration Report TID-20893, Rev 4, July 1974, Standard Nuclear Instrumentation Modules. (Earlier editions date from July 1964.)

Table 4
The Function Codes

Code $F(\ )$	Function	Use of R and W Lines	F16	F8	F4	F2	F1	Code $F(\ )$
0	Read Group 1 register	Functions using the R lines	0	0	0	0	0	0
1	Read Group 2 register		0	0	0	0	1	1
2	Read and Clear Group 1 register		0	0	0	1	0	2
3	Read Complement of Group 1 register		0	0	0	1	1	3
4	Nonstandard		0	0	1	0	0	4
5	Reserved		0	0	1	0	1	5
6	Nonstandard		0	0	1	1	0	6
7	Reserved		0	0	1	1	1	7
8	Test Look-at-Me	Functions not using the R or W lines	0	1	0	0	0	8
9	Clear Group 1 register		0	1	0	0	1	9
10	Clear Look-at-Me		0	1	0	1	0	10
11	Clear Group 2 register		0	1	0	1	1	11
12	Nonstandard		0	1	1	0	0	12
13	Reserved		0	1	1	0	1	13
14	Nonstandard		0	1	1	1	0	14
15	Reserved		0	1	1	1	1	15
16	Overwrite Group 1 register	Functions using the W lines	1	0	0	0	0	16
17	Overwrite Group 2 register		1	0	0	0	1	17
18	Selective Set Group 1 register		1	0	0	1	0	18
19	Selective Set Group 2 register		1	0	0	1	1	19
20	Nonstandard		1	0	1	0	0	20
21	Selective Clear Group 1 register		1	0	1	0	1	21
22	Nonstandard		1	0	1	1	0	22
23	Selective Clear Group 2 register		1	0	1	1	1	23
24	Disable	Functions not using the R or W lines	1	1	0	0	0	24
25	Execute		1	1	0	0	1	25
26	Enable		1	1	0	1	0	26
27	Test Status		1	1	0	1	1	27
28	Nonstandard		1	1	1	0	0	28
29	Reserved		1	1	1	0	1	29
30	Nonstandard		1	1	1	1	0	30
31	Reserved		1	1	1	1	1	31

Since Table 4 is frequently referred to it is repeated here at the end of the Standard for the convenience of users.

# Index

The principal references are given first. Any subsidiary references follow, after a semi-colon (;).

A1, A2, A4, A8 — See Subaddress
ACL, ACN — See Additional Power Lines
Adaptor for NIM units — Fig 7; Section 4.3
Additional Power Lines — Section 8; Section 5.7, Table 1
Address Scan Mode—of block transfer — Section 5.4.3.1
Analog Signals—standards — Section 7.3

B
Bibliography — See Busy
 — Appendix F
Block Transfers — Sections 5; 5.4.1.4, 5.4.3.5, Table K5.4.3.5
 — use of Q in — Section 5.4.3
  — Address Scan Mode — Section 5.4.3.1
  — Repeat Mode — Section 5.4.3.2
  — Stop Mode — Section 5.4.3.3
  — Stop-on-Word Mode — Section 5.4.3.4
Branch Highway—use of — Section 3(7)
Bus Lines — Section 4.4
Busy—signal line — Section 5.4.2; Section 5, Table 1

C
CAMAC categories — See Clear
 — Appendix E
Chamfers
 — on connector socket — Fig 5, (5.6), (5.7), and (5.8); Section 4.1.2
 — on connector plug — Fig 5, (5.3); Section 4.2.2
Clear—signal line — Section 5.5.3; Section 5, Table 1
Clear Group 1, Selective—function code — Section 6.3.5; Table 4
Clear Group 1—function code — Section 6.2.2; Table 4
Clear Group 2, Selective—function code — Section 6.3.6; Table 4
Clear Group 2—function code — Section 6.2.4; Table 4
Clear Look-at-Me—function code — Section 6.2.3; Table 4
Coaxial Connectors — Section 4.2.5
Coaxial Connector Marking — Section 4.2.5; Appendix B
Command
 — definition — Section 6; Section 5.1
 — for access to LAM information — Section 5.4.1.2
 — control, general description — Sections 6.2, 6.4
 — external representation of — Section 6.5
 — read, general description — Section 6.1
 — write, general description — Section 6.3
Command Accepted—signal line — Section 5.4.4; Section 5, Table 1
Command Operations
 — definition — Section 5
 — strobes in — Section 5.2
 — timing of — Section 7.1.3.1, Fig 9
Common Control Signals — Section 5.5; Section 5, Table 1
Complement — See Read Complement of Group 1
Computer—use of CAMAC with — Section 3(5)
Connector Plug, Dataway — Section 4.2.2
 — specification of — Fig 5, (5.1), (5.2), and (5.3); Section 4.2.2
 — chamfers on — Fig 5, (5.3); Section 4.2.2
 — horizontal datum of — Fig 5, (5.1) and (5.3)
Connector Socket, Dataway — Section 4.1.2
 — specification of — Fig 5; Section 4.1.2
 — chamfers on — Fig 5, (5.6), (5.7), and (5.8); Section 4.1.2
 — contacts — Fig 5, (5.4); Section 4.1.2
 — horizontal datum of — Figs 3 and 5, (5.4) and (5.5)
 — use of point contacts in — Section 4.1.2
 — vertical datum of — Figs 3 and 5, (5.5); Section 4.1.2
Connectors, Auxiliary — Section 4.2.5, Appendix C

Connectors, Dataway	Sections 4.1.2, 4.2.2
Connectors, Other	Section 4.2.5
Connectors, Power Supply	Appendix C
Control Commands—general description	Sections 6.2, 6.4
Controller—definition of	Section 5, Appendix A
Control Station	
— Definition of	Section 4.4
— contact allocation	Table 3
Crate	
— specification of	Section 4.1, Figs 1, 2, and 3; Section 4
— horizontal datum of	Fig 3, Section 4.1.1
— power dissipation in	Table 10
Crate Controller Type A—use of N codes	Section 6.5
Cross Coupling—between signal lines	Sections 4.4, 7.1
Current Requirements (and voltage), marking of	Section 8
Data—definition of	Section 5.3
Dataway	
— definition of	Section 4
— construction of	Section 4.4
Dataway Connector	See Connector Plug, Connector Socket
Dataway Lines	
— use of	Table 1; Section 5
— signal standards for	Section 7.1, Tables 5 and 6
Dataway Operations	
— definition of	Section 5, See also Command Operations, Unaddressed Operations
Decoding	
— of function	Section 5.1.3
— of subaddress	Section 5.1.2
Digital Signals	See Signal Standards
Disable—function code	Section 6.4.1; Table 4
Earth—clean	See Additional Power Lines
Enable—function code	Section 6.4.3; Table 4
ESONE Committee	Foreword
Execute—function code	Section 6.4.2; Table 4
$F1$, $F2$, $F4$, $F8$, $F16$	See Function
Fixing Screw—of plug-in unit	Fig 4, Section 4.2.1; Section 4.2.3
Free Bus Lines	Section 5.6.1; Section 4.4
— signal standards for	Section 7.1.4, Table 7
Function—signal lines	Section 5.1.3; Section 5, Table 1
Function Codes	
— summary of	Table 4
— definitions	Section 6
— decoding of	Section 5.1.3
Ground	See Earth
Group 1	
— registers	Section 6
— commands	Sections 6.1.1, 6.1.3, 6.1.4, 6.2.2, 6.3.1, 6.3.3, 6.3.5
Group 2	
— registers	Section 6
— commands	Sections 6.1.2, 6.2.4, 6.3.2, 6.3.4, 6.3.6
Horizontal Datum	
— of connector plug	Fig 5, (5.1) and (5.3)
— of connector socket	Figs 3 and 5, (5.4) and (5.5)
— of crate	Fig 3, Section 4.1.1
— of plug-in unit	Fig 4, Section 4.2.1

DIGITAL INTERFACE SYSTEM (CAMAC)

ANSI/IEEE
Std 583-1982

I
IEC — See International Electrotechnical Commission
Individual Lines—in Dataway — Sections 4.4, 5.1.1, 5.4.1
Inhibit—signal line — Section 5.5.2; Section 5, Table 1
Initialize
  — signal line — Section 5.5.1; Section 5, Table 1
  — use to disable LAM Requests — Section 5.4.1.1
  — use to reset LAM Status — Section 5.4.1.1
Integration—of Z and C in units — Section 7.1.3.2
International Electrotechnical Commission — Foreword, Section 4.1.1
Intrinsic OR — Section 7.1

L
LAM — See Look-at-Me
LAM Mask — Section 5.4.1, Fig 11
  — commands for access — Section 5.4.1.2, Fig 11
LAM Source — Section 5.4.1, Fig 11
LAM Request — Section 5.4.1, Fig 11
  — enabling and disabling — Section 5.4.1.1
  — commands for access — Section 5.4.1.2, Fig 11
LAM Status — Section 5.4.1, Fig 11
  — clearing — Sections 5.4.1.1, 6.2.3, Fig 11
  — commands for access — Section 5.4.1.2, Fig 11
LAM Structure — Section 5.4.1, Fig 11
L Signal — Section 5.4.1, Fig 11
Look-at-Me
  — signal line — Section 5.4.1; Section 5, Table 1
  — gating in modules — Section 5.4.1.3

Mandatory Power Lines — Section 8; 5.7, Table 1
Marking of Voltage and Current Requirements — Section 8
May—permitted practice — Section 2
Module—definition of — Section 5, Appendix A
Module Characteristic—address of — Section 6
Multiple Station Addressing — Sections 5.1.3, 5.1.1, 6.5
Multi width units — Section 4.2
  — dimensions of — Fig 4
  — L signals at — Section 5.4.1
Must—mandatory practice — Section 2

N
N — See Station Number
Negative Logic—convention of signals — Section 7.1
NIM Committee — Foreword
NIM Units
  — use in CAMAC crate — Sections 4.1, 4.3
  — power supplies for — Section 8
  — adaptor for — Fig 7; Section 4.3
Nonstandard
  — function codes — Sections 6.1.5, 6.2.5, 6.3.7, 6.4.5; Section 5.1.3
  — signal lines — Section 5.6
Normal Station
  — definition of — Section 4.4
  — contact allocation — Table 2

Overwrite—function codes — Sections 6.3.1, 6.3.2; Table 4
Overwrite, selective — Sections 6.3.3, 6.3.4

P
P1, P2
  — at normal stations — See Free Bus lines
  — at control station — See Patch Contacts
P3–P7 — See Patch Contacts
Patch Contacts — Section 5.6.2; Section 4.4
  — signal standards for — Section 7.1.4, Table 7

Patch Points	Section 5.6.2; Section 4.4
Plug-In Units	
— specification of	Section 4.2; Sections 4, 5
— power dissipation in	Table 10
— horizontal datum of	Fig 4; Section 4.2.1
— vertical datum of	Fig 4; Section 4.2.1
Power Dissipation—in plug-in unit and crate	Table 10; Section 8
Power Lines	Section 5.7; Section 4.4, Table 1
— standards of	Section 8, Table 10
— mandatory	Section 8
— additional	Section 8
— supplementary 6 V	Sections 10, 5.7, Fig 12
Power Return	Section 4.4
— 0 V line	Section 4.2.2
— width of contacts	Section 8
— resistance of	Section 8; Appendix D
Power Supply	Section 8; Appendix C
Power Supply Connector	Section 4.1.3; Fig 3
Power Supply Units—dimensions of	Fig 8; Sections 4.2, 4.2.4
Printed Wiring Card	
Pull-Up Current Source	Sections 7.1, 7.1.2, Table 6
— Dataway lines	Table 7; Section 7.1.4
— nonstandard connections	

Q

	See Response
$R1$–$R24$	See Read
Read—signal lines	Section 5.3.2; Section 5, Table 1
Read and Clear Group 1—function code	Section 6.1.3; Table 4
Read Commands—general description	Section 6.1
Read Complement of Group 1—function code	Section 6.1.4; Table 4
Read Group 1—function code	Section 6.1.1; Table 4
Read Group 2—function code	Section 6.1.2; Table 4
References	See Bibliography
Register—definition of	Section 6
Repeat Mode—of block transfer	Section 5.4.3.2
Reserved	
— function codes	Sections 6.1.5, 6.2.5, 6.4.5; Section 5.1.3
Response—signal line	Section 5.4.3; Section 5, Table 1

$S1$, $S2$	See Strobes
Selective Clear—function codes	Sections 6.3.5; 6.3.6; Table 4
Selective Overwrite	Sections 6.3.3, 6.3.4; Table 4
Selective Set—function codes	Sections 6.3.3, 6.3.4; Table 4
Should—recommended practice	Section 2
Signal Standards—for Dataway lines	
— voltage standards	Sections 7.1.1, Table 5
— current standards	Section 7.1.2, Table 6
— timing	Section 7.1.3, Figs 9 and 10
Signal Standards	Section 7
— nonstandard connections	Section 7.1.4, Table 7
— for unterminated signals	Section 7.2.1, Table 8
— for terminated signals	Section 7.2.2, Table 9
— classes and markings	Section 4.2.5; Appendix B
Significance—of data bits	Section 5.3
Station—definition of	Section 4, see also Control Station, Normal Station
Station Number—signal lines	Section 5.1.1; Section 5, Table 1
Status Information	Section 5.4
Stop Mode—of block transfer	Section 5.4.3.3
Stop-on-word mode	Section 5.4.3.4
Strobes—signal lines	Sections 5.2, 5.5; Section 5, Table 1
Subaddress	
— signal lines	Section 5.1.2; Section 5, Table 1
— decoding of	Section 5.1.2
Supplementary 6 V Power lines	Sections 10, 5.7

Terminated Signals	Section 7.2.2, Table 9
Test Look-at-Me	
— function code	Section 6.2.1; Section 5.4.1, Fig 11
— mandatory use of	Section 5.4.1.1
— use of Q with	Section 5.4.3
Test Status—function code	Section 6.4.4; Section 5.4.1, Fig 11
Transition Times of Signals	Section 7.1
Unaddressed Operations	Sections 5, 5.2, 5.5
— timing of	Section 7.1.3.2, Fig 10
Unterminated Signals	Section 7.2.1, Table 8
— protection of outputs	Section 7.2.1
Ventilation	
— of crate	Section 4.1.1, 4.1.3, Figs 2 and 6; Section 8
— of plug-in unit	Section 4.2.1
Ventilation, forced air	Section 9
Vertical Datum	
— of connector sockets	Fig 5, (5.5); Section 4.1.2, Fig 3
— of crate	Section 4.1.1, Fig 3
— of plug-in unit	Fig 4; Section 4.2.1
Voltage and Current Requirements, marking of	Section 8
W1–W24	See Write
Write—signal lines	Section 5.3.1; Section 5, Table 1
Write Commands—general description	Section 6.3
Write Data—acceptance at S2	Section 5.2
X	See Command Accepted
Y1, Y2	See Power Lines — Supplementary 6 V
Z	See Initialize

ANSI/IEEE
Std 595-1982
(Revision of ANSI/IEEE
Std 595-1976)

An American National Standard

IEEE Standard Serial Highway Interface System (CAMAC*)

Sponsor

**Instruments and Detectors Committee of the
IEEE Nuclear and Plasma Sciences Society**

Approved September 17, 1981

IEEE Standards Board

Approved December 15, 1981

American National Standards Institute

*Computer Automated Measurement and Control

© Copyright 1982 by

**The Institute of Electrical and Electronics Engineers, Inc
345 East 47th Street, New York, NY 10017**

*No part of this publication may be reproduced in any form,
in an electronic retrieval system or otherwise,
without the prior written permission of the publisher.*

IEEE Standards documents are developed within the Technical Committees of the IEEE Societies and the Standards Coordinating Committees of the IEEE Standards Board. Members of the committees serve voluntarily and without compensation. They are not necessarily members of the Institute. The standards developed within IEEE represent a consensus of the broad expertise on the subject within the Institute as well as those activities outside of IEEE which have expressed an interest in participating in the development of the standard.

Use of an IEEE Standard is wholly voluntary. The existence of an IEEE Standard does not imply that there are no other ways to produce, test, measure, purchase, market, or provide other goods and services related to the scope of the IEEE Standard. Furthermore, the viewpoint expressed at the time a standard is approved and issued is subject to change brought about through developments in the state of the art and comments received from users of the standard. Every IEEE Standard is subjected to review at least once every five years for revision or reaffirmation. When a document is more than five years old, and has not been reaffirmed, it is reasonable to conclude that its contents, although still of some value, do not wholly reflect the present state of the art. Users are cautioned to check to determine that they have the latest edition of any IEEE Standard.

Comments for revision of IEEE Standards are welcome from any interested party, regardless of membership affiliation with IEEE. Suggestions for changes in documents should be in the form of a proposed change of text, together with appropriate supporting comments.

Interpretations: Occasionally questions may arise regarding the meaning of portions of standards as they relate to specific applications. When the need for interpretations is brought to the attention of IEEE, the Institute will initiate action to prepare appropriate responses. Since IEEE Standards represent a consensus of all concerned interests, it is important to ensure that any interpretation has also received the concurrence of a balance of interests. For this reason IEEE and the members of its technical committees are not able to provide an instant response to interpretation requests except in those cases where the matter has previously received formal consideration.

Comments on standards and requests for interpretations should be addressed to:

    Secretary, IEEE Standards Board
    345 East 47th Street
    New York, NY 10017
    USA

Foreword

[This Foreword is not a part of ANSI/IEEE Std 595-1982, IEEE Standard Serial Highway Interface System (CAMAC).]

This standard defines a serial highway interface system for use with CAMAC crate assemblies in accordance with ANSI/IEEE Std 583-1982, Modular Instrumentation and Digital Interface System (CAMAC), and with other controlled devices. It is based on ERDA[1] Report TID-26488, Dec 1973, and ESONE Report ESONE/SH/01, Nov 1974 (modified by subsequently issued supplements), as developed by the NIM Committee of the US Department of Energy,* and the ESONE Committee of European Laboratories.** A parallel highway interface system, also intended for use with ANSI/IEEE Std 583-1982, is defined in ANSI/IEEE Std 596-1982, Parallel Highway Interface System (CAMAC). A report essentially identical to ANSI/IEEE Std 595-1982 has been issued by the ESONE Committee as EUR 6100.

This standard was reviewed and balloted by the Nuclear Instruments and Detectors Committee of the IEEE Nuclear and Plasma Sciences Society. Because of the broad applicability of this standard, coordination was established with numerous IEEE societies, groups, and committees, including the Communications Society, Computer Society, Industry Applications Society, Industrial Electronics and Control Instrumentation Group, Instrumentation and Measurement Group, and the Power Generation and Nuclear Power Engineering Committees of the Power Engineering Society. The revision of this standard was in conjunction with the 1981 review (1982 issue) of the entire family of IEEE CAMAC standards, undertaken to incorporate existing addenda and corrections directly into the standard.

At the time of approval of this standard, the membership of the Nuclear Instruments and Detectors Committee of the IEEE Nuclear and Plasma Sciences Society was as follows:

D. C. Cook, *Chairman*   **Louis Costrell**, *Secretary*

J. A. Coleman	T. R. Kohler	P. L. Phelps
J. F. Detko	H. W. Kraner	J. H. Trainor
F. S. Goulding	W. W. Managan	S. Wagner
F. A. Kirsten	G. L. Miller	F. J. Walter
	D. E. Persyk	

At the time it approved this standard, the American National Standards Committee on Nuclear Instruments, N42, had the following representatives:

Louis Costrell, *Chairman*   **D. C. Cook**, *Secretary*

Organization Represented	*Name of Representative*
American Conference of Governmental Industrial Hygienists	Jesse Lieberman
American Nuclear Society	Frank W. Manning
Health Physics Society	J. B. Horner Kuper
	J. M. Selby *(Alt)*
Institute of Electrical and Electronics Engineers	Louis Costrell
	Julian Forster *(Alt)*
	David C. Cook *(Alt)*
	A. J. Spurgin *(Alt)*
Instrument Society of America	M. T. Slind
	J. E. Kaveckis *(Alt)*
Lawrence Berkeley National Laboratory	Lee J. Wagner
Oak Ridge National Laboratory	Frank W. Manning
	D. J. Knowles *(Alt)*
US Department of the Army, Materiel Command	Abraham E. Cohen
US Department of Commerce, National Bureau of Standards	Louis Costrell
US Department of Energy	Gerald Goldstein
US Federal Emergency Management Agency	Carl R. Siebentritt, Jr
US Nuclear Regulatory Commission	Robert E. Alexander
US Naval Research Laboratory	David C. Cook
Members-at-Large	J. G. Bellian
	O. W. Bilharz, Jr
	John M. Gallagher, Jr
	Voss A. Moore
	R. F. Shea
	E. J. Vallario

---

[1] Energy Research and Development Administration (now part of Department of Energy).

*NIM Committee[1]
L. Costrell, Chairman

C. Akerlof	N. W. Hill	L. B. Mortara
E. J. Barsotti	D. Horelick	V. C. Negro
B. Bertolucci	M. E. Johnson	L. Paffrath
J. A. Biggerstaff	C. Kerns	D. G. Perry
A. E. Brenner	F. A. Kirsten	I. Pizer
R. M. Brown	P. F. Kunz	E. Platner
E. Davey	R. S. Larsen	S. Rankowitz
W. K. Dawson	A. E. Larsh, Jr	S. J. Rudnick
S. R. Deiss	R. A. LaSalle	G. K. Schulze
S. Dhawan	N. Latner	W. P. Sims
R. Downing	F. R. Lenkszus	D. E. Stilwell
T. F. Droege	R. Leong	J. H. Trainor
C. D. Ethridge	C. Logg	K. J. Turner
C. E. L. Gingell	S. C. Loken	H. Verweij
A. Gjovig	D. R. Machen	B. F. Wadsworth
B. Gobbi	D. A. Mack	L. J. Wagner
D. B. Gustavson	J. L. McAlpine	H. V. Walz
D. R. Heywood		D. H. White

NIM Executive Committee for CAMAC
L. Costrell, Chairman

E. J. Barsotti   F. A. Kirsten
J. A. Biggerstaff   R. S. Larsen
S. Dhawan   D. A. Mack
           J. H. Trainor

NIM Dataway Working Group
F. A. Kirsten, Chairman

E. J. Barsotti	C. Kerns
J. A. Biggerstaff	P. F. Kunz
L. Costrell	R. S. Larsen
S. Dhawan	D. R. Machen
A. Gjovig	L. Paffrath
D. R. Heywood	S. Rankowitz
D. Horelick	S. J. Rudnick

NIM Software Working Group
D. B. Gustavson, Jr, Chairman
W. K. Dawson, Secretary

A. E. Brenner	P. A. LaSalle
R. M. Brown	F. B. Lenkszus
S. R. Deiss	C. A. Logg
S. Dhawan	J. McAlpine
C. D. Ethridge	L. B. Mortara
M. E. Johnson	D. G. Perry

NIM Mechanical and Power Supplies Working Group
L. J. Wagner, Chairman

L. Costrell
C. Kerns
S. J. Rudnick
W. P. Sims

NIM Analog Signals Working Group
D. I. Porat, Chairman

L. Costrell
C. E. L. Gingell
N. W. Hill
S. Rankowitz

NIM Serial System Subgroup
D. R. Machen, Chairman

E. J. Barsotti
D. Horelick
F. A. Kirsten
R. G. Martin
L. Paffrath
S. J. Rudnick

NIM Block Transfer Subgroup
E. J. Barsotti, Chairman

W. K. Dawson
F. A. Kirsten
R. A. LaSalle
F. R. Lenkszus
R. G. Martin
R. F. Thomas, Jr

NIM Multiple Controllers Subgroup
P. F. Kunz, Chairman

E. J. Barsotti   D. R. Machen
F. A. Kirsten   R. G. Martin

---

[1] National Instrumentation Methods Committee of the US Department of Energy.

**ESONE Committee[2]

W. Attwenger, Austria, *Chairman* 1980–81; H. Meyer, Belgium, *Secretary*

Representatives of ESONE Member Laboratories

W. Attwenger, Austria
R. Biancastelli, Italy
L. Binard, Belgium
J. Biri, Hungary
B. Bjarland, Finland
D. A. Boyce, England
B. A. Brandt, F. R. Germany
F. Cesaroni, Italy
P. Christensen, Denmark
W. K. Dawson, Canada
M. DeMarsico, Italy
C. A. DeVries, Netherlands
H. Dilcher, F. R. Germany
B. V. Fefilov, USSR
R. A. Hunt, England
W. Kessel, F. R. Germany
R. Klesse, France
E. Kwakkel, Netherlands
J. L. Lecomte, France
J. Lingertat, D. R. Germany
M. Lombardi, Italy
M. Maccioni, Italy
P. Maranesi, Italy
C. H. Mantakas, Greece
D. Marino, Italy
H. Meyer, Belgium
K. D. Muller, F. R. Germany
J. G. Ottes, F. R. Germany
A. D. Overtoom, Netherlands
E. C. G. Owen, England
L. Panaccione, Italy
M. Patrutescu, Roumania
R. Patzelt, Austria
A. C. Peatfield, England
I. C. Pyle, England
B. Rispoli, Italy
M. Sarquiz, France
W. Schoeps, Switzerland
R. Schule, F. R. Germany
P. G. Sjolin, Sweden
L. Stanchi, Italy
R. Trechcinski, Poland
M. Truong, France
P. Uuspaa, Finland
H. Verweij, Switzerland
A. J. Vickers, England
S. Vitale, Italy
M. Vojinovic, Yugoslavia
K. Zander, F. R. Germany
D. Zimmermann, F. R. Germany

ESONE Executive Group (XG)

W. Attwenger, Austria, XG, *Chairman* 1980–81
H. Meyer, Belgium, *Secretary*

P. Christensen, Denmark
M. Dilcher, F. R. Germany
B. Rispoli, Italy
M. Sarquiz, France
R. Trechcinski, Poland
A. Vickers, England
H. Verweij, Switzerland

ESONE Technical Coordination Committee (TCC)

A. C. Peatfield, England, *TCC Chairman*
P. Christensen, Denmark, *TCC Secretary*

W. Attwenger, Austria
R. Biancastelli, Italy
G. Bianchi, France
H. Dilcher, F. R. Germany
P. Gallice, France
W. Kessel, F. R. Germany
J. Lukacs, Hungary
R. Patzelt, Austria
P. J. Ponting, Switzerland
W. Schoeps, Switzerland
S. Vitale, Italy

ECA/ESONE CAMAC Document Maintenance Study Group (DMSG)

P. Gallice, France, *Chairman*

R. C. M. Barnes, England
L. Besse, Switzerland
J. Davis, England
F. Iselin, Switzerland
H. Meyer, Belgium
H. J. Trebst, F. R. Germany

When the IEEE Standards Board approved this standard on September 17, 1981, it had the following membership:

Irvin N. Howell, Jr, *Chairman*   Irving Kolodny, *Vice Chairman*

Sava I. Sherr, *Secretary*

G. Y. R. Allen
J. J. Archambault
James H. Beall
John T. Boettger
Edward Chelotti
Edward J. Cohen
Len S. Corey
Jay Forster
Kurt Greene
Loering M. Johnson
Joseph L. Koepfinger
J. E. May
Donald T. Michael*
J. P. Riganati
F. Rosa
Robert W. Seelbach
Jay A. Stewart
W. E. Vannah
Virginius N. Vaughan, Jr
Art Wall
Robert E. Weiler

*Member emeritus.

---

[2]European Standards on Nuclear Electronics Committee.

Contents

SECTION	PAGE
Abbreviations and symbols	13
CAMAC and NIM Standards and Reports	14
1. Introduction	15
1.1 Aims	15
1.2 Interpretation	15
2. Principles of the Serial Highway System	16
2.1 Configuration	16
2.2 Messages	17
2.3 Transmission of Bytes	17
2.4 System Clock	18
2.5 Serial Highway Ports	18
2.6 Serial Driver	19
2.7 Extended uses of the Serial Highway	19
2.8 Serial Crate Controller	20
3. Message Structure for Serial Crate Controllers	20
3.1 Command Message	20
3.1.1 Complete Command Messages	20
3.1.2 Truncated Command Message	21
3.2 Reply Message	21
3.3 Demand Message	22
3.4 Message Fields	22
3.4.1 Crate Address Field (6 Bits; $SC1$-$SC32$)	22
3.4.2 Station Number Field (5 Bits; $SN1$-$SN16$)	22
3.4.3 Subaddress Field (4 Bits; $SA1$-$SA8$)	22
3.4.4 Function Field (5 Bits; $SF1$-$SF16$)	22
3.4.5 Write-Data Field (24 Bits; $SW1$-$SW24$)	23
3.4.6 Read-Data Field (24 Bits; $SR1$-$SR24$)	23
3.4.7 Message Identification Field (2 Bits; $M1$-$M2$)	23
3.4.8 Status Field (4 Bits; Error, Response, Command Accepted, Delayed Error)	23
3.4.9 Serial Graded-L (SGL) Field (5 Bits; $SGL1$-$SGL5$)	23
3.5 Formatting Bytes	23
3.5.1 Delimiter Bytes	23
3.5.2 Column Parity Field	24
3.5.3 END Byte	24
3.5.4 WAIT Byte	24
3.5.5 ENDSUM Byte	25
3.5.6 SUM Byte	25
3.5.7 SPACE Byte	25
4. Command/Reply Message Sequences	25
4.1 General Requirements	25
4.1.1 Find Header State	29
4.1.2 Receive Command State	29
4.1.3 Execute Command State	29
4.1.4 Send Reply State	29
4.2 Read Operation	30
4.3 Write Operation	30
4.4 Control Operation	30
4.5 Truncation of the Command Message	31

SECTION			PAGE
	4.6	Reply Space	31
		4.6.1 Completion of Dataway Operation	31
		4.6.2 Termination of Command/Reply Transaction	31
		4.6.3 Length of the Reply Space	32
5.		Demand Message Generation	32
	5.1	Control of Demand Message Initiation	34
	5.2	Delay Buffer	34
	5.3	Identification of Demands	35
6.		Identification of Message Type	35
	6.1	Complete Command Message (Minimum Length 8 Bytes; MI=00)	35
	6.2	Truncated Command Message (Length 2 Bytes; MI None)	36
	6.3	Reply Message (Length 3 or 7 Bytes; MI=01)	36
	6.4	Demand Message (Length 3 Bytes; MI=1–)	36
7.		Serial Highway *D* Ports	36
	7.1	*D*-Port Connectors	37
		7.1.1 Mechanical	37
		7.1.2 Contact Assignments	37
	7.2	Data and Clock Signals	38
		7.2.1 Transmission Lines	38
		7.2.2 Logic States	38
		7.2.3 Balanced Transmitter	38
		7.2.4 Balanced Receiver	39
	7.3	Control Signals	39
		7.3.1 Signal Standards	39
		7.3.2 Control Signal Sources and Receivers	39
8.		Timing	41
	8.1	Frequency of System Clock	41
	8.2	Byte Stream	42
	8.3	Signal Timing	42
		8.3.1 Clock Signals	42
		8.3.2 Transmitted Data Signals	43
		8.3.3 Received Data Signals	43
		8.3.4 Data Signal Conditioning	43
	8.4	Propagation Delays	43
9.		Bit-Serial and Byte-Serial Modes	43
	9.1	Byte-Serial Mode	43
	9.2	Bit-Serial Mode	44
		9.2.1 Noncontiguous Byte Frames	44
		9.2.2 Nonuniform Bit Periods	44
		9.2.3 Extraction of Byte Clock	44
10.		Synchronization	45
	10.1	Message Synchronization	45
		10.1.1 Maintenance of Message Synchronism	45
		10.1.2 Loss of Message Synchronism	45
		10.1.3 Establishing Message Synchronism	45
	10.2	Byte Synchronization	46
		10.2.1 Maintenance of Byte Synchronism	46
		10.2.2 Loss of Byte Synchronism	46
		10.2.3 Establishing Byte Synchronism	46
	10.3	Lost Synchronism: Actions by Serial Crate Controller	46

SECTION			PAGE
11.	Access to Registers in Serial Crate Controller		47
	11.1	Status Register	47
	11.2	Other Registers	48
		11.2.1 Look-at-Me Pattern	48
		11.2.2 Reread Data	49
12.	Features of Serial Crate Controller Accessed Via the Status Register		49
	12.1	Dataway Common Controls	49
		12.1.1 Initialize and Clear	49
		12.1.2 Inhibit	50
	12.2	Command/Reply Transaction Status	50
	12.3	Demand Handling	50
		12.3.1 Enable Demand Messages	50
		12.3.2 Demand $L24$	51
		12.3.3 Selected Look-at-Me Present	51
	12.4	Reconfiguration Options	51
		12.4.1 Dataway Off-Line State	51
		12.4.2 Bypass	52
		12.4.3 Loop Collapse	53
13.	Serial Crate Controllers: Front Panel Features		53
	13.1	Manual Controls	53
		13.1.1 Crate Address Switch	53
		13.1.2 Dataway Off-Line Switch	53
		13.1.3 Initialize and Clear Switches	53
	13.2	Indicators	54
		13.2.1 Crate Address	54
		13.2.2 Dataway On-Line	54
		13.2.3 Dataway Inhibit	54
		13.2.4 Unsynchronized State	54
		13.2.5 Crate Addressed	54
		13.2.6 Demand Message	54
		13.2.7 Bypassed	54
	13.3	Connectors	54
	13.4	Other Front Panel Features	54
14.	SGL-Encoder Connector		54
	14.1	Mechanical	54
	14.2	Signals at the SGL-Encoder Connector	55
		14.2.1 Signals $L1$-$L24$	56
		14.2.2 L-sum	56
		14.2.3 Signals SGLE1-SGLE5	56
		14.2.4 Demand Message Initiate	56
		14.2.5 Selected Look-at-Me Present	56
		14.2.6 Demand Busy	57
		14.2.7 External Repeat	57
		14.2.8 Byte Clock	57
		14.2.9 Start Timer	57
		14.2.10 Time Out	58
		14.2.11 Controller Busy	58
		14.2.12 Signals $N1$, $N2$, $N4$, $N8$, $N16$	58
		14.2.13 Auxiliary Controller Lockout	58
		14.2.14 Request Inhibit	59
	14.3	Signal Standards for the SGL-Encoder Connector	59

SECTION			PAGE
	14.4	Hung Demand Time Out	59
		14.4.1 Internal Timer	60
		14.4.2 External Timer	60
	14.5	SGL-Encoder Options	61
		14.5.1 Passive SGL Encoder	61
		14.5.2 Demand Masking	61
		14.5.3 SGL-Field Encoding	61
		14.5.4 False Hung Demand	61
	14.6	Access for Auxiliary Controllers	61
		14.6.1 Access to N Lines	61
		14.6.2 Auxiliary Controller Lockout Signal	61
		14.6.3 Interlock Between Serial Crate Controller and Auxiliary Controller	62
15.	Recovery from Errors		62
	15.1	Transmission-Path Failures	62
		15.1.1 Failures Within Serial Crate Controllers: Bypass Switching	62
		15.1.2 Failures in the Serial Highway: Loop Collapse Switching	63
	15.2	Loss of Synchronism	64
	15.3	Transmission Errors	64
		15.3.1 Principle of Geometric Code	64
		15.3.2 Implementation of Geometric Code	64
		15.3.3 Performance of Geometric Code	65
		15.3.4 Error Detection by Context	65
		15.3.5 Error Detection at Modules	66
	15.4	The Error-Reply Message	66
	15.5	Error Indications in Reply Messages	66
		15.5.1 Error Bit	66
		15.5.2 Command Accepted Bit	66
		15.5.3 Delayed-Error Bit	67
		15.5.4 Delayed Response Bits	67
	15.6	Error Recovery Using the Reread Command	67
16.	Summary: Sequence of Actions in Serial Crate Controller		68
	16.1	Find Header	69
		16.1.1 Exit to Receive Command State	69
		16.1.2 Exit to Pass Message State	69
		16.1.3 Exit to Send Demand State	69
		16.1.4 Error Exit to Pass Message State	69
		16.1.5 Error Exit to Lost Byte Sync State	70
	16.2	Receive Command	70
		16.2.1 Exit to Execute Command State	70
		16.2.2 Error Exit to Send Reply State (ERR = 1)	70
		16.2.3 Error Exit to Send Reply State (Off-Line)	70
		16.2.4 Error Exit to Send Reply State (Bypassed)	70
		16.2.5 Error Exit to Lost Message Sync State	70
		16.2.6 Error Exit to Lost Byte Sync State	70
	16.3	Execute Command	70
		16.3.1 Exit to Send Reply State	70
		16.3.2 Error Exit to Lost Message Sync State	70
		16.3.3 Error Exit to Lost Byte Sync State	70

SECTION			PAGE
16.4	Send Reply		70
	16.4.1	Exit to Find END State	70
	16.4.2	Exit to Find Header State	70
	16.4.3	Error Exit to Lost Message Sync State	70
	16.4.4	Error Exit to Lost Byte Sync State	71
16.5	Find END		71
	16.5.1	Exit to Find Header State	71
	16.5.2	Error Exit to Lost Byte Sync State	71
16.6	Pass Message		71
	16.6.1	Exit to Find Header State	71
	16.6.2	Error Exit to Lost Byte Sync State	71
16.7	Send Demand		71
	16.7.1	Exit to Find Header State	71
	16.7.2	Error Exit to Lost Byte Sync State	71
16.8	Lost Byte Sync		71
	16.8.1	Exit to Lost Message Sync State	71
16.9	Lost Message Sync		71
	16.9.1	Exit to Find Header State	71
	16.9.2	Error Exit to Lost Byte Sync State	71

TABLES

Table 3-1	Length of Command/Reply Transactions	23
Table 3-2	Contents of Message Identification Field	23
Table 7-1	Contact Assignments for D-Port Connectors	37
Table 7-2	Summary of Characteristics of Balanced Transmitter	38
Table 7-3	Summary of Characteristics of Balanced Receiver	39
Table 7-4	Standards for Control Signals at D Ports	40
Table 11-1	Commands Implemented by SCC	47
Table 11-2	Assignment of Status Register Bits	48
Table 11-3	Initial State of Status Register Bits After Power Up	49
Table 12.1	Control of Dataway Off-Line State	52
Table 14-1	Contact Assignments at SGL-Encoder Connector	55
Table 14-2	SGL-Encoder Connector: Signal Standards and Pull-Up Current Sources for All Signals Other Than Coded N	59
Table 15-1	Error Indications in Reply Message	66

FIGURES

Fig. 2-1	CAMAC SH Loop Configuration	16
Fig. 2-2	Basic Message Format	17
Fig. 2-3	Bit-Serial Byte Frame	17
Fig. 2-4	Direct D-Port Interconnection	18
Fig. 2-5	Indirect Connection Via Undefined Standard	19
Fig. 2-6	Compatible Devices	19
Fig. 3-1	Command Message: Bit Assignments	20
Fig. 3-2	Command Message: Field Assignments	21
Fig. 3-3	Truncated Command Message: Bit Assignments	21
Fig. 3-4	Truncated Command Message: Field Assignments	21
Fig. 3-5	Reply Message: Bit Assignments	21
Fig. 3-6	Reply Message: Field Assignments	22

FIGURES		PAGE
Fig. 3-7	Demand Message: Bit Assignments	22
Fig. 3-8	Demand Message: Field Assignments	22
Fig. 4-1	Command/Reply Sequence: Read Operation, Bit-Serial Mode	26
Fig. 4-2	Command/Reply Sequence: Read Operation, Byte-Serial Mode	26
Fig. 4-3	Command/Reply Sequence: Write Operation, Bit-Serial Mode	27
Fig. 4-4	Command/Reply Sequence: Write Operation, Byte-Serial Mode	27
Fig. 4-5	Command/Reply Sequence: Control Operation, Bit-Serial Mode	28
Fig. 4-6	Command/Reply Sequence: Control Operation, Byte-Serial Mode	28
Fig. 5-1	Example of Message Sequence in Loop Having Three SCCs, Showing Demand Message in Three Contexts	33
Fig. 5-2	Demand Message Generation: *A*—Demand Message Directly Displaces WAIT Bytes: *B*—Demand Message Delays Incoming Message	33
Fig. 7-1	Example of Use of Bus 1 and Bus 2 Contacts at *D* Ports	38
Fig. 7-2	Example of Balanced Transmitter	39
Fig. 7-3	Example of Balanced Receiver	39
Fig 7.3a	Example of a Receiver with Hysterisis Added by Use of External Resistors	39
Fig. 7-4	Examples of Circuits for Control Signal Sources and Receivers	40
Fig. 8-1	Timing of Clock and Data Signals at *D* Ports	41
Fig. 14-1	Example of Associated Parts of SCC and SGL Encoder	56
Fig. 14-2	Relationship Between Byte-Clock Signal at SGL-Encoder Connector and Received Bit/Byte-Clock Signals: *A*—Byte-Serial Mode; *B*—Bit Serial Mode	57
Fig. 14-3	Relationship Between Signals at SGL-Encoder Connector Concerned with Demand Message Generation: *A*—Shown for Duration of LAM Less Than and Greater Than Time-Out Period; *B*—Similar to *A* except for Demand Enable After LAM Assertion	58
Fig. 15-1	Bypass Switching for One *D* Port Signal	63
Fig. 15-2	Loop Collapse Switching for One *D*-Port Signal	63
Fig. 15-3	Geometric Error Detection: Basic Principle	64
Fig. 15-4	Geometric Error Detection: As applied to SH	64
Fig. 16-1	Major-State Sequence in SCC	68
Fig. 16-2	Major-State Sequence in SCC: Omitting All Error Conditions	69

APPENDIX

A1.	Specification of CAMAC Serial Crate Controller Type L2		72
	A1.1	Interpretation	72
	A1.2	General Features of Serial Crate Controller Type L2	72
		A1.2.1 Format	72
		A1.2.2 Dataway Connectors	72
	A1.3	Messages for Serial Crate Controller Type L2	72
		A1.3.1 Message Formats	72
		A1.3.2 Message Sequences	72
		A1.3.3 Message Fields	73
		A1.3.4 Demand Message Generation	73
	A1.4	Serial Highway *D* Ports on Serial Crate Controller Type L2	73
	A1.5	Internal Structure of Serial Crate Controller Type L2	73
		A1.5.1 Synchronization	73
		A1.5.2 Status Register	73
		A1.5.3 Commands Executed	73
	A1.6	Front Panel Features of Serial Crate Controller Type L2	74
		A1.6.1 Manual Controls and Indicators	74
		A1.6.2 Other Front Panel Features	74
	A1.7	SGL-Encoder Connector on Serial Crate Controller Type L2	74

APPENDIX			PAGE

A2. Supplementary Information 74
    A2.1 Transition Diagram 74
    A2.2 Flow Chart 74
    A2.3 Block Diagram 74
        A2.3.1 Serial Input and Output 76
        A2.3.2 Command/Reply Transaction 76
        A2.3.3 Onward Transmission of Other Messages 77
        A2.3.4 Demand Message 77
A3. Bibliography 80

APPENDIX FIGURES

Fig A2-1 Transition Diagram for Serial Crate Controller Type L2 75
Fig A2-2 Implementation Independent Flow Chart of SCC Type L2 78
Fig A2-3 SCC Type L2 Block Diagram 79

Index... 81

Abbreviations and Symbols used in this Standard

A	Subaddress (Dataway signal)*
ACL	Auxiliary Controller Lockout
B	Busy (Dataway signal)*
BCK	Byte Clock
C	Clear (Dataway signal)*
CBY	Controller Busy
DSBY	Demand Busy
DERR	Delayed Error
DMI	Demand Message Initiate
DSQ	Delayed Q Response
DSX	Delayed Command Accepted Response
ERPT	External Repeat
ERR	Error bit
I	Inhibit (Dataway signal)*
L	Look-at-Me (Dataway signal)*
LAM	Look-at-Me (Demand)
LSB	Least Significant bit
MI	Message Identification
MSB	Most Significant bit
N	Station Number (Dataway signal)*
NRZL	Non-Return-to-Zero-Level
PH	Parallel Highway of ANSI/IEEE Std 596
Q	Response, Status (Dataway signal)*
S	Prefix for Serial Highway fields and bits
S1	Strobe (Dataway signal)*
S2	Strobe (Dataway signal)*
SA	Subaddress bit
SC	Crate Address bit
SCC	Serial Crate Controller
SCC–L2	Serial Crate Controller, Type L2
SD	Serial Driver
SF	Function bit
SGL	Serial Graded LAM (Demand)
SGLE	Demand Message bits from SGL-Encoder
SH	Serial Highway of ANSI/IEEE Std 595
SLP	Selected LAM Present
SN	Station Number bit
SQ	Q Response bit
SR	Read bit
STIM	Start Timer
SW	Write bit
SX	Command Accepted bit
T	Clock Period
TIMO	Time-Out
X	Command Accepted (Dataway signal)*
Z	Initialize (Dataway signal)*

*ANSI/IEEE Std 583-1982

CAMAC and NIM Standards and Reports

Title	IEEE, ANSI Std No	IEC No	DOE No	EURATOM (EUR) or ESONE No
CAMAC Instrumentation and Interface Standards*	SH08482* (Library of Congress No 8185060)	—	—	—
Modular Instrumentation and Digital Interface System (CAMAC)	ANSI/IEEE Std 583-1982	516	TID-25875† and TID-25877†	EUR 4100e
Serial Highway Interface System (CAMAC)	ANSI/IEEE Std 595-1982	640	TID-26488†	EUR 6100e
Parallel Highway Interface System (CAMAC)	ANSI/IEEE Std 596-1982	552	TID-25876† and TID-25877†	EUR 4600e
Multiple Controllers in a CAMAC Crate	ANSI/IEEE Std 675-1982	729	DOE/EV-0007	EUR 6500e
Block Transfers in CAMAC Systems	ANSI/IEEE Std 683-1976 (Reaff 1981)	677	TID-26616†	EUR 4100 suppl
Amplitude Analog Signals within a 50 Ω System	—	—	TID-26614	EUR 5100e
The Definition of IML A Language for Use in CAMAC Systems	—	—	TID-26615	ESONE/IML/01
CAMAC Tutorial Articles	—	—	TID-26618	—
Real-Time BASIC for CAMAC	ANSI/IEEE Std 726-1982	§	TID-26619†	ESONE/RTB/03
Subroutines for CAMAC	ANSI/IEEE Std 758-1979 (Reaff 1981)	713	DOE/EV-0016†	ESONE/SR/01
Recommendations for CAMAC Serial Highway Drivers and LAM Graders for the SCC-L2	—	—	DOE/EV-0006	ESONE/SD/02
Definitions of CAMAC Terms	Included in SH08482	678	DOE/ER-0104	ESONE/GEN/01
Standard Nuclear Instrument Modules NIM	—	547**	TID-20893 (Rev 4)	—

†Superseded by corresponding IEEE Standard listed.

*This is a hard cover book that contains ANSI/IEEE Std 583-1982, ANSI/IEEE Std 595-1982, ANSI/IEEE Std 596-1982, ANSI/IEEE Std 675-1982, ANSI/IEEE Std 683-1976 (Reaff 1981), ANSI/IEEE Std 726-1982 and IEEE Std 758-1979 (Reaff 1981), plus introductory material and a glossary of CAMAC terms.

**Covers only mechanical features and connector pin assignments.

§ In preparation.

NOTE: *Availability of Documents*
ANSI Sales Department, American National Standards Institute, 1430 Broadway, New York, NY 10018.
IEEE IEEE Service Center, 445 Hoes Lane, Piscataway, New Jersey 08854, USA.
IEC International Electrotechnical Commission, 1, rue de Varembé, CH-1211 Geneva 20, Switzerland.
DOE and TID Reports National Bureau of Standards, Washington, D.C. 20234, USA, Attn: L. Costrell.
EURATOM Office of Official Publications of the European Communities, P.O. Box 1003, Luxembourg.
ESONE Commission of the European Communities, CGR-BCMN, B-2440 GEEL, Belgium, Attn: ESONE Secretariat, H. Meyer.

An American National Standard

IEEE Standard Serial Highway Interface System (CAMAC)

1. Introduction

1.1 Aims. This standard defines a Serial Highway (SH) system using byte-organized messages, and configured as a unidirectional loop to which are connected a system controller and up to sixty-two CAMAC* crate assemblies in accordance with ANSI/IEEE Std 583-1982, [Modular Instrumentation and Digital Interface System (CAMAC*)] or other controlled devices. The highway transfers data and control information in either bit-serial mode (using one data signal and a bit-clock signal) or byte-serial mode (using eight data signals and a byte-clock signal). Clock rates up to 5 MHz may be used, depending on individual system characteristics.

In the primary application the controlled devices are CAMAC-crate assemblies, with Serial Crate Controllers (SCCs) which conform to a defined message structure. In this application the SH is intended to complement the Parallel Highway (PH) defined in ANSI/IEEE Std 596-1982, Parallel Highway Interface System (CAMAC). It will be attractive in certain applications that the PH was not designed to cover, for example, where there are long distances between crates, or where simplicity of interconnections is desirable. However, the time required to perform a complete operation, including a Dataway cycle, will generally be longer in a serial system than on the PH.

In other applications, some or all of the controlled devices connected to the SH can be equipment that conforms to a subset of the full specification, and is not necessarily constructed in CAMAC format or controlled by CAMAC commands. SCCs conforming to the full specification and devices conforming to the subset can coexist on the highway without mutual interference. Other devices and buses, such as that of IEEE Std 488-1978, Digital Interface for Programmable Instrumentation and Related System Components, can be readily incorporated into the system through an interfacing module.

The SH system is defined primarily in terms of the message format and signal standards at the input and output ports of devices connected to the highway. Interconnections between devices may be made directly, using the defined signal standards, or indirectly through communications channels with other signal standards and types of modulation.

A recommended SCC Type L2 is defined and provides facilities for either bit-serial or byte-serial operation.

1.2 Interpretation. This standard is a reference text describing and specifying the CAMAC SH system. It should be read in conjunction with, and is supplementary to ANSI/IEEE Std 583-1982.

No part of this standard is intended to supersede or modify ANSI/IEEE Std 583-1982.

In this standard there are mandatory requirements, recommendations, and examples of permitted practice.

| Mandatory clauses of the standard are enclosed in blocks, as here, and usually include the word *must*. |

Definitions of recommended or preferred practice (to be followed unless there are sound reasons to the contrary) include the word *should*.

Examples of permitted practice generally include the word *may*, and leave freedom of choice to the designer or user.

INTERFACE SYSTEM (CAMAC)

> In order to *conform* with the specification of the CAMAC SH system, an equipment or system must satisfy all the mandatory requirements in this standard, excluding the Appendix. If constructed as a CAMAC plug-in unit, the equipment must also satisfy the mandatory requirements of ANSI/IEEE Std 583-1982.

Section A1 of this standard's appendix defines SCC Type L2 in such a way that operationally interchangeable units can be produced by different manufacturers. The main text of this specification contains a less restrictive definition of SCCs that are not necessarily interchangeable.

> The single letter designations *L*, *M*, *N*, *P*, and *R* (prefixed in some cases by "CC") are reserved for future use in connection with this standard and are not to be used until assigned.

> In order to *conform* with this specification of the CAMAC SCC Type L2, an equipment must satisfy all the mandatory requirements of Section A1 of the appendix to this standard.
>
> In order to be *compatible* with the CAMAC SH system, equipment need not satisfy all the mandatory requirements, but must not interfere with the full operation of all the features of the SH and SCC (including Type L2) as defined in this standard.

No part of this standard is intended to exclude the use of equipment that is compatible in the preceding sense, even if it does not conform fully to the specification or is not constructed as CAMAC plug-in units.

No license or other permission is needed in order to use this standard.

2. Principles of the Serial Highway System

This section summarizes the basic principles that apply to all devices connected to the SH.

All other sections of this standard are concerned with the primary application, where the connected devices are CAMAC crate assemblies with SCCs.

2.1 Configuration. The SH interconnects a master device [the Serial Driver (SD)] and up to sixty-two CAMAC crate assemblies or other controlled devices. At any time there is only one active master device, but the standard does not exclude systems in which more than one device is capable of acting as master. Fig 2-1 shows the basic configuration.

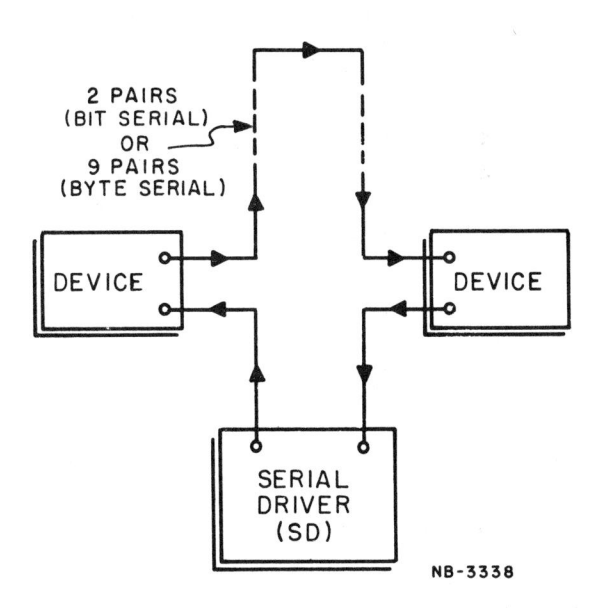

Fig 2-1
CAMAC SH Loop Configuration
(Sixty-Two Addressable Devices)

The addressing scheme allows a maximum of sixty-two controlled devices, whose assigned addresses need not be related to the actual sequence of devices along the highway.

The SH forms a unidirectional loop from the output port of the SD, through each controlled device in turn, and back to the input port of the SD. (When describing conditions with respect to a particular device, it is often convenient to use the term 'upstream' to refer to the part of the SH between the output port of the

SD and the device, and the term 'downstream' to refer to the part between the device and the input port of the SD.)

2.2 Messages. All messages transmitted on the SH consist of sequences of 8-bit bytes as shown in Fig 2-2. All information related to the message is contained within these 8-bit bytes.

The 8 bits constituting a byte are labeled Bit 1 (least significant) to Bit 8 (most significant). In all bytes, Bits 1-6 are available for information fields.

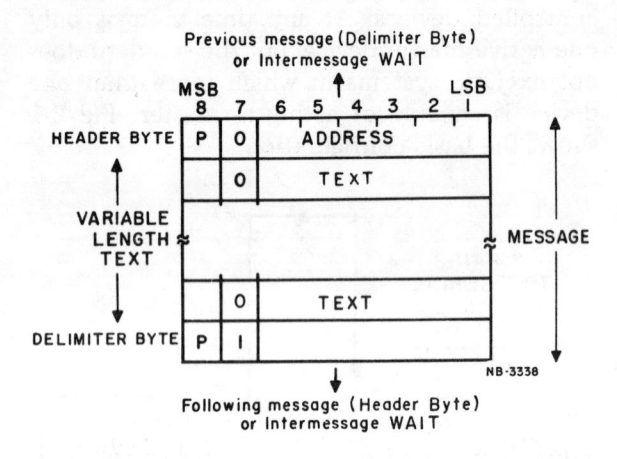

**Fig 2-2
Basic Message Format**

Bit 7 of every byte is a *Delimiter* bit, which allows receiving devices to identify the first and last bytes of each message.

Bit 8 is available for use as an odd-parity bit (with appropriate value so that the byte contains an odd number of bits in the logic 1 state). It is always used in this way in the first and last bytes of a message, and in all bytes of messages associated with CAMAC SCCs.

Every message starts with a *Header* byte. This includes a device address (a Crate Address when the device is an SCC). In a message from the SD the Header byte contains the address of the destination. In a message to the SD it contains the address of the source. Bit 7 of the Header byte is at logic 0 and Bit 8 conserves odd parity over the whole byte.

Every message ends with a *Delimiter* byte, in which Bit 7 is at logic 1, and Bit 8 conserves odd parity.

The length and content of the 'text' between the Header byte and the Delimiter byte of a message can be chosen to suit the needs of the individual device. In principle it need not be uniform for all devices in a system. In each byte between the Header byte and the Delimiter byte, Bit 7 is at logic 0.

If there are any bytes between the Delimiter byte of one message and the Header byte of the next, they are also Delimiter bytes, with Bit 7 at logic 1.

Thus, the Header byte of a message can be identified because, after one or more bytes with Bit 7 at logic 1, it is the first byte with Bit 7 at logic 0. Similarly, the last byte of a message can be identified because, after one or more bytes with Bit 7 at logic 0, it is the first byte with Bit 7 at logic 1.

Error detection over a block of bytes constituting a message or part of a message can be provided by the combination of byte parity in Bit 8 of each byte and a set of column-parity bits in bits 1-6 of the last byte of the block. This *Geometric Error-Detection* code detects all 1 bit, 2 bit, and 3 bit errors, and most errors with 4 bits or more. The scheme offers good protection against bursts of errors and is easy to implement by hardware or software.

2.3 Transmission of Bytes. Bytes are transmitted either in bit-serial mode (using one data signal and an accompanying bit-clock signal) or in byte-serial mode (using eight data signals and an accompanying byte-clock signal). In bit-serial mode the 8-bit byte is transmitted with the least significant bit (Bit 1) first. It is preceded by a Start bit (logic 0) and followed by a Stop bit and optional Pause bits (logic 1), as shown in Fig 2-3. The Start and Stop bits form a 'byte frame' from which receiving devices can recover a byte clock.

In the text of this standard, the bit pattern of an 8-bit byte with least significant bit l and most significant bit m is represented by the bit string $(miiiiiil)_2$. The same byte with Start and Stop bits is represented by the bit string $(1, miiiiiil, 0)_2$.

**Fig. 2-3
Bit-Serial Byte Frame**

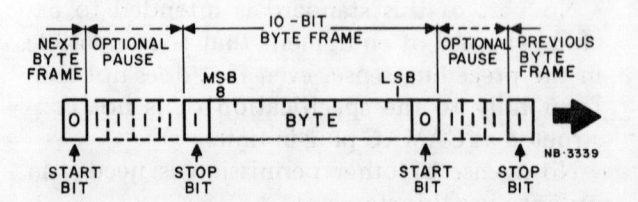

The message structure and protocol of the SH are identical in the two modes of transmission.

Throughout an SH system, bytes are transferred in synchronism with the byte clock, which accompanies the data in byte-serial mode and is derived from the byte framing in bit-serial mode.

In each byte-clock period, each device receives 1 byte and transmits 1 byte, but the contents (Bits 1 to 8) of the received and transmitted bytes are not always identical. Devices normally retransmit the contents of all received bytes, although the contents of a byte received in 1 byte period may be retransmitted in a later byte period. A device can generate a message by interrupting this process of retransmission. The contents of the required number of bytes are generated by the device, and the contents of a corresponding number of received bytes are not retransmitted. The message protocol should ensure that these received bytes do not contain important information. For example, they may be SPACE or WAIT bytes as defined in Section 3.

2.4 System Clock. System-clock signals, at bit rate or byte rate as appropriate, are generated at one point in the system (usually at or in the SD) and are retransmitted by each device connected to the SH.

The clock rate is therefore uniform throughout a system. The absolute maximum clock rate is 5.0 MHz, but the performance of the communications channels or of the connected devices may require a lower clock rate in particular systems.

2.5 Serial Highway Ports. The characteristics of the SH (such as signal standards, timing, message structure, and type of connector) are defined with respect to the ports[3] where the highway enters and leaves each connected device.

Nothing in this standard excludes the use of different standards within the connected devices or within any communications channel used between devices.

All devices connected to the highway have two ports, one for input and one for output. These ports either conform to the defined D-port standards or are related to the standards in such a way that the device could, in principle, be connected to D ports through an appropriate adaptor.

At each D port the data and clock signals are of balanced Non-Return-to-Zero-Level (NRZL) type, compatible with an existing standard (see Section 7) for balanced voltage digital interface circuits. Each signal is carried by a separate pair of wires and is generated by a balanced transmitter and received by a differential receiver.

At each D port there is one pair of contacts for the system clock (at bit rate or byte rate) plus eight pairs for data (only one pair of which is used in bit-serial mode).

A section of the SH can be formed by a direct connection between the output D port of one device and the input D port of the next device, as shown in Fig 2-4, typically using dedicated 100 Ω twisted-pair cable. Alternatively, any section of the SH can include a communications channel whose signal standards and modulation technique are chosen to suit the particular overall system requirements, as shown in Fig 2-5. In this case, signal conditioning units are needed to convert the data and clock signals from the D-port standard to the communications-channel standard, and back to the D-port standard. In the context of this standard any interconnections that do not use the D-port standard are undefined and are described as U-port standards. Although the data and clock signals are on separate wire pairs at the D ports, they may be combined into one unidirectional channel between U ports, for example, by using a suitable modulation technique.

**Fig 2-4
Direct D-Port Interconnection**

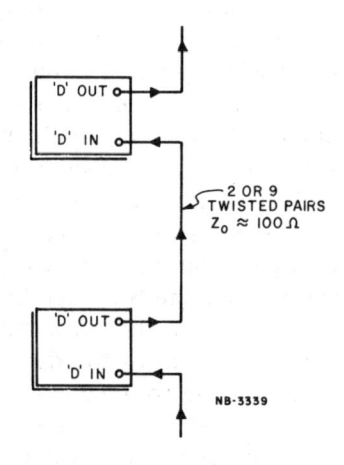

---

[3] The term port implies an entrance or exit of a network, etc.

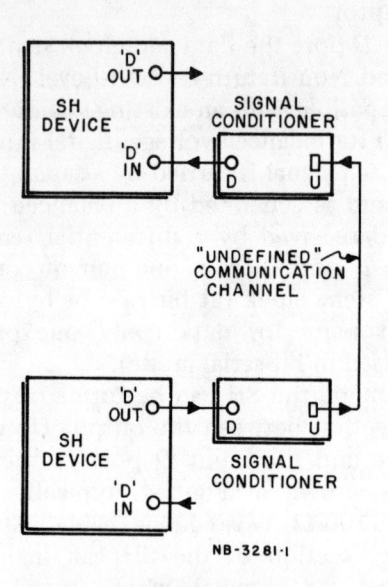

**Fig 2-5
Indirect Connection Via Undefined Standard**

Within an SH system some sections of the highway may use D-port standards while others use various U-port standards, all at the same system-clock rate. For example, a cluster of adjacent devices can have direct interconnections between D ports, with perhaps one signal conditioning unit at the input to the cluster and another at the output from the cluster.

2.6 Serial Driver. The SD is the link between the SH and (directly or indirectly) a computer or other system controller. It consists of a transmitting section associated with a D-output port, and a receiving section associated with a D-input port.

The transmitting section typically accepts commands and data from the computer and assembles them into the appropriate message format with Header byte and Delimiter byte. It transmits the resulting bit or byte stream to the SH, accompanied by the system-clock signals at bit or byte rate. It may generate an Error-Detection code within messages and interpose bytes between successive messages.

The receiving section typically accepts the bit or byte stream from the SH, together with the system-clock signals. It identifies the byte and message format, and passes the data, demands, and status information to the computer. It may check the Error-Detection code, implement error-recovery procedures, and discard any intermessage bytes.

The SD reacts to all messages received at its D-input port, whereas other devices only react to messages that are addressed to them.

This standard defines the SD only in terms of the signals, message structures, and message sequences at its D ports. Many of the activities concerned with generating and receiving messages can be handled either by hardware in the SD or by software in the associated computer.

Asynchronous serial communications ports are available on most modern minicomputers for interfacing to teletypewriters, visual display units, modems, etc. These ports can drive the SH in bit-serial mode through a simple adaptor, which is thus a special case of the SD.

2.7 Extended Uses of the Serial Highway. Every controlled device connected to the SH is transparent to messages that are addressed to other devices, irrespective of the internal structure or length of these messages. The SH can thus support many different types of compatible devices, provided these conform to the basic rules for signal standards, and the use of Header bytes and Delimiter bits to define the beginning and end of messages.

Compatible devices connected to the SH can consist, for example, of those shown in Fig 2-6:

(1) CAMAC crates with recommended SCCs Type L2, conforming to Section A1 of this standard's appendix, and using the CAMAC message structures defined in this standard

**Fig 2-6
Compatible Devices**

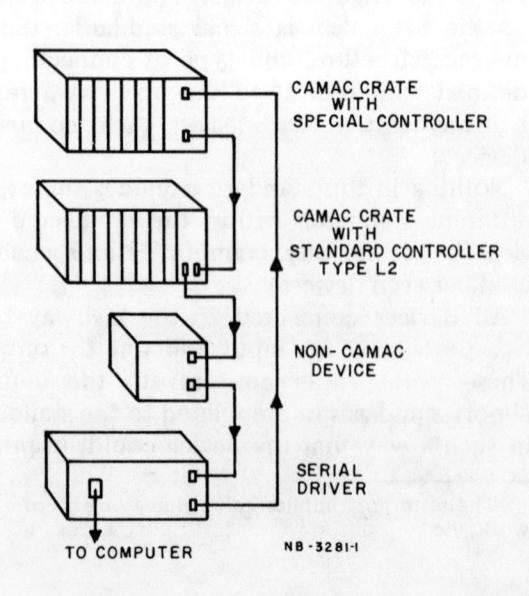

INTERFACE SYSTEM (CAMAC)

ANSI/IEEE
Std 595-1982

3. Message Structure for Serial Crate Controllers

When a CAMAC crate with SCC conforming to this standard is connected to the SH, the message structure includes the following features:

Three types of messages are used by SCCs. *Command messages* are generated by the SD and instruct an addressed SCC to perform a CAMAC operation. The addressed crate controller should transmit a truncated form of the Command message. In response to a Command message the addressed crate controller sends a *Reply message* to the SD. The Command message from the SD to an SCC, and the resulting Reply message from the SCC to the SD, constitute a *Command/Reply transaction*. Any SCC may generate a *Demand message* to indicate that there is a Look-at-Me (LAM) request on the Dataway in the crate.

Bits in SH messages are distinguished from corresponding Dataway signals by the prefix *S*. For example, the bits *SA*1-*SA*8 correspond to Dataway signals *A*1-*A*8.

3.1 Command Message. Command messages can be either complete or truncated.

3.1.1 Complete Command Messages.

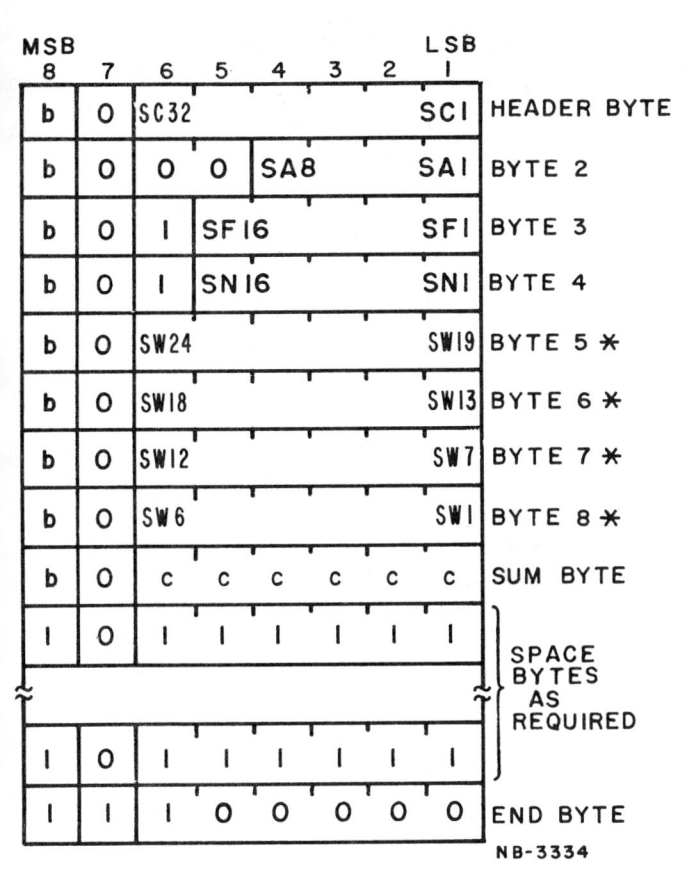

MSB 8	7	6	5	4	3	2	LSB 1	
b	0	SC32					SC1	HEADER BYTE
b	0	0	0	SA8			SA1	BYTE 2
b	0	1	SF16				SF1	BYTE 3
b	0	1	SN16				SN1	BYTE 4
b	0	SW24					SW19	BYTE 5 ✶
b	0	SW18					SW13	BYTE 6 ✶
b	0	SW12					SW7	BYTE 7 ✶
b	0	SW6					SW1	BYTE 8 ✶
b	0	c	c	c	c	c	c	SUM BYTE
1	0	1	1	1	1	1	1	} SPACE
~	~	~	~	~	~	~	~	BYTES AS
1	0	1	1	1	1	1	1	} REQUIRED
1	1	1	0	0	0	0	0	END BYTE

NB-3334

Fig 3-1
Command Message: Bit Assignments

b odd byte-parity bits
c Even column-parity bits
✶ Bytes 5, 6, 7, 8 included if *SF*16 = 1 and *SF*8 = 0

The complete Command message must have the structure shown in Fig 3-1, where the group of bytes 5-8 is included for Write commands (*SF*16=1 and *SF*8=0), but is omitted for Read and Control commands. The message must be transmitted as a consecutive sequence of bytes, starting with the Header byte (Crate Address) and finishing with the END byte.

(2) CAMAC crates with other SCCs, conforming to the main body of this standard, and using messages that are possibly variants or extensions of the standard CAMAC messages

(3) Devices in other mechanical formats or using other message structures

2.8 Serial Crate Controller. When a CAMAC crate assembly is connected to the SH, an SCC is used as the link between the SH and the Dataway highway in the crate. An SCC is a multiwidth CAMAC plug-in unit, occupying two or more stations, with Dataway connectors for the Control station and at least one Normal station. It has front-panel connectors for the two *D* ports.

The complete Command message consists of the following bytes, as shown in Fig. 3-2: a Header byte, in which the Crate Address field indicates the destination of the message; 3 bytes containing the Subaddress, Function, and Station Number fields of the CAMAC command; 4 bytes containing the 24 bit Write data, which are omitted when not required; and a SUM byte. This part of the message allows the addressed crate to assemble, check, and execute the CAMAC command. The message continues with a sequence of SPACE bytes (see Section

3.5.7) providing the opportunity for the SCC to transmit a reply; and concludes with an END Delimiter byte (see Section 3.5.3).

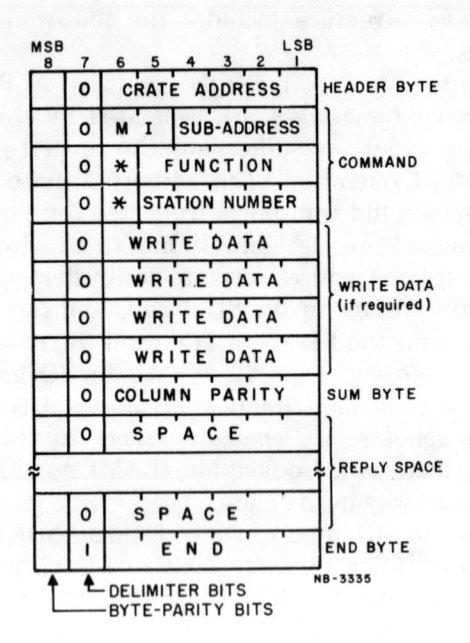

Fig 3-2
Command Message: Field Assignments

* Reserved bits
MI Message Identifier Field

3.1.2 *Truncated Command Message.* The addressed SCC should transmit a truncated form of the Command message, consisting of the Header byte followed by an END byte (see Fig 3-3 and 3-4, and Section 4.5).

Fig 3-3
Truncated Command Message: Bit Assignments

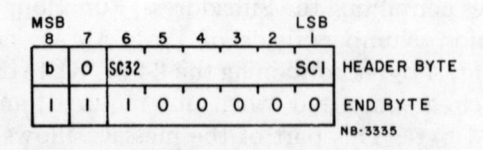

b Odd byte-parity bit

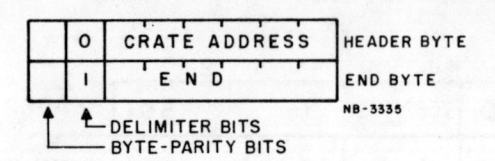

Fig 3-4
Truncated Command Message: Field Assignments

3.2 Reply Message

The Reply message must have the structure shown in Fig 3-5, where the group of bytes 3-6 is included in the reply to a Read command ($SF16=0$ and $SF8=0$) but is omitted in the reply to other commands. The message must be transmitted as a consecutive sequence of bytes, starting with the Header byte (Crate Address) and finishing with the ENDSUM byte.

Thus the Reply message consists of the following bytes, as shown in Fig 3-6: a Header byte, in which the Crate Address field indicates the source of the message; a Status byte; 4 bytes containing 24 bit Read data, which are omitted when not required; and an ENDSUM Delimiter byte.

Fig 3-5
Reply Message: Bit Assignments

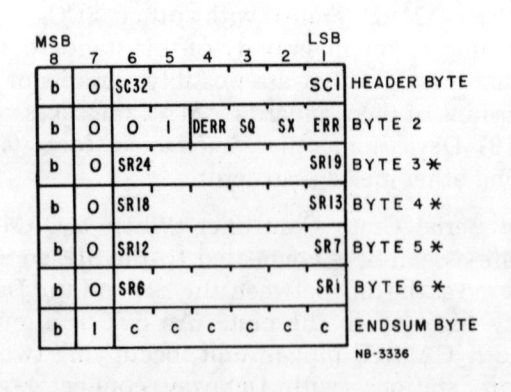

b Odd byte-parity bits
c Even column-parity bits
* Bytes 3, 4, 5, 6, included if $SF16 = 0$ and $SF8 = 0$

INTERFACE SYSTEM (CAMAC)

ANSI/IEEE
Std 595-1982

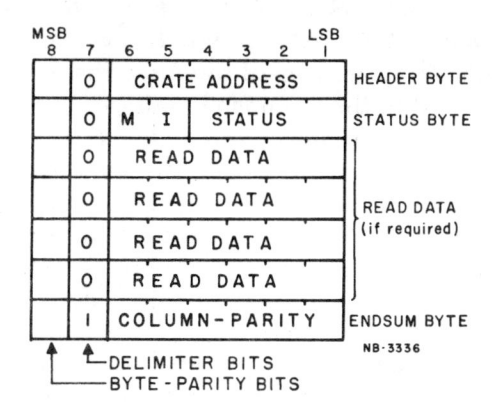

```
MSB                                LSB
 8  7  6  5  4  3  2  1
    0     CRATE ADDRESS         HEADER BYTE
    0  M  I    STATUS           STATUS BYTE
    0     READ DATA          ┐
    0     READ DATA          │ READ DATA
    0     READ DATA          │ (if required)
    0     READ DATA          ┘
    1     COLUMN-PARITY         ENDSUM BYTE
                                   NB-3336
    ↑  ↑
    │  └─ DELIMITER BITS
    └──── BYTE-PARITY BITS
```

Fig 3-6
Reply Message: Field Assignments

MI Message Identifier Field

3.3 Demand Message

> The Demand message must have the structure shown in Fig 3-7. The message must be transmitted as a consecutive sequence of bytes, starting with the Header byte (Crate Address) and finishing with the ENDSUM byte.

Thus the Demand message consists of the following bytes, as shown in Fig 3-8: a Header byte in which the Crate Address field indicates the source of the message; a byte further identifying the demand; and an ENDSUM Delimiter byte.

3.4 Message Fields.
The information in the Command, Reply, and Demand messages is divided into message fields as follows:

3.4.1 *Crate Address Field (6 Bits; SC1-SC32).* This field defines the destination address in Command messages and the source address in Reply and Demand messages.

Fig 3-7
Demand Message: Bit Assignments

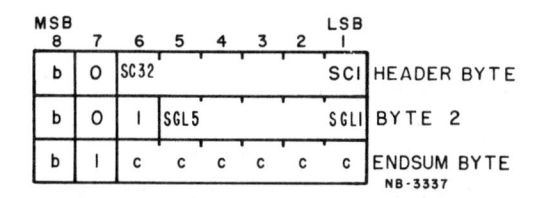

```
MSB                          LSB
 8  7  6  5  4  3  2  1
 b  0  SC32            SC1   HEADER BYTE
 b  0  1  SGL5         SGL1  BYTE 2
 b  1  c  c  c  c  c   c     ENDSUM BYTE
                               NB-3337
```

b Odd byte-parity bits
c Even column-parity bits

> Each SCC must respond to any address assigned to it from the set 01_8 - 76_8 and must not respond to either 00 or 77_8.

The address 00 is reserved for use at the SD. Under certain error conditions a SPACE byte can be falsely identified as a Header byte. The recommended SPACE byte has a bit pattern corresponding to address 77_8. This address is therefore not used, and there are 76_8 (62_{10}) available crate addresses.

3.4.2 *Station Number Field (5 Bits: SN1-SN16).* This field in the Command message defines the Station Number within the CAMAC crate (see Section 5.1.1 of IEEE Std 583-1975).

In general the codes $N(1)$ to $N(23)$ are used as addresses of modules in the CAMAC crate. Internal features of the SCC are addressed by $N(30)$ (see Section 11).

```
MSB                          LSB
 8  7  6  5  4  3  2  1
    0     CRATE ADDRESS        HEADER BYTE
    0  MI  SERIAL GRADED L     SGL BYTE
    1     COLUMN-PARITY        ENDSUM BYTE
                                 NB-3337-1
    ↑  ↑
    │  └─ DELIMITER BITS
    └──── BYTE-PARITY BITS
```

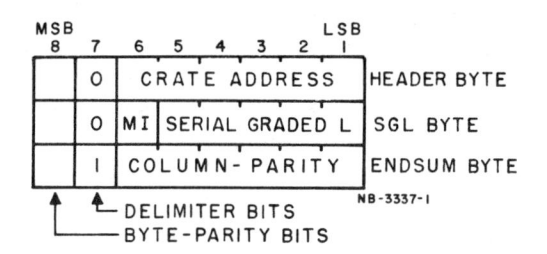

Fig 3-8
Demand Message: Field Assignments

MI Message Identification Field

3.4.3 *Subaddress Field (4 Bits; SA1-SA8).* This field in the Command message defines a subaddress at the selected station in the crate (see Section 5.1.2 of ANSI/IEEE Std 583-1982).

3.4.4 *Function Field (5 Bits; SF1-SF16).* This field in the Command message defines the action to be performed at the selected Station and Subaddress in the crate (see Section 5.1.3 of ANSI/IEEE Std 583-1982).

The values of $SF16$ and $SF8$ in this field distinguish between Read, Write, and Control commands, and hence determine whether a data field is included in the Command or Reply message. The lengths of the Command and Reply messages corresponding to the various values of $SF16$ and $SF8$ are summarized in Table 3-1.

Table 3-1
Length of Command/Reply Transactions

Operation	Function Field F16	Function Field F8	Command from Header to SUM Inclusive	Reply from Header to ENDSUM Inclusive	Command Reply Transaction
Read	0	0	5	7	12*
Control	0	1	5	3	8*
	1	1	5	3	8*
Write	1	0	9	3	12*

*Minimum length, assuming that Reply Header is transmitted by SCC as First SPACE Byte is received and ENDSUM is transmitted as END Byte is received.

3.4.5 *Write-Data Field (24 Bits; SW1-SW24).* This field is included in the Command message if SF16=1 and SF8=0. It contains the data associated with a Write command.

3.4.6 *Read-Data Field (24 Bits; SR1-SR24).* This field is included in the Reply message if the Function field of the Command message had SF16=0 and SF8=0. It contains the data requested by a Read command.

3.4.7 *Message Identification Field (2 Bits; M1-M2).* This field in the Command and Reply messages (2 bits) and in the Demand message (only M2) is used to identify the three types of message.

> The significance of the Message Identification (MI) field must be as shown in Table 3-2.

This field is needed by the SD in order to distinguish between Reply and Demand messages of equal length. It may also be used by SCC to identify Command messages, as additional security against executing false commands.

Table 3-2
Contents of Message Identification Field

	Message Identification Field M2	Message Identification Field M1
Message		
Command	0	0
Reply	0	1
Demand	1	—

3.4.8 *Status Field (4 Bits; Error, Response, Command Accepted, Delayed Error).* This field in the Reply message shows how the SCC has responded to the Command message. The Error (ERR) bit indicates whether the error detection checks on the Command message were satisfactory (see Section 15.5.1). The Delayed Error (DERR) bit provides similar information about the previous command. For a command that has been executed, the SQ and SX bits indicate the Response Q and Command Accepted X status of the feature of the module or controller that has been accessed by the command.

> The contents of the Status field in the Reply message must conform to the requirements of Sections 12.2 and 15.5.

3.4.9 *Serial Graded-L (SGL) Field (5 Bits; SGL1-SGL5).* This field of the Demand message identifies the type of demand, the source of the demand, or the action required by the demand. It may be derived from the L signals on the Dataway of the crate by any process of selection, grouping, priority coding, etc, that is performed by the SCC or by a separate SGL Encoder connected to the SCC (see Section 14).

> The SGL bit pattern 11111_2 must only be used to indicate the Hung Demand state (see Section 14).

3.5 Formatting Bytes. The formatting bytes are an important part of the message structure, but do not contain information fields as defined in the previous section. They indicate the end of each message (END and ENDSUM), provide the column-parity component of the Geometric Error-Detection scheme (SUM and ENDSUM) and maintain the byte clock during intervals within messages (SPACE) and between messages (WAIT).

3.5.1 *Delimiter Bytes*

> A Delimiter byte must have Bit 7 at logic 1, and Bit 8 with the appropriate value to conserve odd byte parity.

All other bytes that do not satisfy these conditions are Nondelimiter bytes. The class of Delimiter bytes includes END, ENDSUM, and WAIT bytes.

Delimiter bytes are used to indicate the last byte of each message and any bytes that occur between messages. Each message consists of a sequence of Nondelimiter bytes terminated by a Delimiter byte. It may possibly be followed by further Delimiter bytes (WAIT bytes). Delimiter bytes cannot occur legitimately elsewhere within messages.

Thus the first byte of a message (the Header byte) can be recognized by its context as the first Nondelimiter byte following one or more Delimiter bytes. The last byte of a message can be recognized by its context as the first Delimiter byte following one or more Nondelimiter bytes.

3.5.2 *Column-parity Field.* The SUM and ENDSUM bytes contain a column-parity field in Bits 1-6. This field provides the column-parity component of the Geometric Error-Detection scheme (see Section 15.3). Each bit of the column-parity field conserves even parity over the corresponding bit position in each byte, from the Header byte to the SUM or ENDSUM byte, inclusive.

The contents of the column-parity field are equivalent to the sum modulo-2 of all preceding bytes of the message, excluding bits 7 and 8 in each byte. (Hence the terms SUM and ENDSUM.)

3.5.3 *END Byte.* The END byte is a Delimiter byte generated by the SD to terminate complete Command messages and by SCCs to terminate truncated Command messages.

In the END byte, the Delimiter bit (Bit 7) is at logic 1, and Bit 8 conserves the odd byte parity.

> An END byte, with bit-pattern 11100000_2, equivalent to 340_8, must be generated by the SD as the last byte of every Command message.

Although the END byte has the same binary pattern as the WAIT byte (see Section 3.5.4), it differs in context and in the requirements concerning its retransmission by SCCs. The END byte is always preceded by a Nondelimiter byte.

> All unaddressed crate controllers must retransmit received END bytes unchanged. An addressed crate controller must either retransmit the received END byte of the Command message or replace it by an ENDSUM byte (see Section 4.1.4).

3.5.4 *WAIT Byte.* The WAIT byte is a Delimiter byte generated by the SD and by addressed SCCs. The SD may generate WAIT bytes between successive Command messages. An addressed SCC generates WAIT bytes, in response to received bytes, between the truncated Command message (see 3.1.2) and the Header byte of the Reply message and also, if required, between the ENDSUM byte of the Reply message and the completion of the Command/Reply transaction.

The WAIT byte has no information field, but the associated byte clock allows messages to propagate around the SH. Thus, although it is not mandatory for the SD to generate bytes during intervals between Command messages, the generation of WAIT bytes is recommended in order to permit the propagation of Reply messages to be completed in a typical SH which includes logical delays.

> Any bytes generated by an SD or SCC during intervals between messages must be WAIT bytes with the bit pattern 11100000_2 equivalent to 340_8.

In this WAIT byte pattern, the Delimiter bit (Bit 7) is set to logic 1, and Bit 8 is at logic 1 to conserve odd byte parity over the whole byte.

The WAIT byte therefore has the same bit pattern as the END byte (see Section 3.5.3) but differs in context and in the requirements concerning its retransmission by SCCs. A WAIT byte always follows another Delimiter byte. Under certain conditions, connected with the generation of Demand messages, an SCC is permitted to transmit another byte in place of a received WAIT byte (see Section 5).

A specific bit pattern has been defined for the WAIT byte in order to assist the process of establishing message synchronism (see Section 10.1.3), which normally takes place while the SD is generating a sequence of WAIT bytes.

The chosen bit pattern is one of several that have the useful property, when they are transmitted repetitively in bit-serial mode, that each byte, including its associated Start and Stop bits, has only one transition from 0 to 1, and one transition from 1 to 0, thus assisting the establishment of byte synchronism.

3.5.5 *ENDSUM Byte.* The ENDSUM byte is a Delimiter byte, and is generated by SCC to terminate each Reply or Demand message. The column-parity field (Bits 1-6) conserves even column parity over all bytes between the Header byte and the ENDSUM byte, inclusive. The Delimiter bit (Bit 7) is at logic 1, and Bit 8 conserves odd byte parity.

3.5.6 *SUM Byte.* The SUM byte is a Nondelimiter byte, and is generated by the SD in the Command message (see Figs 3-1 and 3-2). The column-parity field (Bits 1-6) of this byte conserves even column parity over all bytes of the message between the Header byte and SUM byte, inclusive. The Delimiter bit (Bit 7) is at logic 0, and Bit 8 conserves odd byte parity over the whole of the SUM byte.

The addressed SCC uses the SUM byte in the Geometric Error-Detection test on the Command message, before deciding whether to execute the command.

3.5.7 *SPACE Byte.* The SPACE byte is a Nondelimiter byte. A sequence of SPACE bytes generated by the SD between the SUM and END bytes of the Command message forms the *Reply space*. The addressed SCC generates its Reply message in place of some or all of the SPACE bytes in the *Reply space*.

> In the SPACE byte generated by the SD, the Delimiter bit (Bit 7) must be at logic 0, and Bit 8 must conserve odd byte parity.

The bit pattern of SPACE bytes generated by the SD should be 10111111_2, equivalent to 277_8.

> An SCC that is expecting to receive SPACE bytes in the Reply space of a Command message must accept any Nondelimiter byte as a substitute for a SPACE byte.

Acceptable substitutes for SPACE bytes need not have the recommended bit pattern or conserve odd byte parity.

4. Command/Reply Message Sequences

This section defines the sequences of input and output bytes that occur when an SCC receives a Command message, performs a CAMAC Read, Write, or Control operation, and transmits a Reply message. Some features of the Reply space within the Command message are examined in more detail.

The message sequences for Read, Write, and Control operations in bit-serial mode are shown in Figs 4-1, 4-3, and 4-5, respectively. The corresponding sequences in byte-serial mode are shown in Figs 4-2, 4-4, and 4-6. All these figures show the normal conditions, with no errors and no Delay buffer switched into the message path in the SCC (see Section 5.2).

For a given mode, the sequences for Read, Write, and Control operations differ only in the presence or absence of data fields. For a given type of operation, the sequences for bit-serial and byte-serial modes differ only in the relative timing of received and retransmitted bytes at the beginning and end of the sequence. In bit-serial mode the retransmission of bytes (indicated by arrows in the figures) typically involves a delay of only one bit period, whereas in byte-serial mode there is typically a delay of one byte period.

These figures indicate that extra SPACE bytes can be added to the Reply space of the Command message for two reasons. Essential SPACE bytes, indicated by an asterisk, allow time for the execution of the command. The minimum number of these bytes is, in principle, zero, but a minimum of one byte may be more convenient to implement. Optional excess SPACE bytes, indicated by the number sign #, are used in one method of controlling the length of the Reply space (see Section 4.6.2). The minimum number of these bytes is zero.

4.1 General Requirements. The SCC is normally awaiting a Header byte. The SD sends a Command message (see Section 3.1) consisting of a Header byte, the command and data bytes, the SUM byte, a sequence of SPACE bytes, and

INTERFACE SYSTEM (CAMAC) ANSI/IEEE Std 595-1982

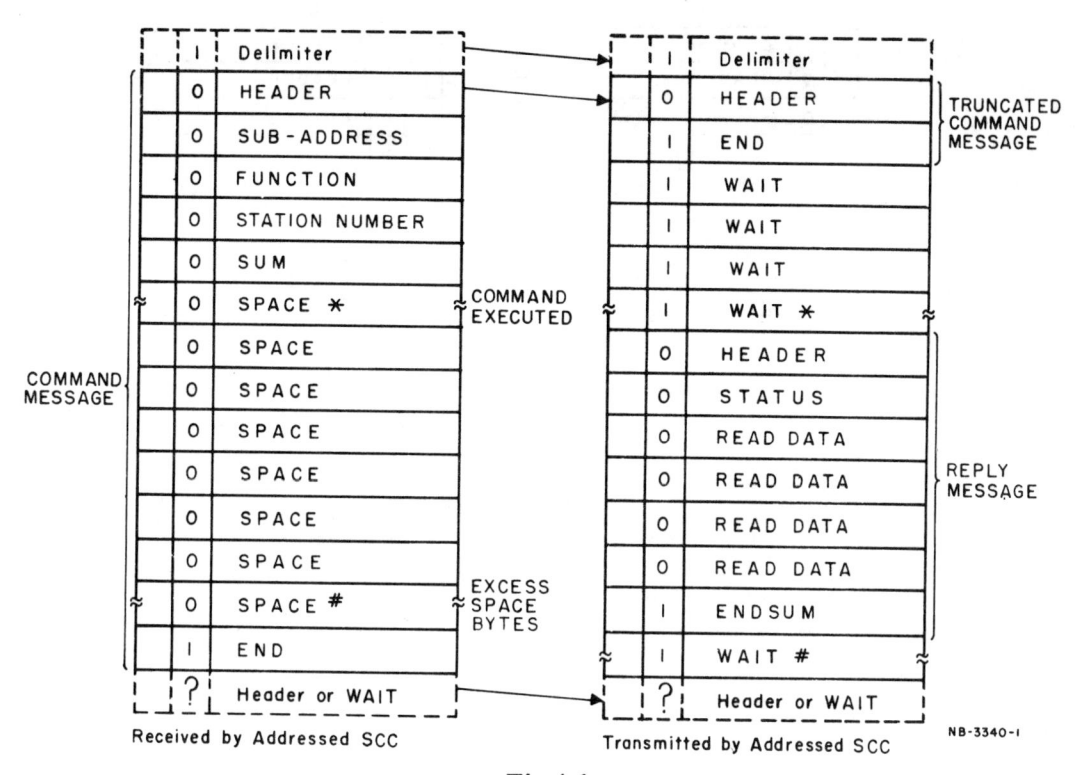

Fig 4-1
Command/Reply Sequence: Read Operation, Bit-Serial Mode

* Number of bytes as required to accommodate execution of command: minimum number 0
\# Number of bytes as required to accommodate excess SPACE bytes: minimum number 0

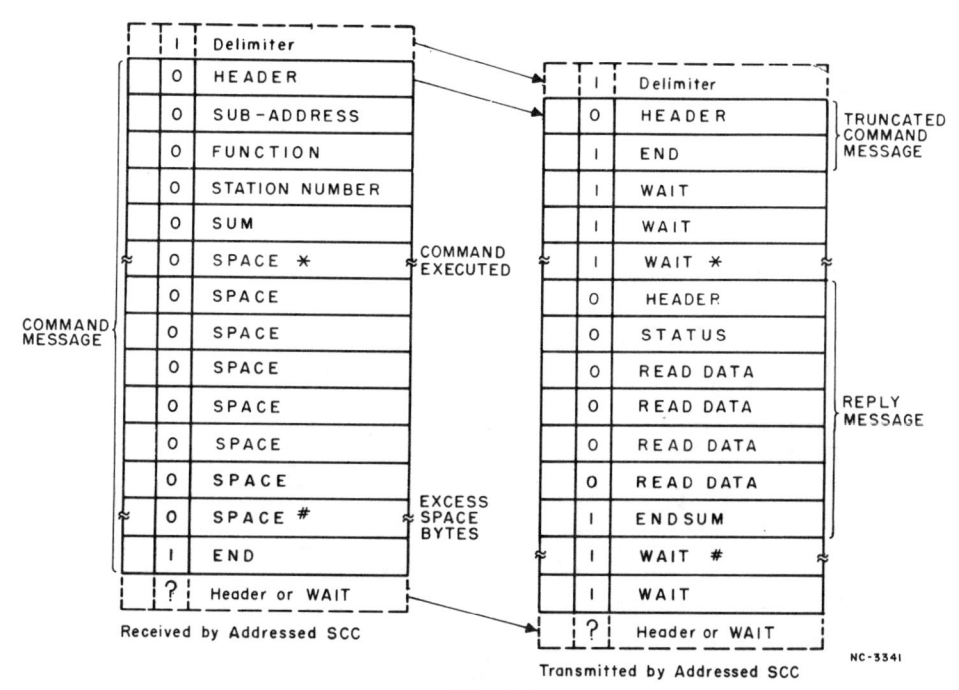

Fig 4-2
Command/Reply Sequence: Read Operation, Byte-Serial Mode

* Number of Bytes as required to accommodate execution of command: minimum number 0
\# Number of Bytes as required to accommodate excess SPACE bytes: minimum number 0

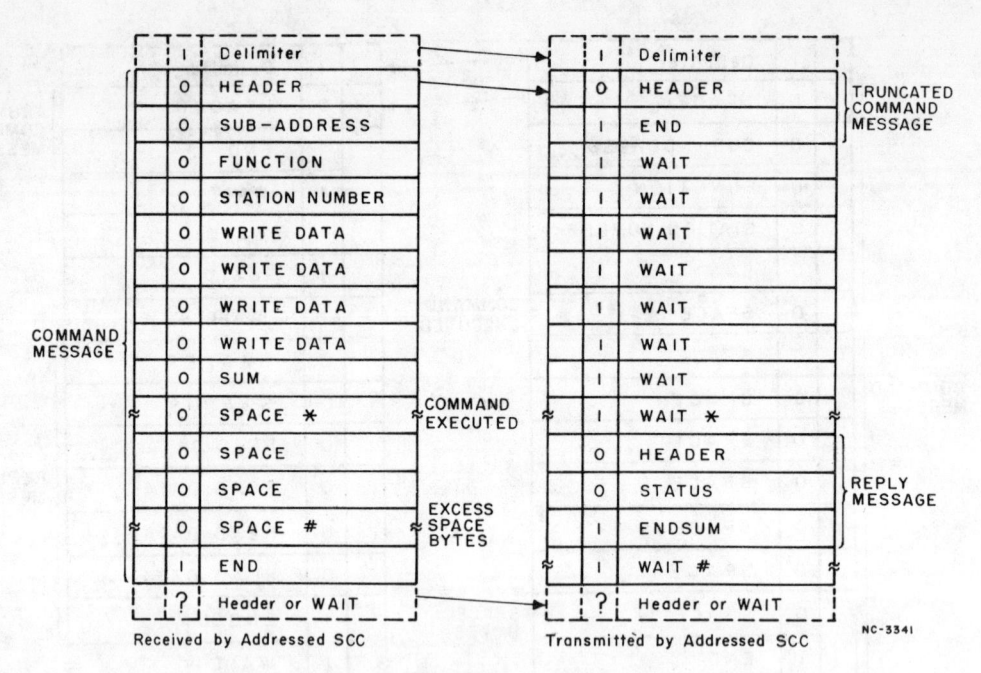

Fig 4-3
Command/Reply Sequence: Write Operation, Bit-Serial Mode

* Number of bytes as required to accommodate execution of command: minimum number 0
\# Number of bytes as required to accommodate excess SPACE bytes: minimum number 0

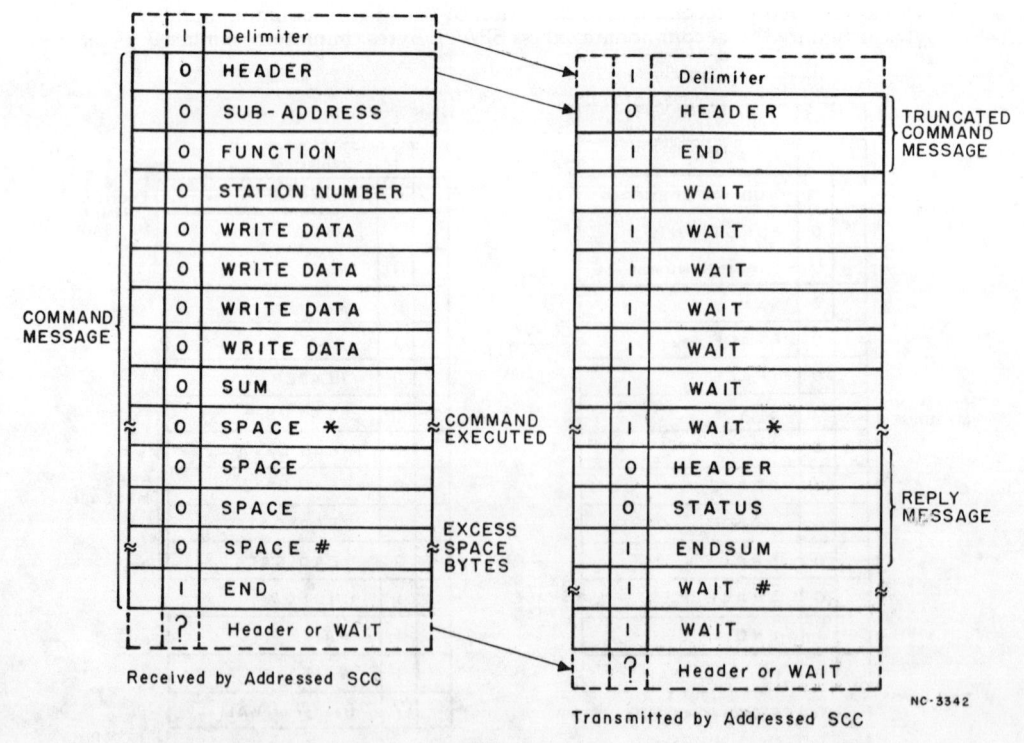

Fig 4-4
Command/Reply Sequence: Write Operation, Byte-Serial Mode

* Number of bytes as required to accommodate execution of command: minimum number 0
\# Number of bytes as required to accommodate excess SPACE bytes: minimum number 0

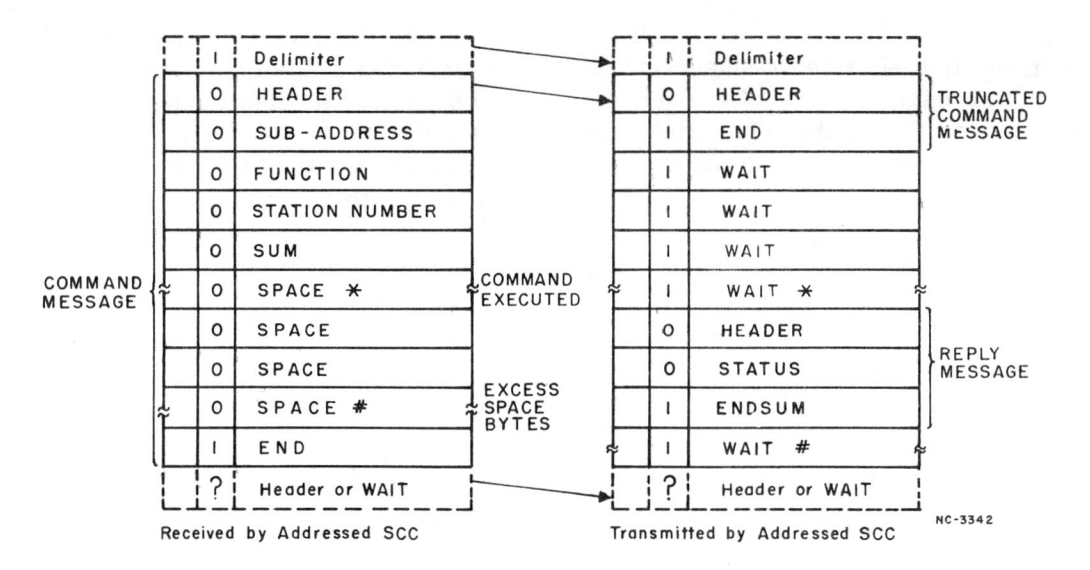

Fig 4-5
Command/Reply Sequence: Control Operation, Bit-Serial Mode

* Number of bytes as required to accommodate execution of command: minimum number 0
\# Number of bytes as required to accommodate excess SPACE bytes: minimum number 0

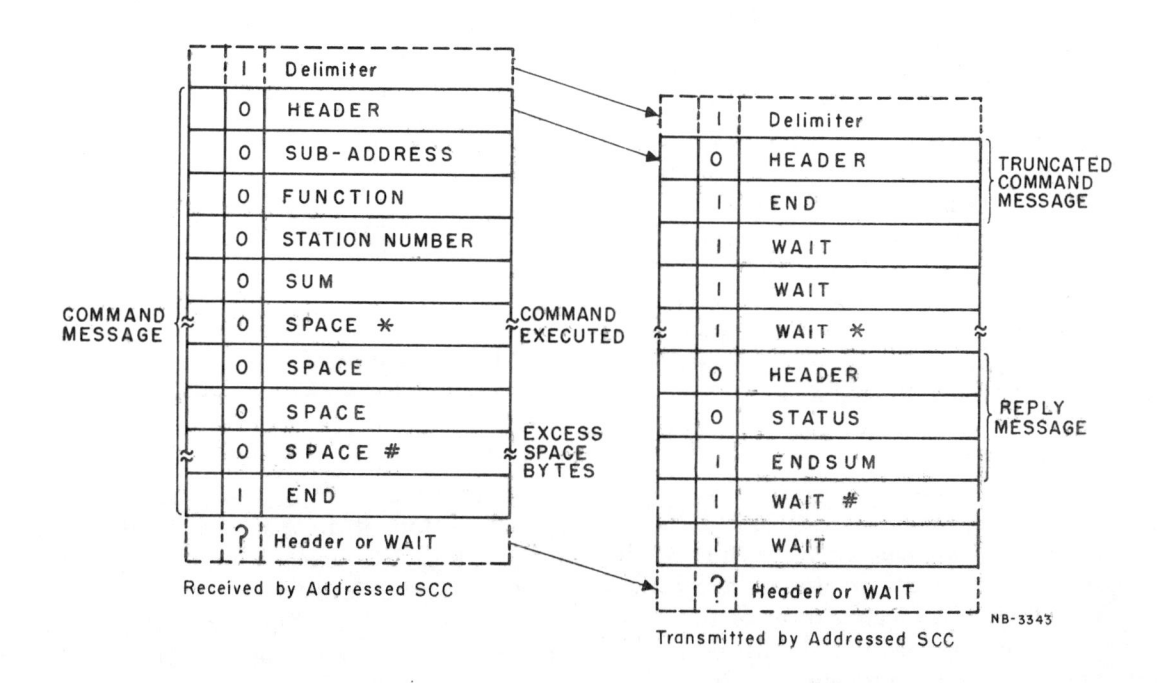

Fig 4-6
Command/Reply Sequence: Control Operation, Byte-Serial Mode

* Number of bytes as required to accommodate execution of command: minimum number 0
\# Number of bytes as required to accommodate excess SPACE bytes: minimum number 0

the END byte. The SCC recognizes the Header byte addressed to itself, receives and checks the command and data, executes the command, and sends a reply during the Reply space and, finally, terminates the transaction when it receives the END byte.

4.1.1 *Find Header State.* These sequences assume that the SCC has previously received 1 or more Delimiter bytes (the END or ENDSUM byte of a previous message, possibly followed by 1 or more WAIT bytes) and is awaiting the Header byte at the beginning of a new message.

> When an SCC is awaiting a Header byte, it must retransmit each byte that it receives and must also examine the contents of each byte in order to take appropriate action as follows:
> If the SCC receives any Delimiter byte (with Bit 7 = 1 and correct byte parity), it must continue to await a Header byte and to permit the initiation of Demand messages.
> If the SCC receives a Nondelimiter byte with Bit 7 = 0, correct byte parity, and Crate Address field matching the 6-bit assigned address, then it must treat this as the Header byte of a Command message addressed to it. It must inhibit the initiation of Demand messages and prepare to accept the remainder of the Command message.
> If the SCC receives any byte other than a Delimiter byte or the Header byte of a message addressed to it, then it must treat this as the Header byte of a message to or from another SCC. It must inhibit the initiation of Demand messages and pass the message unchanged.

4.1.2 *Receive Command State.* While receiving the Command message the addressed SCC should transmit a truncated form of the Command message (see Section 3.1.2), consisting only of the Header byte and an END byte, followed by a sequence of WAIT bytes until it is ready to transmit the Reply message (see Section 3.2).

> The addressed SCC must use the state of the $SF16$ and $SF8$ bits in the Function field of the Command message (see Section 3.4.4) to determine whether the SUM byte is expected as the fifth byte (if $SF16=0$ or $SF8=1$) or as the ninth byte (if $SF16=1$ and $SF8=0$).
> The addressed SCC must check the even column parity of the received Command message from the Header byte to the SUM byte, inclusive. It must also check the odd byte parity over each byte individually.

The addressed SCC may optionally check that the contents of the MI field (see Section 3.4.7) of the received message are **00**.

4.1.3 *Execute Command State*

> The addressed SCC must execute the command and send the appropriate reply if the following conditions are satisfied:
> (1) The byte and column parity are correct
> (2) The MI field (if tested) is correct
> (3) The SCC is in the Dataway On-line condition (see Section 12.4.1) or, if it is Off line, the command is addressed to an internal feature of the SCC
> (4) The SCC is not in the Bypassed condition (see Section 12.4.2), or, if it is bypassed, the command is one that resets the Bypassed state.
> Otherwise, the SCC must not execute the command.

4.1.4 *Send Reply State*

> If the addressed SCC remains in message synchronism (and in byte synchronism, if appropriate) during the Command/Reply transaction, it must send a Reply message.
> If the SCC executes the command, it must send a Reply message with format appropriate to the Function code, and with contents derived from the results of the operation. Transmission of the Reply message must not start until the Q and X responses (and Read data, if appropriate) are established. If execution of the command involves a Dataway operation, transmission of the Reply message must not start earlier than time t_3 in Fig 9 of ANSI/IEEE Std 583-1982.

If the SCC does not execute the command because the parity or MI field test is not correct [see conditions (1) and (2) of Section 4.1.3], it must send a 3-byte Error-reply message (see Section 15.4).

If the SCC does not execute the command because the Dataway Off-line or Bypass conditions are not satisfied [see Conditions (3) and (4) of Section 4.1.3], it must send a Reply message with format appropriate to the Function code and with Status field bit *SX* at logic **0** to indicate that the command has not been accepted by the SCC (see Section 15.5.2).

If the command has not been executed, transmission of the Reply message may start as soon as the presumed SUM byte has been received.

For certain operations, associated with the control signals for Bypass and Loop Collapse devices, the initiation of the Reply message is delayed with respect to the execution of the command (see Sections 12.4.2 and 12.4.3).

Following the SUM byte, the SD transmits, and the SCC receives, a sequence of SPACE bytes constituting the Reply space. Under many typical conditions of Dataway-cycle timing and SH transmission rate, execution of the command can be completed before the first SPACE byte is received by the SCC. If so, the SCC may transmit the Header byte of the Reply message as it receives the first SPACE byte, then transmit the Status byte of the reply as it receives the second SPACE byte, and so on.

Each byte of the Reply message must be transmitted by the addressed SCC in response to a received Nondelimiter byte. After receiving the SUM byte, and before sending the Header byte of the Reply message, the SCC must accept any received bytes and must transmit WAIT bytes.

The SCC must accept any received Nondelimiter byte while it is transmitting each byte of the Reply message up to, but excluding, the ENDSUM byte. When transmitting the ENDSUM byte, the SCC must accept any received byte, including a Delimiter byte.

If the addressed SCC receives a Delimiter byte after it has recognized the Header byte of the Command message and before it begins to transmit the ENDSUM byte of the Reply message, it must abandon the Command/Reply transaction and go into Lost Message Sync state (see Section 10.1.2).

In the following three sections it is assumed, for clarity, that the SCC starts to generate the Reply message as it receives the first SPACE byte, and that it generates the ENDSUM byte as it receives the END byte of the Command message. Section 4.6 deals with other permitted conditions, where SCC receives SPACE bytes before it has finished executing the Command, while it is transmitting the ENDSUM byte, and after it has transmitted the ENDSUM byte.

4.2 Read Operation. The sequence of bytes transmitted by the SD and SCC is shown in Figs 4-1 and 4-2, for bit-serial and byte-serial modes, respectively. The Command message does not include a Write-data field. The Reply message includes a 24-bit Read-data field.

In the Function field of the Command message, bits $SF16=0$ and $SF8=0$ indicate a Read operation. The SCC therefore expects to find the SUM byte as the fifth byte in the received message, and performs the column-parity check over the first 5 bytes of the message.

4.3 Write Operation. The sequence of bytes transmitted by the SD and SCC is shown in Figs 4-3 and 4-4, for bit-serial and byte-serial modes, respectively. The Command message includes a 24-bit Write-data field. The Reply message does not include a Read-data field.

In the Function field of the Command message, bits $SF16=1$ and $SF8=0$ indicate a Write operation. The SCC therefore expects to find the SUM byte as the ninth byte of the received message, and performs the column-parity check over the first 9 bytes of the message.

4.4 Control Operation. The sequence of bytes transmitted by the SD and SCC is shown in Figs 4-5 and 4-6, for bit-serial and byte-serial modes, respectively. Neither the Command message nor the Reply message includes a data field.

In the Function field of the Command message, bit $SF8=1$ indicates a Control operation (neither Read nor Write), and the SCC therefore expects the SUM byte as the fifth byte in the received message and performs the column-parity check over the first 5 bytes of the message.

4.5 Truncation of the Command Message. The addressed SCC should transmit a truncated form of the received Command message, consisting only of the Header byte (containing the Crate Address field) and an END byte.

This recommended practice is followed in the Command/Reply sequence shown in Figs 4-1 to 4-6, and it is a mandatory feature of the SCC Type L2 (see Section A1.3.2). It is recommended for all other SCCs for the following reasons. Only one crate can respond to the Command message even if, for example, the same Crate Address has been assigned to more than one crate. The sequence of Delimiter bytes following the truncated Command message gives a positive indication of the beginning of the Reply message and also provides opportunities for downstream crate controllers to generate Demand messages (see Section 5.1) and to regain byte synchronism (see Section 10.2.3).

However, for some special applications it may be necessary for the addressed SCC to retransmit the Command message (up to, and including the SUM byte). Such applications require appropriate SCCs and SDs, but it is possible for these to coexist on the same SH with other SCCs that generate the recommended truncated Command message.

If the SCC does not transmit a truncated form of the Command message it must transmit at least one Delimiter byte, in response to a received byte, immediately before transmitting the Header byte of the Reply message.

4.6 Reply Space. This section covers some additional aspects of the SPACE bytes generated by SD during the Command message, which are partially or wholly replaced by the Reply message from the addressed SCC.

4.6.1 *Completion of Dataway Operation.* If the byte period is long compared with the command-execution time, the addressed SCC can be ready to send the reply before it receives the first SPACE byte. Under these conditions the first byte of the Reply message can be transmitted by the SCC in place of the first SPACE byte.

Otherwise, the SCC receives 1 or more SPACE bytes before it is ready to send the reply, and the number of extra SPACE bytes indicated by an asterisk in Figs 4-1 to 4-6 is nonzero. The SCC transmits WAIT bytes in place of the received SPACE bytes, until it has executed the command. Then it transmits the first byte of the reply message in place of the next SPACE byte.

4.6.2 *Termination of Command/Reply Transaction.* In one mode of operation, the SD continues to generate excess SPACE bytes until it has received the Reply message. It then generates the END byte to terminate the transaction. This mode has the advantage that it avoids the need for SD to control the number of SPACE bytes precisely. It has also the disadvantages that it reduces the opportunities for Demand message generation upstream from the addressed SCC, and needs precautions to avoid locking up if the command is addressed to a nonexistent crate. The number of extra SPACE bytes indicated by the number symbol # in Figs 4-1 to 4-6 is nonzero, and the SCC transmits WAIT bytes in place of the received SPACE bytes.

In another mode, the SD generates a calculated number of SPACE bytes and then terminates the transaction by sending the END byte. It then continues to generate WAIT bytes until it has received the REPLY message. In this mode there are few excess SPACE bytes, or none.

Both of these modes of terminating the Command/Reply transaction lead to simple recovery from errors (see Section 15.6) because each transaction is completed before the next is started.

In another mode, appropriate to high-performance systems with low error rates, the SD does not wait to receive the reply to one Command message before starting to generate the next. In this mode the SD preferably generates the exact minimum number of SPACE bytes, so that the addressed SCC generates the ENDSUM byte of the Reply message when it receives the END byte of the Command message. Under error-free conditions, an SD operating in this mode transmits a sequence of Command messages and receives a corresponding sequence of Reply messages in the same relative order, but possibly considerably delayed and with interposed Demand messages. The Reread feature (see Section 15.6) cannot be relied upon to assist recovery from transmission errors in this mode.

4.6.3 Length of the Reply Space

> The SD must include within the Command message enough SPACE bytes to provide time for the SCC to execute the command and to transmit the Reply message.

A simple formula for estimating the length of the Reply space is given in the following equation. This equation applies to the recommended mode of operation, in which the addressed SCC transmits a truncated Command message (see 3.1.2). It also covers the worst case condition where the SCC completes the full Dataway operation before transmitting the Header byte of the Reply message. A safe working value for the number of SPACE bytes S in the Reply space is:

$$S = N_{op} + N_{rep} + 1$$

where N_{op} and N_{rep} are the numbers of received SPACE bytes needed to accommodate the execution of the command and the transmission of the Reply message, respectively. N_{op} is the next greater integer above T_{op}/T_{byte}, where T_{op} is the maximum Dataway cycle time of the SCC and T_{byte} is the minimum byte period of the particular SH system. N_{rep} is 2 bytes for Write and Control commands, and 6 bytes for Read commands.

By taking into account the detailed timing relationship between received bytes and the Dataway operation, in a particular SCC, it may be possible to operate with a smaller value of S than that given by this equation. For example, if the byte period T_{byte} is very long compared with the Dataway operation time T_{op}, the value of N_{op} can be replaced by zero. In other systems, the value of N_{op} can be reduced by starting to transmit the Reply message before the full Dataway operation has been completed (see Section 4.1.4). In this latter case it is necessary to ensure that, if two successive Write or Control commands are sent to the same SCC, the number of bytes between the SUM byte of one command and the Subaddress byte of the next is adequate to maintain the Dataway command signals for the full duration of the Dataway operation.

5. Demand Message Generation

The format of Demand messages is defined in Section 3.3. Any SCC may transmit a Demand message, typically in response to a Dataway L signal, by interposing it between any two messages in the incoming SH message sequence.

Demand message generation is controlled by bits in the Status Register of the SCC (see Section 12.3) and by the Delimiter bits in the bytes received by the SCC. Each byte with the Delimiter bit at logic 1 permits the initiation of Demand messages, and each byte with the Delimiter bit at logic 0 inhibits the initiation of Demand messages. An SCC is therefore unable to generate a Demand message while it is receiving a Command message addressed to itself, or while it is retransmitting either a Command message addressed to another crate or a Reply or Demand message generated by an upstream crate.

A Delimiter byte that allows the SCC to start generating a 3-byte Demand message may be followed by Nondelimiter bytes of a new incoming message. The SCC therefore needs a buffer memory so that it can preserve incoming bytes while the Demand message is being transmitted.

If the SCC addressed by a Command message follows the preferred practice of transmitting a truncated Command message (see Section 3.1.2), there are differences in the opportunities available for Demand message generation at crates upstream and downstream of the addressed crate. Demand message generation is inhibited at upstream crates for the complete duration of the Command/Reply transaction, from the Header byte to the END byte of the Command message. At downstream crates, the generation of Demand messages is inhibited during the 2-byte truncated Command message. The END byte of this message, and the following WAIT bytes, permit Demand generation in the interval between the truncated

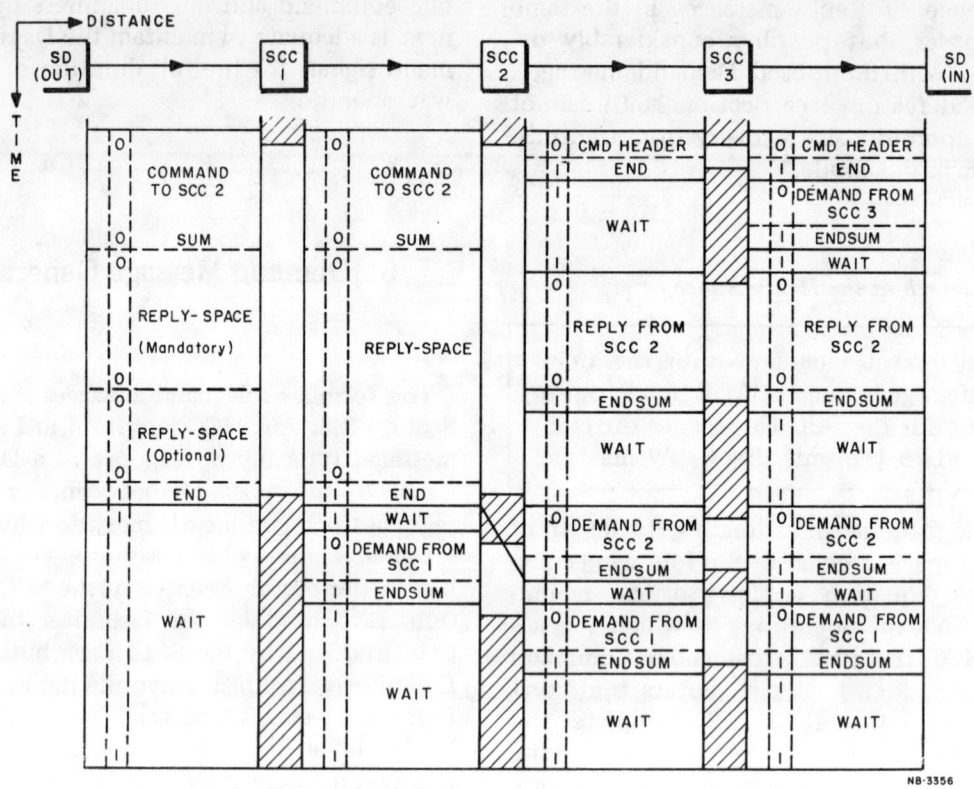

**Fig 5-1
Example of Message Sequence in Loop Having Three SCCs,
Showing Demand Message in Three Contexts**
(See Section 5)

NOTES:
1. Propagation delays assumed to be zero
2. Cross-hatched areas indicate SCC Demand and Command response Enabled
3. Demand from SCC1 is delayed by 3-byte Delay in SCC2

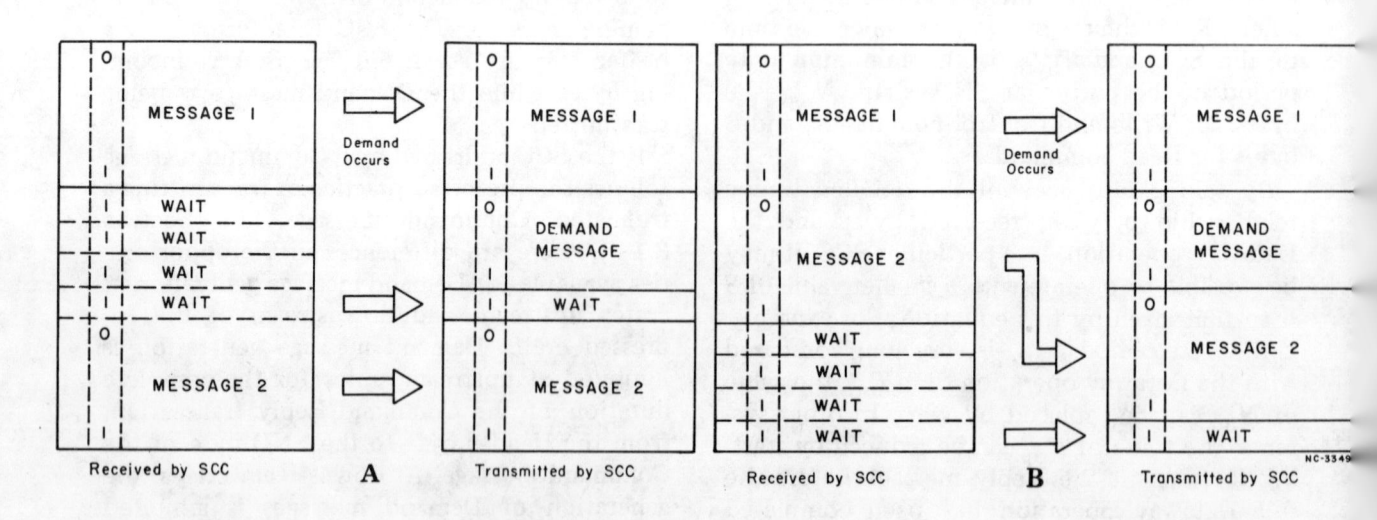

**Fig 5-2
Demand Message Generation
A Demand Message Directly Displaces WAIT Bytes
B Demand Message Delays Incoming Message**

Command message and the Reply message. Demand message generation is again inhibited by the Header byte of the Reply message and is permitted by the ENDSUM byte of this message.

A typical message sequence including Demand messages is shown in Fig 5-1. Here a demand occurs in SCC1 shortly after the Command message to SCC2 has passed. A demand occurs in SCC2 during the Command/Reply transaction, and the Demand message is generated immediately afterward, so that the Demand message from SCC2 is interposed ahead of the Demand message from SCC1, and delays it. A Demand message is generated by SCC3 in the interval between the truncated Command message and the Reply message from SCC2.

5.1 Control of Demand Message Initiation

> The following conditions must be satisfied before an SCC initiates the generation of a Demand message:
>
> (1) Demand message generation has been enabled by the appropriate bit of the Status Register of the SCC (see Section 12.3.1)
>
> (2) A demand is present, which has either appeared since the last Demand message was transmitted by the SCC, or was present when the SCC changed to the Demand Enabled state
>
> (3) The SCC is able to accept three incoming bytes while generating a Demand message (see Section 5.2)
>
> (4) The previous byte transmitted at the output port was a Delimiter byte (see Section 3.5.1).

An SCC with an SGL Encoder connector (see Section 14.1) should interpret Condition (2) as requiring that the Demand Message Initiate (DMI) signal is at logic 1, and has changed from logic 0 to 1 since the SCC assembled and transmitted the last Demand message.

5.2 Delay Buffer.
An SCC is not permitted to transmit more bytes than it receives (see Section 8.2). Therefore, when an SCC generates a 3 byte Demand message this has to displace 3 WAIT bytes from the stream of bytes passing along the SH.

If the byte stream happens to contain 3 WAIT bytes at the appropriate time, each byte of the Demand message directly replaces 1 WAIT byte. Otherwise, the SCC transmits the Demand message and then displaces 3 WAIT bytes that occur later in the byte stream. In the meantime the byte stream passing through the SCC has to be delayed by up to 3 byte periods (see Fig 5-2).

> While an SCC generates a 3 byte Demand message, it must continue to accept bytes from the SH at its input port. The Demand message must displace 3 WAIT bytes from the byte stream passing through the SCC, and all other bytes must be transmitted in the correct sequence after the Demand message has been transmitted.
>
> In order to meet these requirements, any SCC that can generate 3 byte Demand messages must be able to delay the SH byte stream by an appropriate number of byte periods. This delay must not be less than the actual number of non-WAIT bytes received while the Demand message is being transmitted. The delay must be applied to the SH byte stream when the SCC begins to transmit the Demand message and must continue until a total of 3 WAIT bytes have been displaced from the SH byte stream (unless loss of byte synchronism occurs, see Section 10.3). The delay must be switched in and out between messages, and in synchronism with the byte clock.

The recommended implementation of this requirement is a 3 byte Delay buffer. When the SCC begins to generate a Demand message, incoming bytes are delayed by being passed through this buffer. The recommended SCC Type L2, defined in Section A1 of the appendix, switches in a fixed delay of 3 bytes, regardless of the contents of the byte stream. In another permitted implementation, more complex but giving somewhat better system performance, an SCC that is transmitting a Demand message switches in 1 byte-unit of delay each time that it receives a non-WAIT byte.

After generating a 3-byte Demand message, the SCC continues to route the incoming byte stream through the Delay buffer until such time as it is able to remove 3 WAIT bytes from the byte stream. The Delay buffer is then switched out of the SH message path.

The recommended SCC Type L2, defined in Section A1 of the appendix, switches out the fixed delay of 3 bytes when the buffer contains 3 consecutive WAIT bytes (and a preceding END or ENDSUM byte has passed through the buffer). In another permitted implementation, more complex but giving somewhat better system performance, the SCC switches out 1 byte-unit of delay each time a WAIT byte is received (distinguishing between a genuine WAIT byte and any preceding END byte).

After an SCC has generated a Demand message, it cannot initiate another Demand message until it has received the appropriate number of WAIT bytes to allow it to switch out the Delay buffer [see Condition (3) in Section 5.1].

> When an SCC is operating in bit-serial mode and loses byte synchronism, the delay associated with Demand message generation must be removed completely from the path by which the SCC retransmits the received byte stream (see Section 10.3).

For further information concerning the means of enabling and disabling Demand message generation and of testing the Demand status, see Section 12.3. For information concerning coding the SGL field of the Demand message, and dealing with unserviced demands, see Section 14.

5.3 Identification of Demands. The Header byte of the Demand message indicates the crate in which the demand has occurred. The 5-bit SGL field in the second byte of the message can be used to identify the demand in more detail, for example, by a code indicating either the station from which the L signal originated or the action that is required.

This information can be supplemented by using the Read LAM-pattern command. Each bit of the data word read by this command indicates the state of the corresponding Dataway L line (see Section 11.2.1).

6. Identification of Message Type

Under normal operating conditions the messages on the SH consist of complete Command messages (with or without Write-data field), truncated Command messages, Reply messages (with or without Read-data field), and Demand messages. Under error conditions there may also be various forms of incomplete and spurious messages due, for example, to loss of synchronism, or abandoned operations, or corruption of WAIT bytes. This section summarizes the information that is available to an SCC or SD in order to distinguish these types of messages and to identify the byte containing the column-parity field.

The three main pieces of information available to an SCC or SD are the *Message Identification* field MI, the Function field *SF*, and the message length. The MI field (see Section 3.4.7) provides the basic means of distinguishing between Reply, Demand, and complete Command messages. The message length (number of bytes from the Header byte to the first Delimiter byte, inclusive) provides a means of distinguishing between complete and truncated Command messages, and between Reply messages with and without a data field. The *SF* field of the complete Command message distinguishes between Command messages with and without a data field.

The SD may also use information about the types of messages that are expected at its input in response to various conditions existing at its output. For example, when no Command/Reply transaction is in progress, the SD expects to receive only Demand messages. When Command/Reply transactions are in progress it expects the Crate Address fields of any received Command and Reply messages to match those of the transmitted Command messages.

6.1 Complete Command Message (Minimum Length 8 Bytes, MI=OO). Under normal conditions complete Command messages are only received by SCCs (if the addressed SCC truncates the Command message as recommended in Section 3.1.2), but under error conditions complete Command messages may also be received by the SD (for example, if the Crate Address field is not recognized by any SCC in the system).

For most purposes an adequate identification of complete Command messages can be made at the SCC by comparing the Crate Address field of the message with the assigned address of the SCC. If the Crate Address field of a received message matches the assigned address of the SCC, then this message is normally a complete Command message. Other messages might erroneously satisfy this condition, but should ultimately be rejected by tests on the column-parity field or message length. As an additional precaution the SCC may check that the contents of the MI field in the second byte are MI=00. (The recommended SCC Type L2 defined in Section A1 of the appendix does not check the MI field.)

At the SCC the location of the SUM byte is determined by reference to the *SF* field. If $(\overline{SF16} + SF8) = 1$, indicating a Read or Control command, then the column-parity field is in the fifth byte. If $SF16 \cdot \overline{SF8} = 1$, indicating a Write command, then the column-parity field is in the ninth byte.

At the SD this message may be distinguished from a truncated Command message by its length (more than 2 bytes), and from all other genuine messages by its MI field. The column-parity field of an untruncated Command message is generally irrelevant at the SD, but it could be identified by reference to the *SF* field.

6.2 Truncated Command Message (Length 2 Bytes; MI None). An SCC cannot legitimately receive a truncated Command message addressed to itself. If this occurs erroneously, the message length will cause the SCC to reject the message (Delimiter in second byte).

At the SD this message can only be distinguished by its length. This identification may be supported by comparing the Crate Address fields in the received message and the transmitted Command message. There is no column-parity field in the truncated Command message.

6.3 Reply Message (Length 3 or 7 Bytes; MI=01). An SCC cannot legitimately receive a Reply message whose Crate Address field matches the assigned address of the SCC. If this occurs erroneously, the SCC will treat it as a Command message and ultimately reject it as a result of the column-parity field, message length, or (if the test is implemented) the MI field.

At the SD, a Reply message without Read-data field has the same length as a Demand message, from which it can only be distinguished by the MI field. Reply messages with and without data fields can only be distinguished by their message length. The ENDSUM byte containing the column-parity field can therefore only be identified by the fact that it is the first Delimiter byte in the message.

6.4 Demand Message (Length 3 bytes; MI=1-). An SCC cannot legitimately receive a Demand message whose Crate Address field matches the assigned address of the SCC. If this occurs erroneously, the SCC will treat it as a Command message and ultimately reject it as a result of the column-parity field, message length, or (if this test is implemented) the MI field.

At the SD, a Demand message can be distinguished by the MI field. It has the same message length as a Reply message without data field. The ENDSUM byte containing the column-parity field can be identified as the first Delimiter byte or as the third byte of the message.

7. Serial Highway *D* Ports

> All SCCs and SDs must have input and output *D*-port connectors and must implement via these connectors either the bit-serial mode, or the byte-serial mode, or both modes.
>
> Any SCC that implements both bit-serial and byte-serial mode must include a means of selecting the required mode.

This mode selection may be made internally. It should not be accessible (and need not be indicated) at the front panel of the SCC.

An SCC or SD may also have additional ports to other standards (*U* ports), unless these are specifically excluded, as in the case of the recommended SCC Type L2.

The *D*-port connectors carry data and clock signals on separate wire pairs, using Non-Return-to-Zero-Level (NRZL) signals. They also carry control signal outputs for external

7.1 D-Port Connectors

7.1.1 Mechanical

> All *D*-port input connectors must be twenty-five way Cannon Type DBC-25P single-density fixed members with pins, or equivalent connectors fully compatible with the Cannon DBC-25S. All *D*-port output connectors must be twenty-five way Cannon Type DBC-25S single-density fixed members with sockets, or equivalent connectors fully compatible with the Cannon DBC-25P.

Equivalent connectors include the AMP-Minrac 17-Series and Cinch-D*SM series.

> The *D*-port connectors must have Cannon Type D20418-2 screw-lock retainers (fixed members) or equivalent retainers that are fully compatible with Cannon Type D20419-16 free members.

The arrangement of *D*-port connectors on the front panel of the SCC is defined in Section 13.3.

7.1.2 Contact Assignments

> The contacts of the *D*-port connectors must be assigned as shown in Table 7-1.

Contacts 22 and 23 are used for the bit clock in bit-serial mode, and for the byte clock in byte-serial mode.

Contacts 4 and 5 are used for the serial bit stream in bit-serial mode, and for the least significant bit of the byte in byte-serial mode. Seven pairs of contacts, from contacts 6 and 7 to contacts 18 and 19, are used in byte-serial mode only for Bits 2 to 8 of the byte.

Contact 24 of the *D*-input port is connected to the corresponding contact of the *D*-output port and also to the Bypass Control signal source within SCC.

Contact 25 of the *D*-output port is connected to the Loop Collapse Control signal source within the SCC. Contact 25 of the *D*-input port is reserved for use with a third control signal, if this is required in special cases.

Contacts 2 and 3 are assigned as Bus 1 and contacts 20 and 21 as Bus 2. These contacts on the *D*-input connector of SCC are linked directly to the corresponding contacts on the *D*-output connector. Bus 1 and Bus 2 are Free-use connections, with no defined use or direction of signal transfer.

As an example of the use of Bus 1 and Bus 2, Fig 7-1 shows an arrangement by which an SCC operating in bit-serial mode can be linked to the SH by only one twenty-five-way cable assembly, connected to the *D*-output port. Bus 1 and Bus 2 are used to bring the bit-serial data and bit-clock signals to the SCC. Turn-around cross-connections at the *D*-input port of the

Table 7-1
Contact Assignments for *D*-Port Connectors

D-Input Connector	Contact	*D*-Output Connector
Circuit ground (earth)	1	Circuit ground (earth)
Bus 1 (Free use)	2 3	Bus 1 (Free use)
Bit serial data in or Byte serial LSB* in	4 5	Bit serial data out or Byte serial LSB* out
Byte serial Bit 2 in	6 7	Byte serial Bit 2 out
Byte serial Bit 3 in	8 9	Byte serial Bit 3 out
Byte serial Bit 4 in	10 11	Byte serial Bit 4 out
Byte serial Bit 5 in	12 13	Byte serial Bit 5 out
Byte serial Bit 6 in	14 15	Byte serial Bit 6 out
Byte serial Bit 7 in	16 17	Byte serial Bit 7 out
Byte serial MSB† in	18 19	Byte serial MSB† out
Bus 2 (Free use)	20 21	Bus 2 (Free use)
Bit/Byte clock in	22 23	Bit/Byte clock out
Bypass Control	24	Bypass Control
Reserved for Control Signal	25	Loop Collapse Control

*Least significant bit.
†Most significant bit.
NOTE: Each Balanced signal input or output occupies a pair of contacts.
The even-numbered contact is Terminal *A*, carrying Signal
The odd-numbered contact is Terminal *B*, carrying $\overline{\text{Signal}}$

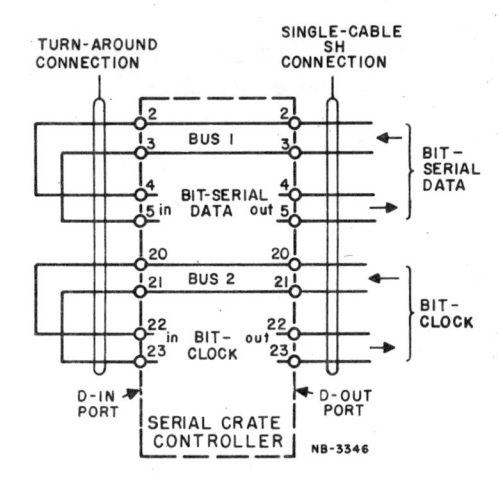

Fig 7-1
Example of Use of Bus 1 and Bus 2
Contacts at D Ports

SCC then link Bus 1 to the data input and Bus 2 to the clock-signal input. The loop to the SD is completed in the usual way from the D-output port.

7.2 Data and Clock Signals. The standards for data and clock signals at D-ports are based on Standard RS-422 of the Electronic Industries Association.[4]

7.2.1 *Transmission Lines*. All D ports transmit or receive data and clock signals over two conductor transmission lines, terminated at least at the receiving end. Each transmission line should be of balanced construction, such as a twisted pair. The nominal characteristic impedance should be 100 Ω. (For conservative line-length limitations see Standard RS-422.)[4]

One conductor of each transmission line is designated 'Signal' (corresponding to Terminal A in RS-422[4]) and occupies an even numbered contact at D ports. The other conductor is designated '$\overline{\text{Signal}}$' (corresponding to Terminal B) and occupies an odd-numbered contact.

---

[4] Standard RS-422, Electrical Characteristics of Balanced Voltage Digital Interface Circuits, Apr, 1975, Electronic Industries Association, 2001 Eye Street, Washington DC 20006.

This standard is closely related to the following provisional recommendation: Provisional Recommendation X27: Electrical Characteristics for Balanced Double-Current Interchange Circuits for General Use with Integrated Circuit Equipment in the Field of Data Communications, International Telegraph and Telephone Consultative Committee, Geneva, Switzerland.

7.2.2 *Logic States*

For all data and clock signals transmitted and received through D ports the voltage level of the Signal conductor with respect to the $\overline{\text{Signal}}$ conductor must be negative in the logic **1** state and positive in the logic **0** state.

7.2.3 *Balanced Transmitter*

The balanced transmitter must conform to the specification of the generator in Standard RS-422.[4]

The characteristics of this transmitter are summarized for easy reference in Table 7-2. An example of a balanced transmitter is given in Fig 7-2.

Table 7-2
Summary of Characteristics of
Balanced Transmitter*

Output resistance, line to line	100 Ω or less
Magnitude of open circuit voltage, line to line [V_o]	6 V or less
Magnitude of open circuit voltage, line to ground [V_{oa}, V_{ob}]	6 V or less
Magnitude of output voltage, terminated in 100 Ω, line to line [V_t]	Not less than 2 V or 50% V_o whichever is greater
Magnitude of offset voltage [V_{os}]‡	3 V or less
Magnitude of difference in [V_t] for two logic states	Less than 0.4 V
Magnitude of difference in [V_{os}] for two logic states	Less than 0.4 V
Magnitude of short-circuit current, line to ground	150 mA or less

*The maximum values of the 10 to 90 percent rise and fall times of signals generated by the transmitter, when applied to a 100 Ω resistive load, are as follows (where T_{min} is the bit or byte period, as defined in Section 8.3)- for clock signals, less than 20 ns or 0.05 T_{min}, whichever is greater; for data signals, less than 0.1 T_{min}.

Where a magnitude is defined, the parameter may be positive or negative.

‡The offset voltage is measured between the center point of a 100 Ω test load consisting of two resistors, 50 Ω ±1 percent each, and the generator circuit ground.

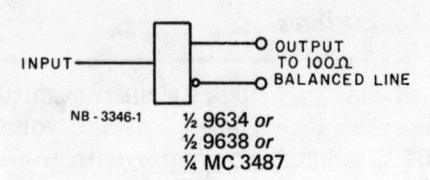

[Fig 7-2 diagram: Balanced Transmitter with INPUT to NB-3346-1, ½ 9634 or ½ 9638 or ¼ MC 3487, OUTPUT TO 100Ω BALANCED LINE]

Fig 7-2
Example of Balanced Transmitter

7.2.4 Balanced Receiver

> The balanced receiver must conform to the specification of the receiver in Standard RS-422[4] and must include a cable termination resistance.

The characteristics of this receiver are summarized for easy reference in Table 7-3. An example of a balanced receiver is given in Fig 7-3.

If the input to an SCC becomes disconnected, such as could occur in certain by-passing configurations, the SCC may receive erroneous messages due to noise at the receiver inputs. It is therefore recommended that line receivers with input hysteresis be used at least for the clock receiver. Examples of such line receivers conforming to Standard RS-422 are shown in Fig 7-3. Some types of receivers may require external hysteresis added by means of resistors as shown in Fig 7-3a.

7.3 Control Signals.
The D-input and D-output ports carry signals for controlling an external Bypass device (see Section 15.1.1), and the D-output port carries a signal for controlling an external Loop Collapse device (see Section 15.1.2). A contact at the D-input port is reserved for a third Control signal, if required. These signals are suitable for operating electromechanical relays in the external devices, and allow the external device to assume an appro-

Fig 7-3
Example of Balanced Receiver

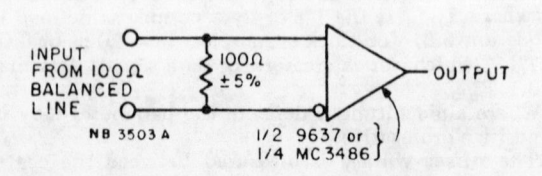

[Fig 7-3 diagram: INPUT FROM 100Ω BALANCED LINE, 100Ω ±5%, NB 3503A, ½ 9637 or ¼ MC 3486, OUTPUT]

Table 7-3
Summary of Characteristics of Balanced Receiver

Input resistance line to line $[R_t]$*	100 Ω ± 10 percent
Input impedance line to ground with 100 Ω termination removed	⩾ 4000 Ω
Magnitude of input voltage, line to line, at which receiver must operate correctly $[V_i]$	⩾ 0.2 V ⩽ 6.0 V
Magnitude of common mode voltage at which receiver must operate correctly $[V_{cm}]$†	⩽ 7.0 V
Maximum magnitude of input voltage, line to ground	⩽ 10.0 V
Magnitude of input voltage, line to line, without damaging the receiver (The termination of 100 Ω may be removed for this test)	

*In Standard RS-422 (see Section 7.2, footnote[4]) the use of a cable termination at the balanced receiver is optional, depending on the specific environment in which the receiver is used, but is here specified as a mandatory feature for the balanced receiver used at the SH D-port.
†The common mode voltage is defined as the algebraic mean of the two voltages line to ground at the receiver input terminals.
NOTE: Where a magnitude is defined, the parameter may be positive or negative

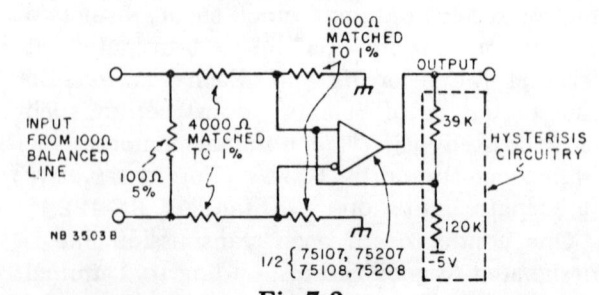

[Fig 7-3a diagram: INPUT FROM 100Ω BALANCED LINE, 100Ω 5%, 4000Ω MATCHED TO 1%, 1000Ω MATCHED TO 1%, op-amp, 39K, 120k, -5V, HYSTERISIS CIRCUITRY, OUTPUT, NB 3503B, ½ {75107, 75207 / 75108, 75208}]

Fig 7-3a
Example of a Receiver with Hysteresis Added by Use of External Resistors

priate fail-safe state if the power supply to the SCC or the external device fails.

Each control signal occupies only one contact at the D port. The signals share the common circuit ground connection.

7.3.1 Signal Standards

> The sources and receivers for the Bypass and Loop Collapse Control signals must conform to the signal standards shown in Table 7-4.

INTERFACE SYSTEM (CAMAC)

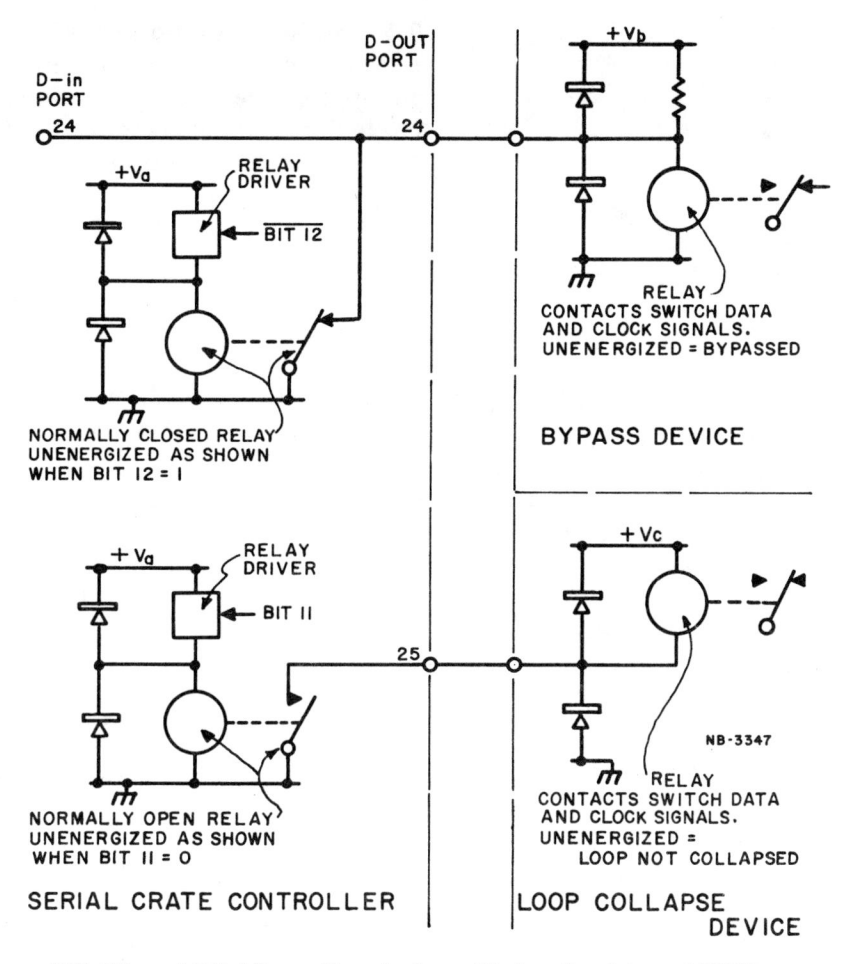

(Bit 11 and Bit 12 are Signals from Status Register of SCC)

Fig 7-4
Examples of Circuits for Control Signal Sources and Receivers
(See also Figs 15-1 and 15-2)

Table 7-4
Standards for Control Signals at D Ports

Logic State	State of Control Line	Current Drawn from Control Line by Source	Receiver Must Respond Correctly to Control Signal in Range*
0	'Free'	Magnitude of current not more than 100 µA for Control Line between 0 and +25 V	+10 to +24 V
1	'Grounded'	Minimum Current sinking capability 115 mA for Control Line at 0.5 V	0 to +3 V

*The receiving device must respond within 80 ms.

7.3.2 *Control Signal Sources and Receivers.* Examples of circuits for generating the Bypass and Loop Collapse Control signals in the SCC and for receiving them in the external devices are shown in Fig 7-4.

The circuits for the Bypass and Loop Collapse Control signals are different because of the requirements that the external devices assume the Bypassed and not Collapsed states, respectively, if there is a power failure in the SCC or the external device.

If the controlled device presents an inductive load, it must include transient suppressors to prevent the voltage on the control line going outside the range −5 V to +30 V.

8. Timing

The timing of data transfers on all segments of the SH is determined by a system clock at the SD. In a serial system based entirely on direct interconnections between D ports, there is a separate path throughout the SH for system-clock signals at bit or byte rate. In a system with U-standard interconnections using synchronous transmission, the transmitting signal converters may modulate the system-clock signals onto the same communications channel as the data. The synchronous receiving signal converters will then reconstitute the separate clock signal for use at the next D port. In a system with U-standard interconnections using asynchronous transmission, each asynchronous receiving signal converter will usually reconstitute the system clock by using a local clock set to the appropriate nominal frequency.

8.1 Frequency of System Clock. The overall design of each serial system involves choosing an appropriate system-clock frequency that meets the operational requirements and is within the operating limits of system components such as the communications channels and SCCs.

> The maximum instantaneous system-clock rate in any SH system must not be more than 5.0 MHz.
> Each SCC and SD must be capable of operation via its D ports at any clock rate up to a stated maximum rate (which may be less than 5.0 MHz).

For the recommended SCC Type L2 (see Section A1 of the appendix) the maximum bit or byte clock rate is defined as 5 MHz.

Thus the SCCs in a serial system place no restrictions on the frequency or regularity of the system clock, except perhaps to set an upper limit of clock rate less than the absolute

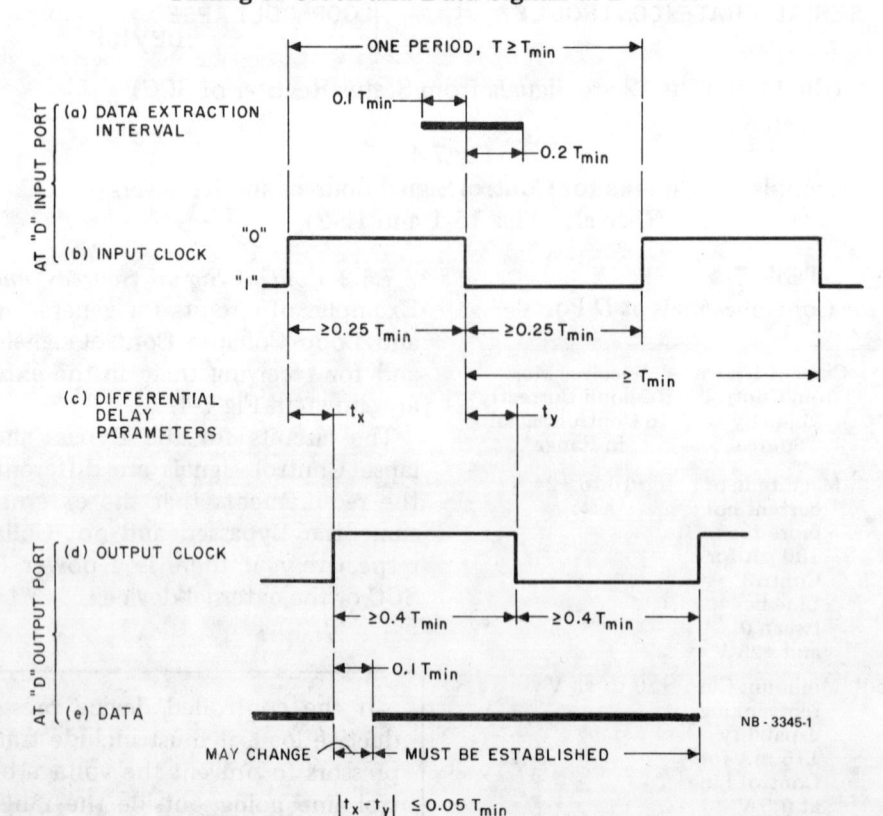

Fig 8-1
Timing of Clock and Data Signals at D Ports

DIFFERENTIAL DELAY: $|t_x - t_y| \leq 0.05\ T_{min}$, IF THIS IS CONSISTENT WITH CONDITIONS AT (d)

maximum of 5 MHz. However, other components of the system, such as the communications channels and signal converters, may demand a particular clock rate and set limits on the permitted nonuniformity of the clock periods.

In any particular system, the clock frequency and communications channels should be chosen so that the data signals received at D-port inputs are established and stable during the data-sampling period shown in Fig 8-1.

8.2 Byte Stream. The SD generates at its output port a series of bytes timed by the system clock. This byte stream propagates along the SH.

> Each time an SCC receives one byte it must transmit one, and only one, byte.

The contents of the bytes received and transmitted by an SCC in a particular byte period are not always identical.

Bytes cannot propagate through SCC in the absence of system-clock signals. Therefore, the system-clock signals should be generated continuously, particularly in systems that rely on the generation of Demand messages.

Messages, and the bytes within messages, are not necessarily contiguous, but may be separated by pauses of variable length. At a given clock rate, the maximum system performance is obtained when the byte stream is a contiguous series of bytes. The performance of auxiliary controllers that respond to the Auxiliary Controller Lockout (ACL) signal (see Section 14.6.2) can be adversely affected by excessive pauses within Command messages.

8.3 Signal Timing. The timing relationship between data signals and clock signals at D ports is shown in Fig 8-1.

The significant times on the balanced-signal waveforms at the D-input ports are measured with respect to the final crossing of the plus or minus 200 mV differential signal level (whichever is applicable). Times at the D-output ports are measured with respect to the final crossing of the plus or minus $V_t/2$ differential signal level (whichever is applicable).

Each 1-to-0 transition of the clock defines the beginning of a new clock period. The 0-to-1 transition of the received clock is the reference for a defined *Data Extraction Interval* during which the received data signals at the D-input port are assumed to be steady. At the D-output port the data signals are established soon after the 1-to-0 transition of the transmitted clock and are maintained until the next 1-to-0 transition.

In Fig 8-1, T is a parameter of the system and is the actual duration of a particular clock period, which may consist of the nominal clock period extended by a pause. T_{min} is a parameter of the SCC or SD and is its minimum rated clock period. For example, T_{min} = 200 ns for the SCC Type L2 defined in Section A1 of the appendix. The relationship between these parameters is that T is not less than T_{min}.

8.3.1 *Clock Signals*

> Each SCC or SD must be capable of correct operation when the duration of the logic 0 or logic 1 state of the received clock signal has the minimum value shown in Fig 8-1.
>
> Each SCC or SD must generate at its D-output port a clock signal in which the duration of each logic 0 and logic 1 state is not less than the minimum value shown in Fig 8-1.
>
> The duration of each logic 0 and logic 1 state of the clock signal transmitted by an SCC at the D-output port must reproduce the corresponding duration of the received clock signal at the D-input port, within the limits of distortion set by the differential delay condition shown in Fig 8-1, whenever this is consistent with the absolute requirements for the duration of the logic states.

Cumulative deterioration of the clock-signal timing can occur in passing through successive sections of the SH (particularly if the clock rate is near the limit set by the bandwidth of the transmission medium), and in passing through successive SCCs (particularly if the clock rate is near the limit set by the circuits that receive, shape, and retransmit the clock signals). The definition of the clock signals therefore allows a margin between the timing of the transmitted and received signals in order to accommodate deterioration within each section of the SH. Distortion of the clock signal within each SCC is limited by the differential delay requirement. Each SCC restores the timing of the clock signal if it has deteriorated beyond acceptable limits.

These properties of the SCC are defined with respect to its own value of T_{min}. System design problems can arise if SCCs with widely differing values of T_{min} are used in the same system. For example, 0.4 T_{min} at the output of one SCC could be less than 0.25 T_{min} at the input to the next SCC.

8.3.2 Transmitted Data Signals

> The data signals transmitted from each D-output port must be established for that portion of the clock period shown in Fig 8-1, where T_{min} is the period corresponding to the stated maximum clock rate of the transmitting device.

8.3.3 Received Data Signals

> The data signals received at each D-input port must be sampled (strobed) within the portion of the received clock period shown as the Data Extraction Interval in Fig 8-1. Here T_{min} is the period corresponding to the stated maximum clock rate for the receiving device. Data signals outside the Data Extraction Interval must be ignored.

8.3.4 Data Signal Conditioning. When an SCC is retransmitting received data, the timing requirements shown in Fig 8-1 involve sampling the received data at or near the 0-to-1 transitions of the received clock signal and establishing the retransmitted data for most of the period between successive 1-to-0 transitions of the transmitted clock.

This implies that there is a temporary buffer store in the data path between the input and output ports, and that the data signals are delayed by up to one clock period in passing through the SCC.

8.4 Propagation Delays. The message stream on the SH is subject to delays due to various factors, including the following:

(1) Signal transmission in the communications channels and signal converters

(2) Signal propagation delays within SCCs

(3) Logical reshaping in SCCs [This can impose a delay of up to 1 clock period. This delay may be either 1 bit period in bit-serial mode, or 1 byte period in byte-serial mode (see Section 8.3.4).]

(4) Generation of Demand messages [Downstream from an SCC that generates a demand, the following message or messages may be delayed by up to 3 byte periods (see Section 5.2).]

Thus the total delay encountered by messages propagating around the SH loop in a small bit-serial system may be only a few bit periods. In a large byte-serial system the normal delay excluding the transmission and propagation delays may be as much as sixty-two byte periods, with a possible worst case of 248 byte periods in the unlikely event of simultaneous demands from all crates.

Good management at the SD can prevent the accumulation of 3-byte delays, due to demand generation, by transmitting an adequate number of WAIT bytes whenever a Demand message is received by the SD.

9. Bit-Serial and Byte-Serial Modes

All messages on the SH are structured as 8 bit bytes. The bytes are transmitted on the SH in one of the two modes, either bit serial or byte serial. All SCCs and SDs have D ports that can handle bytes in one or both of these modes.

9.1 Byte-Serial Mode. In this mode the 8 bits of a byte are transferred in parallel through the D ports, with each bit on a separate contact pair. One byte is transferred during each period of the system clock, as defined in Fig 8-1. The 8 bits are established by the transmitting D port soon after the 1-to-0 transition of the clock and are maintained until the next 1-to-0 transition.

> The SD is permitted to generate system-clock signals in which the nominal clock period is followed by a byte pause of arbitrary length. During this pause the clock signal is maintained in the 1 state and is extended to a total clock period T. The SD and all SCCs must accept at their D ports a system clock of this form, with nonuniform byte periods.

In specific systems the characteristics of communications channels used between D ports or U ports may require that the system clock has uniform periods, or that the maximum duration of byte pauses is limited.

9.2 Bit-Serial Mode. In this mode the 8 bits of a byte are transferred serially through the D ports, on one contact pair. The 8 bits are preceded by a Start bit and followed by a Stop bit, making up a byte frame of 10 bits. One bit is transferred during each period of the system clock.

> In each 10-bit byte frame, the Start bit must be transferred first, and must always be at logic **0**. The 8 data bits must follow, with the least significant (Bit 1) first and the most significant (Bit 8) last. The Stop bit must be transferred last and must always be at logic **1**. The Start bit of each byte frame must always be preceded by a bit at logic **1**.

A data signal is established on the data contact pair by the transmitting D port soon after the 1-to-0 transition of the clock and is maintained until the next 1-to-0 transition. The state of the remaining 7 data contact pairs at the D port is of no significance.

9.2.1 *Noncontiguous Byte Frames.* Successive byte frames may be contiguous or noncontiguous. If two byte frames are contiguous, the Stop bit of the first frame is followed immediately by the Start bit of the next frame. If two frames are noncontiguous, the Stop bit of the first is followed, as shown in Fig 2-3, by a pause consisting of an arbitrary number of bit-clock periods during which the data signal is held at logic **1**. These pause bits are followed by the Start bit of the next frame.

Hence the context of the Start bit is such that it is always preceded by a bit at logic **1**, which is either the Stop bit of a preceding contiguous frame or a pause bit following a noncontiguous frame.

> The SD is permitted to generate either contiguous or noncontiguous byte frames. All SCCs must accept contiguous or noncontiguous byte frames at their D ports.

In specific systems the characteristics of communications channels used between D ports or U ports may require that the number of pause bits is limited, particularly if asynchronous transmission is used.

> At the bit-serial D output port of an SCC, the duration of the pause between any two byte frames must reproduce the duration of the pause received at the same time at the D-input port.

Some communications channels may be based on 11-bit frames (with two Stop bits). These are equivalent to a 10-bit frame followed by at least one pause bit.

Owing to logical delays within the SCC (see Section 5.2), the pause between two particular bytes of the message stream is not necessarily reproduced between the same pair of bytes at the output of an SCC. However, the requirement is sufficient to ensure that if, for example, the SD generates 11-bit frames, these will be correctly reproduced by all SCCs.

9.2.2 *Nonuniform Bit Periods*

> The SD is permitted to generate system clock signals in which the actual bit periods are arbitrarily longer than the bit period corresponding to the nominal bit rate. The SD and all SCCs must accept at their D ports a system clock of this form, with nonuniform bit periods.

9.2.3 *Extraction of Byte Clock*

> Every SCC with a bit-serial D-port input must be capable of deriving a byte clock from received 10-bit byte frames that are either contiguous or separated by pauses of any arbitrary number of bit periods.

In order to derive the byte clock it is necessary to identify the received byte frames, either by the logic **0** state of the Start bits or by their 0-after-1 context (see Section 9.2.1). The derived byte clock at the SGL-Encoder connector is defined in Section 14.2.8.

The Stop bit of a received byte frame initiates any processing of the byte within the SCC, but the Start bit initiates bit-serial transmission of a byte at the *D*-output port. Thus, under some conditions (for example, when receiving a Header byte) a received byte is retransmitted before its contents have been examined by the SCC.

10. Synchronization

The SD and each SCC need to establish synchronism with the message format of the byte stream received from the SH. In addition, when operating in bit-serial mode they need to establish synchronism with the byte-frame format of the bit stream. This section deals with the means by which these two levels of synchronism are established, maintained, and verified.

10.1 Message Synchronization. In bit-serial and byte-serial modes each message is transmitted with a defined sequence of Delimiter and Nondelimiter bytes. In order to extract the message from the serial byte stream, each receiving device has to identify this sequence correctly, and thus it achieves and maintains message synchronism.

10.1.1 *Maintenance of Message Synchronism*

> When message synchronism has been established, a receiving device must, after having received a message terminated by a Delimiter byte, treat the next Nondelimiter byte as the Header byte of a new message. Having identified the Header byte of a message it must treat the next Delimiter byte as the end of the message.

10.1.2 *Loss of Message Synchronism.*
Loss of message synchronism occurs when a receiving device is unable to identify the Header and terminating bytes of a message correctly. This can occur if, for example, any Nondelimiter byte is corrupted into a Delimiter byte or if the information field determining the length of a Command or Reply message is corrupted.

When operating in bit-serial mode it is very probable that loss of byte synchronism will also result in loss of message synchronism. Message synchronism is therefore always reestablished after byte synchronism has been reestablished.

Detection of loss of message synchronism is based on checking whether Delimiter bytes occur in appropriate contexts. An SCC is only required to test message synchronism during Command/Reply transactions addressed to it. In general, loss of message synchronism will also result in failure of the column parity (see Section 15.3.2).

> An SCC must assume the Lost Message Sync state under either of the following conditions:
> (1) The SCC has just reestablished byte synchronism
> (2) The SCC receives a Delimiter byte at any time after accepting the Header byte of a Command message addressed to it and before transmitting the ENDSUM byte of the Reply message.

10.1.3 *Establishing Message Synchronism*

> An SCC with Lost Message Sync status, operating in bit-serial mode, must be in byte synchronism (see Section 10.2.3) before attempting to establish message synchronism.
>
> When byte synchronism has been established, the SCC must search the incoming byte stream for at least one Delimiter byte (typically END, ENDSUM, or WAIT bytes) before asserting message synchronism. Following this the next Nondelimiter byte must be treated as the Header byte of a message.

Increased assurance of true message synchronism is given by requiring that more than one Delimiter byte is recognized before message synchronism is asserted.

10.2 Byte Synchronization.

In byte-serial mode, the system-clock signals indicate the occurrence of successive bytes and are used by SCCs and the SD as a direct means of byte synchronization.

In bit-serial mode, each 8 bit byte is transmitted within a frame of Start and Stop bits. In order to extract the 8-bit byte from the serial bit stream, each receiving device has to identify the framing bits correctly and thus achieve and maintain byte synchronism.

10.2.1 Maintenance of Byte Synchronism

> When byte synchronism has been established, an SCC or SD operating in bit-serial mode must, after having received a byte frame, treat the next logic 0 bit in the received bit stream as the start of a 10-bit byte frame.

In the context of a correctly framed sequence of bytes this is equivalent to recognizing the first 0-after-1 sequence.

10.2.2 Loss of Byte Synchronism. Loss of byte synchronism occurs when an SCC or SD is unable to identify the framing bits (Stop bit and Start bit) in the received bit stream.

When the SCC or SD receives a logic 0 bit in the appropriate context, this is assumed to be the beginning of a 10-bit framed byte. The validity of this assumption is tested by examining the tenth bit of the byte.

Under error conditions the SCC or SD can falsely identify some other logic 0 bit as being a Start bit. In a correctly framed byte the tenth bit is the Stop bit, and is logic 1. The tenth bit in a falsely framed byte can be either logic 0 or logic 1. If it is logic 0, this is a clear indication of loss of byte synchronism. If it is logic 1, there is no clear indication of either loss or maintenance of byte synchronism. Thus, a number of incorrectly framed bytes can be received before loss of byte synchronism is detected.

> When operating in bit-serial mode an SCC must perform a byte-framing test on the tenth bit of each received byte frame. If this bit is in the logic 0 state, the SCC must adopt Lost Byte Sync status.

10.2.3 Establishing Byte Synchronism

> An SCC with Lost Byte Sync status must search the received bit stream for the bit pattern $1,11100000,0_2$, corresponding to a WAIT byte with Start and Stop bits, before reasserting byte-synchronism status.
>
> The SCC must perform this search for a WAIT-byte pattern by either:
>
> (1) In each bit-period, comparing the last 10 bits that have been received and the required bit-pattern (This is the preferred method, which is mandatory for SCC Type L2.)
>
> (2) In each received byte frame (identified by the 0-after-1 context of the Start bit and confirmed by the logic 1 state of the Stop bit) comparing the 8-bit contents of the byte and the 8 bits of the WAIT byte pattern (This is an alternative method, not permitted in SCC Type L2)

After byte synchronism has been established in this way, it is maintained as defined in Section 10.2.1.

10.3 Lost Synchronism: Actions by Serial Crate Controller

> An SCC that has lost byte synchronism or message synchronism must retransmit at its output port the bit or byte stream received at its input port.
>
> While an SCC is in the Lost Byte Sync or Lost Message Sync state it must not accept Command messages addressed to it, or generate Reply or Demand Messages. If the SCC has already accepted the Header byte of a Command message addressed to it and then loses synchronism, it must neither start to generate a Reply message nor continue the transmission of a Reply message that it has already started.
>
> If an SCC loses byte synchronism, it must switch out the Delay buffer. Otherwise, an SCC that has lost message synchronism must neither switch in nor switch out the Delay buffer.

Typical implementations of the SCC (including that described in Section A2 of the appendix) have two types of internal data path. One is a direct bit-serial path that transfers all bits from the input port to the output port and is independent of byte synchronism. The other is a basically byte-serial path that transfers the 8-bit contents of byte frames and is dependent on correct byte synchronism. Corruption of the serial bit stream is minimized by switching out the Delay buffer when byte synchronism is lost, so that the direct bit-serial internal path is used.

11. Access to Registers in Serial Crate Controller

Commands accepted by an SCC are addressed either to modules in the CAMAC crate, or to internal features of the SCC itself. In the first case the command is executed by means of a Dataway addressed operation. In the second case the execution of the command does not involve a Dataway addressed operation.

Internal registers of the SCC are accessed through Subaddresses at Station $N(30)$. Single-bit internal features, without an associated data word, are controlled and interrogated by accessing the appropriate bits of a Status Register.

> Each SCC must accept and execute all commands addressed to its internal registers, except when the SCC is in the Bypassed state. The operation cycle time for executing these commands must not be more than that for a Dataway operation, but in certain specified operations (see Sections 12.4.2 and 12.4.3) the initiation of the Reply message must be delayed. When the commands shown in Table 11-1 are executed successfully, the SQ and SX bits in the Status field must be as shown in the table.
>
> For any other commands addressed to the SCC [at $N(24)$-$N(31)$ inclusive] the SX bit in the Reply message must be at logic 1 if the command can be executed by the SCC and at logic 0 if it cannot be executed.

Table 11-1
Commands Implemented by SCC

	Command			Response	
Operation	N	A	F	SQ	SX
Status Register					
Read	30	0	1	1	1
Write	30	0	17	1	1
Selective set	30	0	19	1	1
Selective clear	30	0	23	1	1
Reread data field	30	1	0	DSQ	1
Read LAM pattern	30	12	1	1	1

11.1 Status Register. The Status Register comprises a collection of single-bit features of the SCC, with the common property that they are accessed at Station $N(30)$, Group 2, Subaddress $A(0)$. Some bits of the register control features of the SCC and others indicate the status of features of the SCC. The features that are accessed via the Status Register are defined in detail in Section 12.

The assignment of bits in the Status Register is shown in Table 11-2, with separate columns for the control exercized through Write operations and the indications accessed by Read operations.

> If an SCC provides access to any feature shown in Table 11-2, it must do so via the appropriate bit or bits of a Status Register as shown in the table. The bit positions that are shown as Reserved must not be used.

The Free bits of the Status Register may be used to control and interrogate any other features of the SCC, unless specifically prohibited (as in the case of SCC Type L2). This specification does not define or coordinate the use of the Free bits.

The bits that control features of the SCC can be set to logic 1 and, in general, reset to logic 0 by operations such as Write $F(17)$, Selective Set $F(19)$, and Selective Clear $F(23)$. Each control bit in the Status Register is associated with a corresponding SW bit in the Write-data field of the Command message.

The bits that indicate the status of internal features can be interrogated by operations such

Table 11-2
Assignment of Status Register Bits

Status Register Bit	Write Operations — Logic 1 into Register Controls	Read Operations — Logic 1 from Register Indicates	Comment
1	Generate Z	(Always 0)	Automatic reset to Logic 0
2	Generate C	(Always 0)	Automatic reset to Logic 0
3	Set $I = 1$	$I_{out} = 1$	
4	—	DERR = 1	Previous Reply status
5	—	DSX = 1	Previous Reply status
6	—	DSQ = 1	Previous Reply status
7	—	Dataway $I = 1$	
8	(Reserved)	(Always 0)*	Reserved
9	Enable Demands	Demands Enabled	
10	Set Internal Demand $L24$	Internal Demand $L24 = 1$	
11	Collapse Loop	Loop Collapsed	Indicates Control Signal
12	Apply Bypass	(Always 0)	
13	Dataway Off line	Dataway Off line	
14	—	Switch Off line	
15	(Reserved)	(Always 0)*	Reserved
16	—	Selected LAM present	
17	(Reserved)	(Always 0)*	Reserved
18	(Reserved)	(Always 0)*	Reserved
19	(Reserved)	(Always 0)*	Reserved
20	(Reserved)	(Always 0)*	Reserved
21	As required	As required	Free use
22	As required	As required	Free use
23	As required	As required	Free use
24	As required	As required	Free use

*Applies while bit has Reserved status

as Read $F(1)$. Each indicating bit in the Status Register is associated with a corresponding SR bit in the Read-data field of the Reply message.

> When power is applied to an SCC, certain bits in the Status Register must assume the states shown in Table 11-3.

11.2 Other Registers. All other addressable registers in the SCC should be accessed through Subaddresses, preferably at station $N(30)$.

11.2.1 Look-at-Me Pattern. The demand handling features of an SCC include the ability to interrogate the status of the twenty-four Dataway LAM signals $L1$ to $L24$, which form a virtual register accessed at Station $N(30)$, Group 2 Subaddress $A(12)$. Thus, some systems may respond to a Demand message from a particular crate by reading the LAM pattern from that crate in order to identify the modules that require servicing.

> In response to the Read LAM pattern command defined in Table 11-1 the SCC must reply with a Read-data field indicating the logic state of the Dataway L lines $L1$ to $L24$, including any simulated demand equivalent to $L24$ (see Section 12.3.2). The logic state of $L1$ must be indicated by Bit $SR1$, etc.
>
> The LAM pattern read by this operation must be independent of the status of the Demand Enable bit of the Status Register and of any selection or grading made by an associated SGL Encoder.
>
> While executing the Read LAM pattern command the SCC must not generate the Dataway Busy signal $B=1$.

Table 11-3
Initial State of Status Register Bits after Power Up

Status Register Bit	State After Power Up	Condition
3	1	Inhibit set ($I = 1$)
9	0	Demands disabled
10	0	Internal Demand $L24 = 0$
11	0	Loop not Collapsed
12	1	SCC Bypassed
13	1	Dataway Off Line

11.2.2 *Reread Data.* This recommended feature of an SCC allows recovery from an unsuccessful Read operation during which data has been read destructively from a module but not received correctly by the SD. Whenever an SCC with this feature executes a Read command, it stores the contents of the Read-data field and *SQ* status bit that it has assembled for the Reply message. Thus, if the SD fails to receive the Reply message correctly, it can generate a Re-read command to access this stored Read data from the SCC. The DERR bit in the Status field of Reply messages has an important role in this recovery process (see Section 15.5.3 and 15.6).

> All SCCs must respond to the Reread command defined in Table 11-1 by generating a Reply message with Read-data field. The *SX* bit in the Status field of the Reply must be *SX*=1 if the Reread command is implemented by the SCC, and *SX*=0 if it is not.
>
> If the SCC implements the Reread feature, the contents of the Read-data field in the Reply message must consist of the Read-data word resulting from the previous operation addressed to the SCC, if this was a correctly executed Read operation. The *SQ* bit in the Status field of the reply must correspond to the DSQ bit in the Status Register (*SQ*=DSQ).

If an SCC has just executed a Write or Control command, or has not executed the previous command, or does not implement the Reread command, then the contents of the Read-data field are undefined.

Thus, the Read data resulting from an executed Read command can be recovered, even if transmission of the Reply message was not completed. This applies equally to Read commands accessing modules and internal features of the SCC.

12. Features of Serial Crate Controller Accessed Via the Status Register

This section defines the features of an SCC whose status can be controlled or interrogated or both by operations addressed to the Status Register.

12.1 Dataway Common Controls. Generation of the Dataway Common Control signals Initialize *Z*, Clear *C*, and Inhibit *I* is controlled by assigned bits of the Status Register.

12.1.1 *Initialize and Clear.* Bit 1 of the Status Register controls the generation of Initialize, and Bit 2 controls Clear, as single-shot operations.

> If the SCC is in the Dataway On-line state (see Section 12.4.1) and is not bypassed (see Section 12.4.2) it must respond to commands that set Bit 1 or Bit 2 of the Status Register to logic 1 by generating a Dataway unaddressed operation (conforming to Section 7.1.3.2 of IEEE Std 583-1975). The operation must be accompanied by the *Z* signal if Bit 1=1 or by the *C* signal if Bit 2=1. The SCC must reset Bit 1 or Bit 2, as appropriate, to logic 0 not later than the end of the Dataway operation.
>
> If the SCC is in the Dataway On-line state and is not bypassed, it must allow Bits 1 and 2 of the Status Register to be set to logic 1 by operations that write into or selectively set the Status Register and have the appropriate value of bit *SW1* or *SW2*, respectively, in the Write-data field of the Command message.
>
> If the SCC is in the Dataway Off-line state or is bypassed (Status Register Bit 12=1), it must maintain Bits 1 and 2 of the Status Register in the 0 state. It must not

generate a Dataway operation in response to commands that attempt to set these bits to logic **1**.

Bits *SR*1 and *SR*2 of the data word read from the Status Register must always be logic **0**.

12.1.2 *Inhibit.* The Dataway Inhibit signal *I* generated by the SCC is controlled by Bit 3 of the Status Register.

If the SCC is in the Dataway On-line state and is not bypassed (Status Register Bit 12=0), it must generate an output to the Dataway Inhibit line corresponding to the logic state of Bit 3 of the Status Register (I_{out} = Bit 3).

If the SCC is either in the Dataway Off-line state (see Section 12.4.1) or is bypassed (Status Register Bit 12=1) it must generate a logic **0** output to the Dataway Inhibit line, irrespective of the logic state of Bit 3 of the Status Register (I_{out} = 0). The Inhibit line is thus free to take up the **0** state if no other is driving it to the **1** state.

Under all conditions, Bit *SR*3 of the Read-data field of the reply to commands that read the Status Register must indicate the state of Bit 3 of the Register, and Bit *SR*7 must indicate the state of the Dataway Inhibit line (Bit 7=I).

If there is another source of the Inhibit signal, the state of Bit 7 is not necessarily the same as that of Bit 3.

The SCC must allow Bit 3 of the Status Register to be set or reset by Write operations, using the appropriate value of Bit *SW*3 in the Write-data field. It must not allow Bit 7 to be set or reset by Write operations.

Bit 3 of the Status Register must be set to logic **1** when the SCC performs a Dataway Initialize operation.

12.2 Command/Reply Transaction Status. Three bits of the Status Register indicate the status of the previous transaction handled by the SCC, and are particularly associated with the means of recovery from certain types of error (see Section 15.6).

Bit 4 is the Delayed Error (DERR) bit, indicating the error status of the previous transaction.

Bit 5 is the Delayed Command Accepted (DSX) bit, indicating the *X* response resulting from the previous transaction.

Bit 6 is the Delayed *Q* response (DSQ) bit, indicating the *Q* response resulting from the previous transaction.

In the reply to commands that read the contents of the Status Register, Bit *SR*4, *SR*5, and *SR*6 in the Read-data field must indicate the current state of Bits, 4, 5, and 6, respectively, of the Status Register. It must not be possible to write into these bits of the Register.

Before completing or abandoning a Command/Reply transaction, an SCC that implements these features must set the DERR, DSX, and DSQ bits of the Status Register as follows:

The DSX and DSQ bits must be set to correspond with the current state of the *SX* and *SQ* responses, respectively, at the end of the operation.

The DERR bit must be set to logic **1** if the transaction is abandoned before the SCC has accepted valid *Q* and *X* responses (and Read data, if appropriate), or if the transaction results in the error response ERR=1 or the Dataway response *X*=0.

If the transaction is abandoned before executing the command, so that there are no Dataway *X* and *Q* responses related to the transaction, the DSX and DSQ bits should be set to logic **0**.

12.3 Demand Handling. Bits 9, 10 and 16 of the Status Register (see Section 11.1) are concerned with the control, testing and monitoring of the Demand handling activities in the SCC.

12.3.1 *Enable Demand Messages.* Bit 9 of the Status Register controls the initiation of Demand messages by the SCC in response to Dataway *L* signals, or to the simulated demand on $L24$, or to any internal source of demands within the SCC.

> All initiation of Demand messages by the SCC must be disabled when Bit 9 of the Status Register is in the logic 0 state. When Bit 9 is in the logic 1 state, Demand messages can be initiated, subject to the conditions defined in Section 5.1.
>
> The SCC must allow Bit 9 of the Status Register to be set or reset by Write Status Register operations, using the appropriate values of bit $SW9$ in the Write-data field. Bit $SR9$ of the data word read from the Status Register must indicate the current state of Bit 9 of the register.

12.3.2 *Demand L24.* As a test facility, the SCC includes a means of simulating a LAM signal from Dataway line $L24$. The initiation of a Demand message and the contents of its SGL field will depend on the appropriate conditions being present at the SGL-Encoder connector and on the generation of Demand messages having been enabled (Status Register Bit 9 at logic 1).

> When Bit 10 of the Status Register is in the 1 state, the SCC must simulate an input from Dataway $L24$. The simulated demand must be available at the SGL-Encoder connector for selection and encoding.
>
> The SCC must allow Bit 10 of the Status Register to be set or reset by Write Status Register operations, using the appropriate value of Bit $SW10$ in the Write-data field of the Command message.
>
> Bit $SR10$ of the Read-data field of the reply to commands that read the Status Register must indicate the current state of Bit 10 of the Register.

12.3.3 *Selected Look-at-Me Present.* Bit 16 of the Status Register indicates whether any Dataway L signal (including $L24$) selected by the external SGL-Encoder unit is presenting an active demand. A simple SGL Encoder may derive this condition directly from the L-sum signal (see 14.2.2) which is the OR combination of the twenty-four L signals. A more complex encoder with facilities for masking out unwanted L signals may derive the condition from the OR combination of the selected L signals.

> Bit $SR16$ of the Read-data field in the reply to commands that read the Status Register must indicate the current state of Bit 16 of the Register. It must not be possible to write into Bit 16.

12.4 Reconfiguration Options. In the Status Register 3 bits control the Dataway Off-line state of the SCC and any Bypass or Loop Collapse switching devices used to change the configuration of the SH.

12.4.1 *Dataway Off-Line State.* This test and maintenance feature of an SCC is controlled jointly by Bit 13 in the Status Register and a front-panel manual control (see Section 13.1.2). The current state of the manual control is indicated by Bit 14 of the Status Register. When an SCC is in the Dataway Off-line state, the Dataway and SH should be isolated from each other, so that operations on both can continue independently. For example, this isolation allows an auxiliary controller to perform Dataway operations independently of any operations taking place on the SH. [The isolation should therefore not prevent access by the auxiliary controller to the N lines (see Section 14.6.1).]

> When an SCC is in the Dataway Off-line state it must receive and transmit all SH messages. It must not execute any command addressed to a module at Stations $N(1)$ to $N(23)$, but must send a Reply message with format appropriate to the Function code and with $SQ=0$ and $SX=0$ in the Status field. It must execute commands addressed to the Status Register, but must not set Bit 1 or Bit 2 of the Status Register to the logic 1 state.
>
> While executing commands addressed to any feature of the SCC, an Off-line SCC must not generate Dataway Strobe signals S1 and S2. If the recommended isolation of the Dataway from the SH is not implemented, then the SCC must generate Busy=1 on the Dataway while executing all commands addressed to features of the SCC, except the Read LAM pattern command.

An Off-line SCC must not generate Demand messages in response to Dataway L signals but may do so in response to internal demands.

If the SCC has any internal demand sources, other than $L24$, that can generate Demand messages it must be able to execute the Read LAM-pattern command (see Section 11.2.1) while it is Off line.

The SCC must assume the Off-line state when either the front panel manual control is in the Off-line position or Bit 13 of the Status Register is at logic 1. It must assume the On-line state only when both the front panel manual control is On line and Bit 13 of the Status Register is at logic 0 (see Table 12-1).

Thus, the front panel manual control can force the Off-line state but requires the consent of the System Controller (via Bit 13) in order to select the On-line state.

An SCC must defer switching from On line to Off line, or from Off line to On line, until the completion of any Command/Reply transaction in which it is engaged.

An SCC that implements this feature must allow Bit 13 of the Status Register to be set or reset by Write Status Register operations with the appropriate value of Bit $SW13$ in the Write-data field of the Command message.

Bits $SR13$ and $SR14$ in the Read-data field of the reply to commands that read the Status Register must indicate the current states of Bit 13 of the Register and the manual Off-line control, respectively.

The Off-line status of the SCC is indicated by the combination of Bit $SR13$ and Bit $SR14$ of the Read-data field (See Table 12-1).

Table 12-1
Control of Dataway Off-Line State

Manual Switch	Status Register Bit 13	Dataway State
On line	0	On line
Off line	0	Off line
On line	1	Off line
Off line	1	Off line

12.4.2 *Bypass*. This feature of an SCC is associated with Bit 12 of the Status Register and with a signal at the D ports to control an optional external Bypass device (see Section 15.1.1).

Each SCC must provide at its output and input D ports a Bypass Control signal conforming to Section 7.3.1. This signal must be controlled by Bit 12 of the Status Register; and in response to any command that sets this bit to the 1 state, the control signal must be driven to the 1 state after the Reply message has been transmitted. In response to any command that resets Bit 12 to the 0 state, the control signal output must immediately be free to go to the 0 state if no other unit connected to the Bypass Control bus is driving it to the 1 state, but initiation of the Reply message must be delayed for 100 ms ± 10 percent.

The SD must generate a Reply space in the Command message (see Section 4.6) of appropriate length to accommodate the delayed reply.

Thus, the reply to a command that bypasses the crate or cluster is transmitted to the SH before the Bypass device operates, and the reply to commands that unbypass the crate or cluster is delayed until the Bypass device has restored the normal route.

The Bypass Control signal output from an unpowered SCC must be in the logic 1 state. When power is restored to the SCC, Bit 12 of the Status Register must remain in the logic 1 state until it is specifically reset by command.

In the Read-data field of the reply to commands that read the contents of the Status Register, Bit $SR12$ must always be logic 0.

While an SCC is in the Bypassed state (Bit 12 of the Status Register in logic 1 state), it must only execute commands that operate on the Status Register to restore the Unbypassed state by resetting Bit 12 to the logic 0 state. Such commands are permitted to operate on other bits of the Status Register.

Any other command received while an SCC is in the Bypassed state must not be executed. The SCC must transmit a Reply message appropriate to the Function field of the Command message, and with the responses $SX = 0$ and $SQ = 1$ in the Status field.

If the SCC is bypassed by an external Bypass device, the reply to an unexecuted command is not propagated through downstream crates to the SD. The external bypass routes the full received Command message to the downstream SH.

12.4.3 *Loop Collapse.* This feature of the SCC is associated with Bit 11 of the Status Register and with a signal at the D-output port to control an optional external Loop Collapse switching device (see Section 15.1.2).

Each SCC must provide at its D-output port, a Loop Collapse Control signal conforming to Section 7.3.1. This signal must be controlled by Bit 11 of the Status Register. In response to any command that sets Bit 11 to the 1 state, the control signal must be driven to the 1 state immediately, and initiation of the Reply message must be delayed for 100 ms ± 10 percent. In response to any operation that resets Bit 11 to the 0 state, after the Reply message has been transmitted the control signal output must become free to go to the 0 state.

The SD must generate a Reply space in the Command message of appropriate length to accommodate the delayed reply.

The Loop Collapse Control signal output from an unpowered SCC must be in the logic 0 state. When power is restored to the SCC, Bit 11 of the Status Register must remain in the 0 state until it is specifically set by command.

Commands to control an external Loop Collapse device should be addressed to the last crate in a cluster. Thus, the reply to a command that collapses the loop (for example, to cut out a faulty SCC or section of SH) is delayed until it can be transmitted over the shortened loop. The reply to a command that extends the loop (thus bringing into use SCCs that may have lost synchronism) is transmitted over the shortened loop before the loop is extended.

In the Read-data field of the reply to commands that read the contents of the Status Register, Bit $SR11$ must indicate the state of Bit 11 of the register.

13. Serial Crate Controllers: Front Panel Features

13.1 Manual Controls. The following manual controls should be provided on the front panel of each SCC.

13.1.1 *Crate Address Switch.* Each SCC should have a Crate Address switch, by which any address in the range 01_8 to 76_8 can be assigned (see Section 3.4.1), preferably without removing the controller from the crate. Special conditions of use may require other means of assignment, a restricted range of addresses, or the assignment of more than one address. These are permitted unless specifically excluded (as in the case of the recommended SCC Type L2).

Precautions should be taken to minimize the risk that the address assignment is changed accidentally. For example, the switch could be screwdriver adjustable, with access through the front panel.

13.1.2 *Dataway Off-Line Switch.* This switch (see Section 12.4.1) has two positions, preferably labeled On line and Off line. The construction or positioning of the switch should minimize the risk that its state can be changed accidentally. A locking toggle switch is recommended.

13.1.3 *Initialize and Clear Switches.* These two pushbuttons, or similar momentary-contact controls, initiate Dataway unaddressed operations with Initialize and Clear, respectively. They should only be effective when the SCC is in the Dataway Off-line state. The front panel layout or markings should indicate this.

13.2 Indicators. The following visual indicators should be provided on the front panel of each SCC.

13.2.1 *Crate Address.* An indication of the address selected by the Crate Address switch. This indication may be an integral part of the switch mechanism, for example, visible through a window in the front panel.

13.2.2 *Dataway On-Line.* An indication that the SCC is in the On-line state. (This is derived from the Dataway Off-line switch and Bit 13 of the Status Register.)

13.2.3 *Dataway Inhibit.* An indication of the state of the Dataway Inhibit signal I.

13.2.4 *Unsynchronized State.* A monostable[3] indication that the SCC is in the Lost Message Sync or Lost Byte Sync state (see Fig 16-1 and Section 16).

13.2.5 *Crate Addressed.* A monostable[5] indication derived from the Controller Busy signal (see Section 14.2.11), and indicating that the SCC has received a Header byte addressed to it.

13.2.6 *Demand Message.* A monostable[5] indication of the state of the DMI signal (see Section 14.2.4). This indicates that the SCC has sent a Demand message, or is actually sending it, or is awaiting an opportunity to send it.

13.2.7 *Bypassed.* An indication that the SCC is in the Bypassed state, derived from Bit 12 of the Status Register. This indication should be independent of the presence or actual condition of any external Bypass device.

13.3 Connectors

> Each SCC must have on its front panel two *D*-port connectors as defined in Section 7-1. Identifying markings must be provided on the front panel adjacent to the connectors.
>
> The connectors must be mounted with the major axis vertical and with contact 1 lowermost.

The preferred markings are *D* Input and *D* Output. The *D*-input connector should be mounted either above or to the left of the *D*-Output connector when viewed from the front.

13.4 Other Front Panel Features. The requirements in this section do not exclude the use of additional controls, indicators, connectors, etc, on the front panel of an SCC, except where this is specifically prohibited. (For example, additional front panel features that could affect operational interchangeability are prohibited on the recommended SCC Type L2.)

14. SGL-Encoder Connector

Demand messages are initiated in response to LAM L signals from the Dataway. The Crate Address field of the Demand message indicates the source crate, and the SGL field is available for further identification of demands, if required. Together, these fields can provide a direct vector to a LAM source or software routine.

Basic facilities for Demand message initiation and SGL-field encoding are recommended for inclusion in every SCC that can generate Demand messages. A wide range of additional facilities can be provided, either within the SCC (but not within the recommended SCC Type L2), or in separate units coupled to the SCC via an SGL-Encoder connector. This arrangement is somewhat analogous to the Crate Controller Type A1 and separate LAM Grader used with the Parallel Highway (see ANSI/IEEE Std 596-1982).

There is a risk that a Demand message can be corrupted in transmission between the source SCC and the SD so that the demand is not serviced. A time-out feature is specified (see Section 14.4) in order to detect unserviced demands (Hung Demands), and to initiate a Demand message with a distinctive SGL field. For other aspects of Demand handling see Sections 5 and 12.3.

14.1 Mechanical. Each SCC should have an SGL-Encoder connector. This connector carries input and output signals that permit an external SGL-Encoder unit to access the Dataway L signals, derive from them the SGL field of the Demand message, detect unserviced demands,

---

[5] A monostable indication has an 'on' state of defined minimum duration (chosen for visibility) when the input signal is of short duration, and a continuous 'on' state when the input signal is maintained. It is typically generated by a retriggerable monostable element with minimum output pulse length of 10 ms.

and initiate the generation of Demand messages for normal and Hung demand conditions. The connector also carries signals that allow an auxiliary controller to supply coded Station Number information.

> The SGL-Encoder connector, if provided, must be wired in accordance with the contact assignments shown in Table 14-1. The SCC must generate, or respond to, signals at the SGL-Encoder connector as defined in this section.

Any SGL-Encoder connector used on an SCC must be a fifty-two way Cannon Type 2DB52P double-density fixed member with pins, or equivalent connector fully compatible with Cannon 2DB52S. It must be mounted at the rear of the SCC, above the Dataway connectors, within the area designated for free access in Fig 3 of ANSI/IEEE Std 583-1982. Contact 1 must be lowermost. The connectors must have Cannon Type D20418-2 screw-lock retainers (fixed members), or equivalent retainers that are fully compatible with Cannon Type D20419-16 free members. (See Table C2, Preferred Auxiliary Connectors, in Appendix C of ANSI/IEEE Std 583-1982).

14.2 Signals at the SGL-Encoder Connector. The significance of the signals passing through the SGL-Encoder connector is indicated in the following text. To assist this explanation Fig 14-1 is a block diagram giving an example of

Table 14-1
Contact Assignments at SGL Encoder Connector

Contact	Signal	Direction*	Contact	Signal	Direction*
1	Demand Busy (DBSY)	out	2	$L1$	out
3	Graded L SGLE1	in	4	$L2$	out
5	Graded L SGLE2	in	6	$L3$	out
7	Graded L SGLE3	in	8	$L4$	out
9	Graded L SGLE4	in	10	$L5$	out
11	Graded L SGLE5	in	12	$L6$	out
13	External Repeat (ERPT)	in	14	$L7$	out
15	(Reserved)		16	$L8$	out
17	Request Inhibit	out**	18	$L9$	out
19	Time Out (TIMO)	out	20	$L10$	out
21	Demand Message Initiate (DMI)	in	22	$L11$	out
23	Start Timer (STIM)	in	24	$L12$	out
25	Selected LAM Present (SLP)	in	26	$L13$	out
27	(Reserved)		28	$L14$	out
29	Aux. Controller Lockout (ACL)	out	30	$L15$	out
31	Byte Clock	out	32	$L16$	out
33	Free use	in or out	34	$L17$	out
35	Free use	in or out	36	$L18$	out
37	Free use	in or out	38	$L19$	out
39	Free use	in or out	40	$L20$	out
41	Controller Busy (CBY)	out	42	$L21$	out
43	Station Number N1	in	44	$L22$	out
45	Station Number N2	in	46	$L23$	out
47	Station Number N4	in	48	$L24$	in or out
49	Station Number N8	in	50	SUM	out
51	Station Number N16	in	52	OV (ground)	—

*Out indicates signal generated by SCC; In indicates signal received by SCC.
**The crate controller needs only a pull-up on the Request Inhibit (contact 17) to be compatible with Auxiliary Controller Bus of ANSI/IEEE Std 675-1982.

INTERFACE SYSTEM (CAMAC)

ANSI/IEEE
Std 595-1982

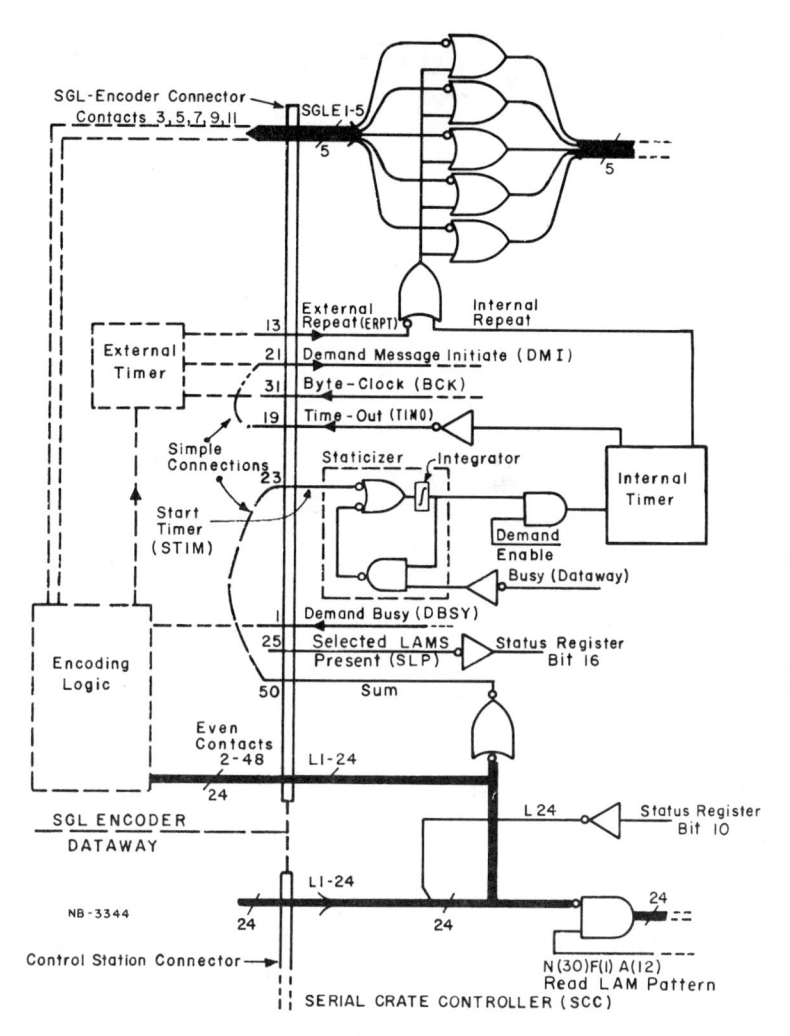

**Fig 14-1
Example of Associated Parts of SCC and SGL Encoder**

parts of the SCC and SGL Encoder closely associated with the interconnection.

14.2.1 *Signals L1-L24.* Signals $L1$-$L24$ are the LAM L signals from Dataway stations 1-24, respectively. Signal $L24$ may be used as a LAM request generated within the SCC or the SGL Encoder. In particular, $L24$ may be controlled by Bit 10 of the Status Register (see Section 12.3.2) as a means of testing the demand-handling process.

14.2.2 *L-sum.* L-sum is a signal from the SCC to the SGL Encoder. It is the inclusive OR of signals $L1$-$L24$. It indicates the presence of one or more demands, independently of any subsequent masking process.

14.2.3 *Signals SGLE1-SGLE5.* Signals SGLE1-SGLE5, from the SGL Encoder to the SCC, provide the contents for the 5 bit SGL field of the Demand message.

14.2.4 *Demand Message Initiate.* The DMI is a signal from the SGL Encoder to the SCC. It indicates the presence of one or more demands after any masking process in the SGL Encoder. The 0-to-1 transition of this signal initiates the generation of a Demand message, after all the conditions listed in Section 5.1 have been satisfied.

14.2.5 *Selected Look-at-Me Present.* The selected LAM present (SLP) signal controls Bit 16 of the Status Register (see Section 12.3.3). It may be derived simply by a cross-connection to the L-sum signal, in which case it merely indicates that one or more of the Dataway L sig-

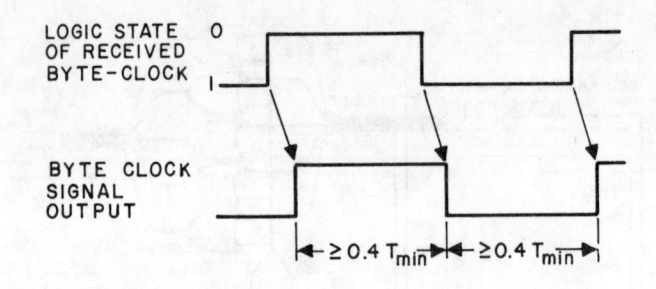

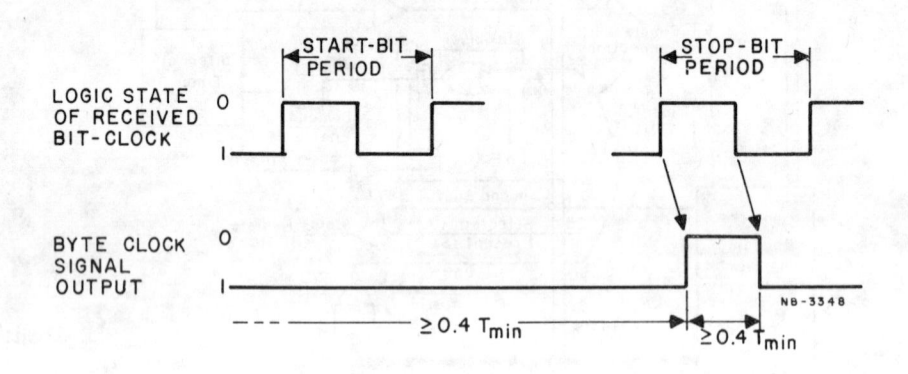

**Fig 14-2
Relationship Between Byte-Clock Signal at SGL-Encoder Connector
and Received Bit/Byte-Clock Signals
A   Byte-Serial Mode; B   Bit-Serial Mode**

nals are active. A more complex SGL Encoder may derive the SLP signal from the L signals after they have been selectively enabled by some form of masking.

14.2.6 *Demand Busy.* The Demand-Busy (DBSY) signal from the SCC indicates that a Demand message is being assembled. When DBSY=1, the SGL Encoder should staticize SGLE1 - SGLE5, so that the SGLE code does not change while the SGL field of the Demand message is being transmitted.

14.2.7 *External Repeat.* The External Repeat (ERPT) signal from the SGL Encoder indicates the presence of a Hung demand. When ERPT=1, the SCC generates a 5-bit SGL field containing the bit pattern 11111_2 in any Demand message that it transmits.

14.2.8 *Byte Clock.* The Byte-Clock (BCK) signal from the SCC to the SGL Encoder is derived by the SCC from the incoming byte stream. It is provided primarily for use by an external unserviced-demand timer in the SGL Encoder (see Section 14.4.2).

When the SCC is operating in byte-serial mode, the 0-to-1 and 1-to-0 transitions of the byte-clock signal must be derived from the corresponding transitions of the clock signal received at the D-input port (Fig 14-2A).

When the SCC is operating in bit-serial mode, the 0-to-1 transition of the byte-clock signal must be derived from the 0-to-1 transition of the bit-clock signal received at the D-input port during the bit period in which the Stop bit is strobed. The 1-to-0 transition of the byte-clock signal must be derived from the 1-to-0 transition of the bit-clock signal at the beginning of the Stop bit period (Fig 14-2B). In both modes the duration of each logic 0 and logic 1 state of the byte-clock signal must not be less than 0.4 T_{min}, as defined in Section 8.3.

14.2.9 *Start Timer.* The Start Timer (STIM) signal from the SGL Encoder is used to control

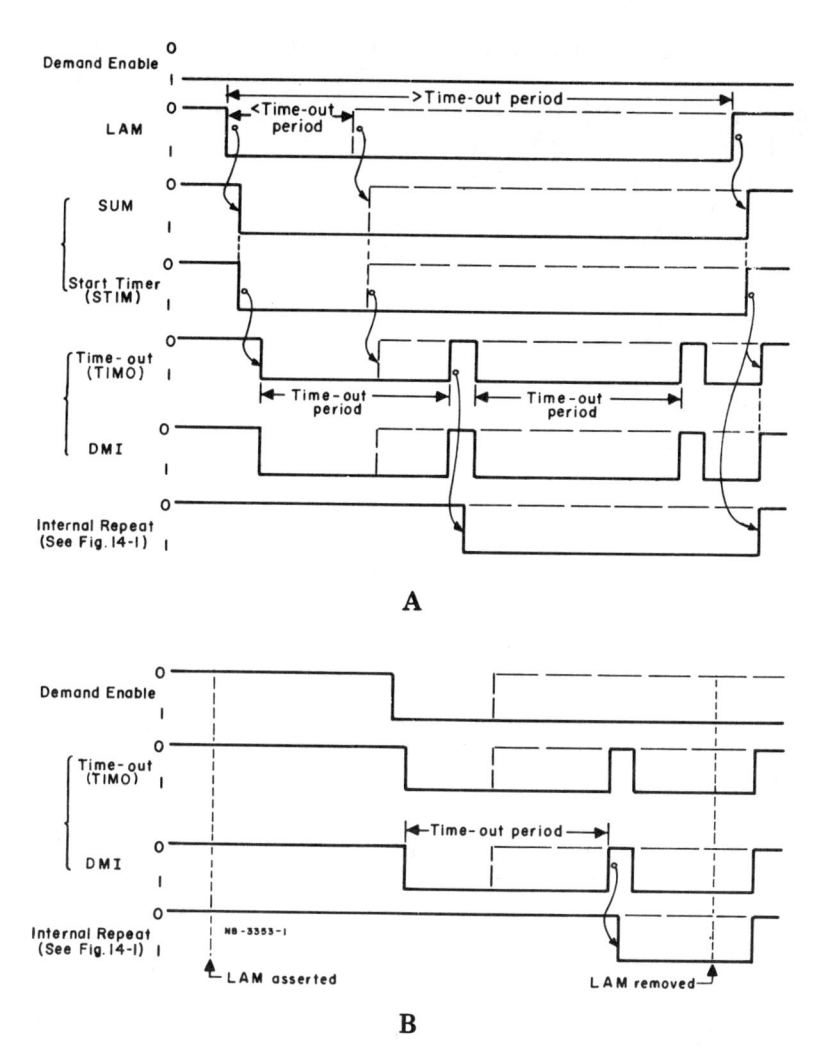

Fig 14-3
Relationship Between Signals at SGL-Encoder Connector
Concerned with Demand Message Generation
A — Shown for Duration of LAM Less Than and Greater Than Time-Out Period;
B — Similar to A except for Demand Enable After LAM Assertion

the internal timer in the SCC.

14.2.10 *Time Out.* The Time-Out (TIMO) signal is the output from the internal timer in SCC (see Section 14.4.1). The STIM and TIMO signals are routed via the SGL Encoder connector so that a choice can be made between the internal timer in the SCC and an external timer in the SGL Encoder. An example of the time relationship of these signals is shown in Fig 14-3.

14.2.11 *Controller Busy.* The Controller Busy (CBY) signal indicates that the SCC is engaged in a Command/Reply transaction. The SCC generates CBY when it receives a Header byte addressed to it and maintains CBY until the transaction is terminated by the END byte or is abandoned.

14.2.12 *Signals N1, N2, N4, N8, N16.* Signals $N1, N2, N4, N8, N16$ allow a separate auxiliary controller to provide the N field of a Dataway command (see Section 14.6.1).

14.2.13 *Auxiliary Controller Lockout.* The ACL signal indicates that the SCC requires use of the Dataway. The SCC generates ACL when it receives a Header byte addressed to it and maintains ACL until execution of the command is completed or is abandoned. This signal is used to control access to the Dataway by an auxiliary controller (see Section 14.6).

14.2.14 *Request Inhibit.* A Request Inhibit pull-up, conforming to the specifications for "Signals from SCC" in Table 14-2, must be provided on pin 17 of the SGL Encoder connector on the Serial Crate Controller (Table 14-1).

The Request Inhibit pull-up is for use with the Request Inhibit line of the Auxiliary Control Bus (ACB) required for auxiliary controllers. (See, for example, "Multiple Controllers in a CAMAC Crate," ANSI/IEEE Std 675-1982, 4.1.3.)

14.3 Signal Standards for the SGL-Encoder Connector

> All signals at the SGL-Encoder connector on an SCC must conform to the signal voltage standards shown in Table V of IEEE Std 583-1975 and the standards for pull-up current sources in Table 14-2.

These signal standards are derived from those for Dataway Read and Write signals but with some additional features associated with certain passive interconnections permitted at the SGL-Encoder connector (see Section 14.5.1).

This specification allows direct links within the SCC between the incoming Dataway L signals $L1$ - $L24$ at the Control station and the outgoing signals $L1$ - $L24$ at the SGL-Encoder connector.

It also allows passive interconnections between certain outputs from an SCC (including $L1$ - $L24$) and certain inputs at the SGL-Encoder connector. These inputs can therefore be connected to either a Dataway L signal, via the link within SCC, or the output from a logic element in the Encoder.

In order to ensure that each such connection has one, and only one, standard pull-up current source, the relevant inputs to SCC have special low-current pull-up sources, and the outputs from logic elements in the Encoder have standard pull-up current sources.

14.4 Hung Demand Time Out.
An internal timer in the SCC provides a basic means of detecting unserviced demands. Alternatively, the SGL Encoder can include an external

Table 14-2
SGL-Encoder Connector: Signal Standards and Pull-Up Current Sources for all Signals Other than Coded-N

Signal Standards at Connector	Signals from SCC	Signals to SCC
Line in 1 State at +0.5 V Minimum current sinking capability (current drawn from line by unit generating the signal)	3.2 mA* L signals 6.4 mA* other signals (from line by SCC)	16 mA* (from line by Encoder)
Line in 1 State at +0.5 V Maximum load current (current fed into line by unit receiving the signal)	3.2 mA* each unit (max 6.4 mA*) (into line by Encoder)	3.2 mA (into line by SCC)
Line in 0 State at +3.5 V Minimum pull-up capability (current fed into line by the SCC)	2.3 mA (for L signals this is common to Encoder and Dataway)	200 uA
Line in 0 State at +3.5 V Maximum current drawn from line by the Encoder	200 uA	200 uA
Pull-Up Current Sources in SCC	**Signals from SCC**	**Signals to SCC**
Internal pull-up current source (Ip) at +0.5 V	6.0 mA $\leqslant$ Ip $\leqslant$ 9.6 mA*	0.8 mA $\leqslant$ Ip $\leqslant$ 1.6 mA
Internal pull-up current source (Ip) at +3.5 V	2.5 mA $\leqslant$ Ip*	300 uA $\leqslant$ Ip
Pull-Up Current Sources in SGL Encoder	**Signals from SCC**	**Signals to SCC (Outputs only)**
Internal pull-up current source (Ip) at +0.5 V	—	6.0 mA $\leqslant$ Ip $\leqslant$ 9.6 mA*
Internal pull-up current		

*Values derived from ANSI/IEEE Std 583-1982.

timer, for example, to provide more complex facilities. The appropriate connections via the SGL-Encoder connector determine whether the internal timer, an external timer, or neither, is used.

If a Demand message is initiated, and then the Demand condition persists for more than a predetermined time-out period, the internal or external timer asserts the Hung Demand state and initiates a Demand message with a unique SGL field. If the demand is still present after a further time-out period the Hung Demand message is repeated, and so on. The time-out period should be considerably longer than the time normally needed to service a demand in the particular system.

The Hung Demand message is not an absolute indication that the original demand is still unserviced. For example, a simple SGL Encoder may not be able to distinguish between the original demand and a new demand that has occurred in the same crate. Also, owing to various delays in the SH message path, the SD may receive a Hung Demand message when it has already sent the command or commands to service the demand.

14.4.1 *Internal Timer*

> Each SCC that is capable of generating Demand messages must have an internal timer that provides a choice of time-out periods between 1 ms and 10 s.
>
> The internal timer must start to operate when the STIM signal from the SGL-Encoder connector is at logic 1 and demands are enabled. It must continue to operate until both STIM and Dataway Busy are at logic 0 or until demands are disabled. It must not be affected by temporary removal of the STIM signal during Dataway operations, for example, due to the source module removing its *L* signal while addressed by a command (see ANSI/IEEE Std 583-1982, Section 5.4.1.3.).
>
> The internal timer must produce TIMO signal at the SGL-Encoder connector. This signal output must be at logic 0 when the timer is not operating. When the timer starts to operate, the TIMO signal must go to logic 1 for the duration of the time-out period. It must then go to logic 0 for a period of not less than 200 ns. This sequence of logic 1 for one time-out period, followed by logic 0 for a short time, must be repeated while the timer continues to operate. (See Fig 14.3). The rise and fall times of the TIMO and Internal Repeat signals must be less than 100 ns.
>
> At the end of the first time-out period the timer must assert the Hung Demand state, maintain this state while the timer continues to operate, and cause the SGL field of any subsequent messages to contain the bit-pattern 11111_2.

In order to bring the internal timer into use, appropriate connections are made to the STIM and TIMO signals at the SGL-Encoder connector. The STIM input to the SCC may be derived from some demand-masking feature in an SGL Encoder. More simply it may be provided by connecting the L-sum output from the SCC to the STIM input.

The TIMO output from the SCC may be connected to the DMI input. Successive 0-to-1 transitions of this signal (at the beginning of each time-out period) initiate Demand messages. The contents of the SGL field in the first message are determined by the SGL Encoder. Thereafter, with the Hung Demand status asserted, the timer provides the special Hung Demand SGL field.

14.4.2 *External Timer*

> If an external timer, connected via the SGL Encoder, is used in place of the internal timer, then it must generate a DMI signal with 0-to-1 transitions at appropriate times to initiate the original Demand message and any subsequent Hung Demand messages. When it detects an unserviced demand, it must either generate the ERPT signal to control the contents of the SGL field or generate logic 1 on the lines SGLE1-5.

The internal timer has its time-out period expressed in real time (1 ms to 10 s) and therefore has to be adjusted to suit the clock rate of the system. The byte-clock signal at the SGL-Encoder connector allows the time-out period of an external timer to be expressed in byte-clock periods, and thus be less dependent on the real-time clock rate of the system.

14.5 SGL-Encoder Options. The simplest form of SGL Encoder is merely a free connector member with passive interconnections between certain contacts. At the other extreme, an SGL Encoder can provide enhanced facilities for masking the L demands, coding the SGL field, and detecting unserviced demands. Examples of such additional facilities are given in the following text.

14.5.1 *Passive SGL Encoder.* Simple interconnections between the SUM output from the SCC and the STIM input, and between the TIMO output and the DMI input are needed in order to use the internal timer (see Fig 14-1).

With only these connections the SGL field for all original demand messages is 00000_2 and for Hung Demand messages is 11111_2.

Passive interconnections, each consisting of a link between one L signal output from SCC and one of the five SGLE inputs to SCC, may be used to provide limited control over the contents of the SGL field of Demand messages.

14.5.2 *Demand Masking.* A simple connection from L-sum to STIM allows any L signal to initiate a Demand message. More complex SGL Encoders may include a means of masking the L signals, so that only those L signals passed by the mask are able to initiate Demand messages. The mask may be varied dynamically by commands sent to the SGL Encoder via the Dataway.

14.5.3 *SGL-Field Encoding.* As an example, the SGL Encoder may assign relative priorities to the L signals, identify the current demand that has highest priority, and encode the SGL field accordingly, via the SGLE lines.

After a simple SGL Encoder has initiated a Demand message, it may not be able to generate another Demand message (other than a Hung Demand message) even if a new demand with high priority occurs. A more complex encoder, with the ability to assign priorities to the L signals, may be able to initiate a new Demand message before a previous demand of lower priority has been serviced.

14.5.4 *False Hung Demand.* A simple SGL Encoder may indicate a false Hung Demand state if the original demand has been serviced but a new demand has occurred in the meantime, and so maintained the STIM signal. A more complex SGL Encoder may, however, associate the STIM condition with a particular demand, and so be able to reset the timer when that demand is cleared.

14.6 Access for Auxiliary Controllers. An auxiliary controller situated at one or more Normal stations of the CAMAC crate, and requiring to generate a command, has direct access to the Subaddress A and Function F lines of the Dataway, but not to the Station Number N lines, which are only accessible at the Control station occupied by the SCC.

Two additional features, not associated with the basic demand handling scheme, are therefore provided at the SGL-Encoder connector in order to give access to the N lines and ensure that the SCC and an auxiliary controller cannot both generate commands on the Dataway at the same time.

14.6.1 *Access to N Lines.* The lines $N1$, $N2$, $N4$, $N8$, $N16$ via the SGL-Encoder connector allow an auxiliary controller to present a coded Station Number address. This is decoded by the N decoder in the SCC, in order to put the appropriate Dataway N line into the logic 1 state.

> Each SCC that can be used in conjunction with an auxiliary Dataway controller must accept coded Station Number addresses at the appropriate contacts of the SGL-Encoder connector (see Table 14-1).
>
> Coded Station Number addresses presented to the SCC via the SGL-Encoder connector must be generated from open collector sources.

14.6.2 *Auxiliary Controller Lockout Signal*

> Each SCC that can be used in conjunction with an auxiliary Dataway controller must generate an ACL signal at the appropriate contact of the SGL Encoder (see Table 14-1).
>
> The ACL signal must go to logic 1 when the SCC recognizes a Header byte addressed to it, and must remain in the 1 state until completion of the Dataway operation cycle (t_9 in Fig 9 of ANSI/IEEE Std 583-1982) or earlier abandonment of the Command/Reply transaction.

14.6.3 Interlock Between SCC and Auxiliary Controller

> While the ACL signal is in the logic 1 state, an auxiliary controller used in conjunction with SCC must complete or abandon any current Dataway operation before the SCC requires use of the Dataway.

In order to satisfy this condition, an auxiliary controller should respond to the ACL signal in the logic 1 state by abandoning any Dataway operation that has been started if it has not yet generated Strobe $S1$ (stage t_3 of Fig 9 of ANSI/IEEE Std 583-1982).

The duration of any Dataway operation controlled by an auxiliary controller associated with SCC should not be more than 1.2 µs nor less than 1.0 µs (between t_0 and t_9 in Fig 9 of ANSI/IEEE Std 583-1982).

15. Recovery from Errors

The SH relies on continuity of the transmission paths for the data and clock signals, and is intended for use in environments that may expose these signals to noise. This section reviews the various features associated with detecting errors and failures, and with recovery from them.

15.1 Transmission-Path Failures. The transmission paths for data and clock signals pass through successive SCCs and the intervening sections of the SH (which may include cables, signal conditioners, or modems).

15.1.1 Failures Within Serial Crate Controllers: Bypass Switching.
In each SCC the data and clock signals are received and transmitted by active components. The effect of failure of these components or their power supplies can be limited by using the SCC in conjunction with a separate Bypass device, so that continuity of the signal paths can be restored by switching to an alternative route that bypasses the faulty SCC.

The Bypass device is controlled by the Bypass Control signal, which is available at both D ports of the SCC and is derived from Bit 12 of the Status Register (see Section 12.4.2).

If an external Bypass device is used it should be connected between the U or D ports of a single crate. When the SCC is bypassed the external device routes all incoming SH signals to the outgoing SH without passing them through the SCC. Under these conditions the Bypassed SCC monitors incoming messages from the SH, in order to receive any Command message requiring the removal of the Bypass. While an SCC is bypassed by the external device it cannot transmit messages to the downstream SH.

In the Bypassed state a Bypass device should link the incoming and outgoing SH lines in a way that is not dependent on power supplies and does not impose abnormal termination loading on any section of the SH transmission lines. For a Bypass device connected to U ports, this latter requirement may involve a means of disconnecting the normal termination impedances of the signal receiver. For a Bypass device connected to D ports (which have internal terminations on all data and clock receivers) it may involve interposing high input-impedance buffers in order to allow the Bypassed SCC to monitor the SH traffic.

The Bypass device can use electromechanical relays to switch the message path. It can therefore have a response time that is long compared with the Command/Reply transaction time of the SH. The switching operation is likely to cause an asynchronous interruption of the SH message path, and so cause loss of synchronism at downstream SCCs.

An example of part of a Bypass device, in Fig 15-1, shows an arrangement of relay contacts for switching one D-port signal. Multiple termination loads on the SH transmission line are avoided in this example by having a switched termination in the Bypass device and a balanced receiver and transmitter to isolate the mandatory termination at the D-input port of the SCC. In a complete Bypass device for bit-serial D-port signals there would be two such switching arrangements, one for the clock signal and one for the data signal. In the power-off state the Bypass switching should be as shown, with the SCC bypassed and the termination disconnected.

A Bypass device may also be connected across the U or D ports of a cluster of several crates, but there are certain risks associated

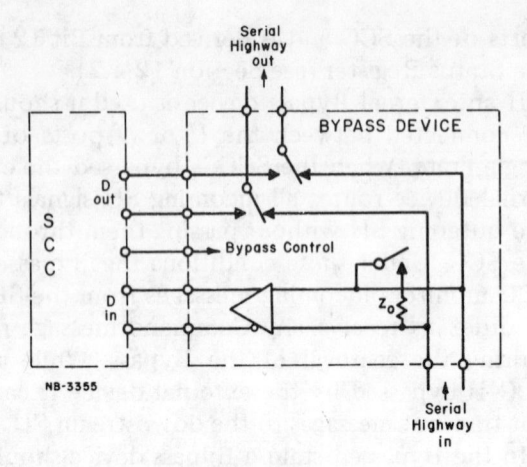

**Fig 15-1
Example of Bypass Switching for One
D-Port Signal**

[In a Bypass Device for Bit-Serial D-Port
Signals This Arrangement is Duplicated for
Data and Clock Signals]

with this arrangement. A command to one crate in the cluster will cause the cluster to be bypassed and will inhibit Dataway operations in that one crate. Other crates in the cluster do not necessarily sense that they have been bypassed. Therefore there is a risk that these other crates may respond to commands involving Dataway operations, although any replies that they generate are unable to reach the SD.

System configurations should take into account the worst case conditions that can be created by bypassing. It is possible for the SD or an SCC to be called upon to transmit or receive signals over the majority of the SH loop.

As an additional safeguard against failure of the SH message path within SCC, the Bypass device could monitor the SH activity at the output port of the SCC and switch automatically to the Bypassed state if this activity ceases.

15.1.2 *Failures in the Serial Highway: Loop Collapse Switching.* Between successive SCCs the data and clock signals are transmitted over segments of the SH. The effect of failure of the cables and other equipment can be limited by using the SCC in conjunction with a separate Loop Collapse device, so that continuity of the signal paths can be restored by switching to an alternative route that avoids the faulty segment of the SH.

The Loop Collapse device is controlled by the Loop Collapse Control signal, which is available at the D-output port of the SCC and is derived from Bit 11 of the Status Register (see Section 12.4.3). The SCC that controls the Loop Collapse device is not within the part of the SH that is disconnected when the device operates.

The Loop Collapse device can use electro-mechanical relays to switch the message path and can therefore have a response time that is long compared with the Command/Reply transaction time of the SH. The switching operation is likely to occur asynchronously with respect to the system clock and so cause loss of synchronism at downstream crates.

The Loop Collapse devices can be associated with individual crates or with clusters of crates, in which case they are controlled by the last crate in the cluster. They can be used to switch U-port signals (preferred) or D-port signals.

An example of part of a Loop Collapse device, in Fig 15-2, shows an arrangement of relay contacts for switching one D-port signal. In a complete Loop Collapse device for bit-serial D-port signals there would be two such switching arrangements, one for the data signal and one for the clock signal. This example assumes that the SH loop passes through all SCCs and then returns to the SD by a parallel route (but not necessarily passing through the SCC,

**Fig 15-2
Example of Loop Collapse Switching for One
D-Port Signal**

[In a Loop Collapse Device for Bit-Serial
D-Port Signals This Arrangement is Duplicated
for Data and Clock Signals]

[The Clock Signal Output to the Disconnected Part of the SH Loop is Held in a Fixed Logic State by the Conditions at *-*]

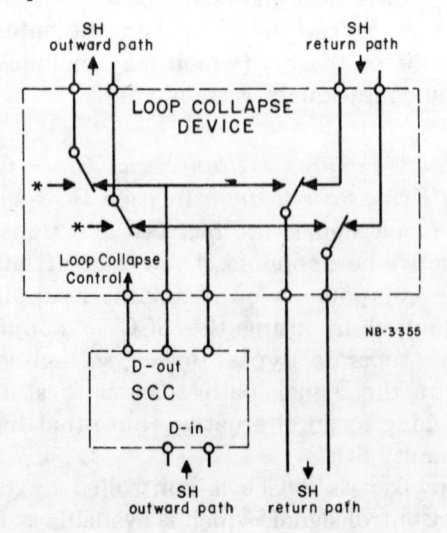

although the Bus 1 and Bus 2 connections could be used, see Section 7.1.2) so that the outward and inward sides of the loop are available to the Loop Collapse device. Operation of the Loop Collapse device connects the outward and inward sides of the loop, thus forming a shortened loop and disconnecting the remainder of the loop and all SCCs connected to it. The clock signals to the disconnected part of the loop are held in a fixed state, so that the disconnected SCCs can neither receive nor transmit messages. In the power-off state the Loop Collapse switching should be as shown, with the SH extended through the device.

Other implementations are possible, for example, using duplicated segments of the SH, so that the whole loop can be restored if there is only one fault per segment. Contact 25 on the *D*-input port of each SCC is reserved for a third control signal and can be relevant to some more complex forms of Loop Collapse switching.

15.2 Loss of Synchronism. Message synchronism depends on the ability of SCCs and the SD to identify the Header and Delimiter bytes of each message. Errors in the clock or data signals (particularly the Delimiter bits) can result in loss of message synchronism. An SCC or SD recognizes that this has occurred because it receives Delimiter bytes in illegal contexts. It can recover message synchronism by the means described in Section 10.1.3. Byte synchronism depends on the ability of SCCs and the SD, operating in bit-serial mode, to identify the Start and Stop bits of byte frames. Errors in the clock or data signals (particularly the Start and Stop bits) can result in loss of byte synchronism. An SCC or SD recognizes that this has occurred because it receives a logic 0 bit at Stop-bit time. It can recover byte synchronism by the means described in Section 10.2.3.

15.3 Transmission Errors. The main means of detecting errors due to corrupted clock or data signals is a Geometric code, with byte (row) and column parity codes. A subsidiary means is provided by the message formats; in particular the contexts in which Delimiter bytes are permitted.

15.3.1 *Principle of Geometric Code.* The Geometric Error Detection code is a simple, but powerful, scheme for detecting transmission errors in serial communication links such as those used for the SH. It is readily adapted to messages of varying length and uses parity codes that are easily generated and checked.

The basic principle of the Geometric code used by the SH is that a block of data is arranged in a matrix of m rows by n columns (see Fig 15-3). Each row has an extra row-parity bit to conserve odd parity, and each column has an extra column-parity bit to conserve even parity, forming an enlarged block of $m+1$ rows by $n+1$ columns.

15.3.2 *Implementation of Geometric Code.* The practical implementation of this principle for use by the SH is shown in Fig 15-4.

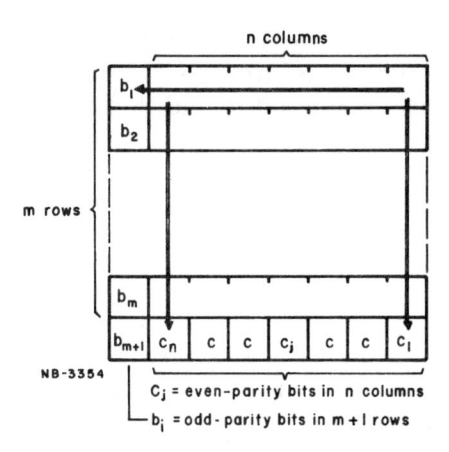

Fig 15-3
Geometric Error Detection: Basic Principle
(See Fig 15-4 for Application to SH)

c_j = even-parity bits in n columns
b_i = odd-parity bits in m+1 rows

Fig 15-4
Geometric Error Detection: As Applied to SH

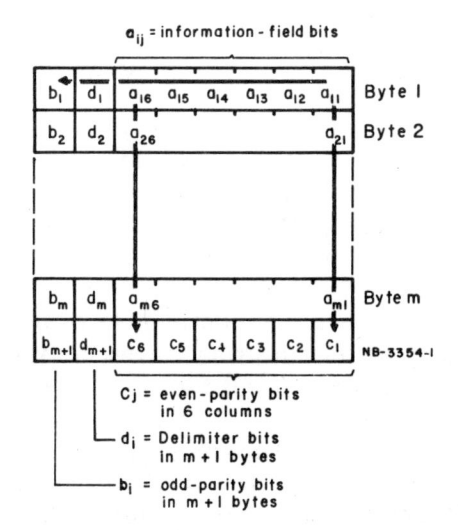

a_{ij} = information-field bits
c_j = even-parity bits in 6 columns
d_i = Delimiter bits in m+1 bytes
b_i = odd-parity bits in m+1 bytes

Here, the byte-parity bit b_i in byte i conserves odd parity over the whole byte, including the Delimiter bit d_i and the 6 information-field bits a_{i_1} to a_{i_6}.
Thus
$$b_i = \overline{a_{i_1} \oplus a_{i_2} \oplus \cdots \oplus a_{i_6} \oplus d_i}$$
where the symbol $\oplus$ indicates the exclusive OR function or modulo-2 sum. The byte always contains an odd number of bits in the logic 1 state.

In the final byte (the SUM byte in Command messages and the ENDSUM byte in Reply and Demand messages) each of the 6 column-parity bits c_1 to c_6 conserves even parity over a column of information-field bits a_{ij} to a_{mj}.
Thus, for column j
$$c_j = a_{1j} \oplus a_{2j} \oplus \cdots \oplus a_{mj}$$
and the column always contains an even number of bits in the logic 1 state.

Bits 7 and 8 in this final byte are not column-parity bits but are, respectively, the Delimiter bit d_{m+1} and the byte-parity bit b_{m+1}.
Thus
$$b_{m+1} = \overline{c_1 \oplus c_2 \oplus \cdots \oplus c_6 \oplus d_{m+1}}$$

> The SD must generate the correct Geometric Error Detection code in all Command messages sent to SCC.
>
> Each SCC must check the Geometric Error Detection code in all Command messages addressed to it.
>
> Each SCC must generate the correct Geometric Error Detection code in all Reply and Demand messages that it transmits.

The SD should check the Geometric Error Detection code in all Reply and Demand messages that it receives from SCCs.

15.3.3 *Performance of Geometric Code.* The Geometric Error-Detection code covering a data block of m rows by n columns requires the transmission of $m+1$ rows by $n+1$ columns and has the following properties:

(1) It detects all occurrences of one, two or three errors in the enlarged block of $m+1$ rows by $n+1$ columns

(2) It detects any odd number of errors, regardless of their distribution within the block

(3) It detects all burst errors $\leqslant n+2$ bits in length

(4) It detects a large proportion of the errors not included in (1) to (3) (For example, of all 4-bit error patterns in the block, it fails to detect only those arranged in a rectangular format)

The performance of Geometric codes can be calculated for random bit-error rates. In practice the errors due to telephone channels are not characterised by random independent noise and, in the implementation used for the SH, any errors in the Delimiter bit are detected partly by context rather than by column parity. However, some approximate performance data for an eight-column Geometric code operating under random noise conditions are quoted in the following text as an indication of the power of the scheme.

If the communications channel has a random error rate such that the probability that a bit is in error is 10^{-4} then, after applying the Geometric Error-Detection procedure, the probability that a block contains an undetected error is approximately 10^{-13} for 9-byte blocks and 10^{-14} for 3-byte blocks. If the probability of bit error is 10^{-5} then the probability that a block contains an undetected error is 10^{-17} and 10^{-18} for 9-byte and 3-byte blocks, respectively. As a practical example, if 9-byte messages are transmitted at 2000 bits per second with a random error rate of 10^{-4}, the average rate of errors on the channel is one bit error in 5 s, but the average rate of occurrence of blocks with undetected errors is reduced to one block in 10^4 years.

15.3.4 *Error Detection by Context.* In the rare cases where multiple errors are not detected by the Geometric code, some additional protection against errors is provided by the message structure, which requires that certain bits and fields occur in the appropriate context.

Errors in the Delimiter bit result in bytes occurring in the wrong context, and so lead to loss of message synchronism (see Section 10.1.2).

Undetected errors in bits $SF8$ and $SF16$ of the Function field of a Command message can result in the SCC expecting a Write-data field when there is none, or vice versa. If so, the SCC does not identify the column-parity field correctly, and thus it is likely that the column parity will be violated.

Undetected errors in the MI field of the Command message could, if the SCC implements the appropriate test, result in the SCC rejecting the corrupted Command message.

15.3.5 *Error Detection at Modules.* A typical CAMAC module implements few of the full range of Subaddresses and Functions. There is, therefore, some additional protection against errors because a corrupted command will probably result in the response $X=0$, either from the intended module, or from some other module or unoccupied station.

In special cases where maximum security against false operations in a module is essential, the module can include additional precautions. For example, the module could be constructed so that it uses extra redundancy, in commands or data, to protect against false execution of some particularly critical operation.

15.4 The Error-Reply Message. When an SCC detects a violation of byte or column parity in a received Command message, it cannot rely on the contents of the Function field to determine the length of the Reply message. The SCC therefore generates an Error-Reply message with a fixed length of 3 bytes in which the ERR bit in the Status field is at logic 1. An Error-Reply message is also sent if, as a result of the optional MI-field test, the SCC detects nonzero contents in the MI field of any message addressed to it (see Section 3.4.7).

15.5 Error Indications in Reply Messages. The Status field in the Reply message contains error indications relating to the execution of the current command and the previous command to the SCC.

15.5.1 *Error Bit.* The ERR bit, in bit-position 1 of the Status field of the Reply message, indicates whether the SCC has detected errors in the current Command message as a result of testing the byte and column parity and (optionally) the MI field. If ERR=0, the command has been executed by the SCC (although not necessarily accepted by the addressed module). If ERR=1, the SCC has detected a parity error (see Section 15.3.2), or incorrect MI field (see Section 3.4.7), and has therefore not executed the command. Reply messages in which the ERR bit is logic 1 are always 3-byte messages without a Read-data field. (See Table 15-1).

> If an SCC detects a violation of byte or column parity in a received Command message, it must set the ERR bit in the Status field of the Reply message to logic 1.
>
> If an SCC tests the MI field of received messages and detects that the contents of the field are not 00, it must set the ERR bit in the Status field of the Reply message to logic 1. (Optional test)

15.5.2 *Command Accepted Bit.* The SX bit is in bit-position 2 of the Status field and corresponds to the Command Accepted X response from the Dataway (for commands to modules) or from some feature of the SCC (for commands addressed to the SCC). When ERR=1, indicating that the command has not been executed, the SX bit has no significance and is arbitrarily defined as SX=0. (see Table 15-1).

> After executing a command, the SCC must generate the SX bit in the Status field of the Reply message to correspond with the Command Accepted X response from the addressed module or feature of the SCC.

An unsuccessful operation, indicated by ERR=0 and SX=0, may be due either to the command not having been accepted by the module or feature to which it is addressed, or to the command requiring a Dataway operation when the SCC is in the Off-line or Bypassed state. These conditions can be distinguished by means of the SQ bit in the Status field of the Reply message and bit 13 in the Status Register (see Section 11.1) of the SCC.

Modules conforming to earlier specifications may give the response $X=0$ to all commands

Table 15-1
Error Indications in Reply Message

Execution of Command	ERR	SX	SQ	Length (bytes)
Successful	0	1	Q**	3 or 7
Unsuccessful				
*Dataway Off-line	0	0	0	3 or 7
*Command not accepted	0	0	0	3 or 7
Bypassed	0	0	1	3 or 7
Not Executed (Parity or MI Error)				

*These can be distinguished by reading the contents of the Status Resister. Bit 13 = 1 indicates Dataway Off-line.

**Response from the addressed feature of the module or controller.

but can usually be modified to give $X=N$ in order to allow recognition of successful operations as required by Table 15-1 (see Section 5.5.4 of ANSI/IEEE Std 583-1982).

15.5.3 *Delayed-Error Bit.* The DERR bit, in bit-position 4 of the Status field, indicates whether the previous command to the SCC was executed successfully. The DERR bit is at logic 1 if the previous Command message received by the SCC contained a parity error or (optionally) an incorrect MI field, or if the Command/Reply transaction was abandoned before executing the command, or if the command was executed but was not accepted by the addressed module or feature of SCC. It is primarily associated with recovery from situations in which an assembled Reply message has been lost, either by corruption of the Reply in transmission or by the Command/Reply transaction having been abandoned prematurely.

Before completing or abandoning a Command/Reply transaction the SCC sets the DERR bit of the Status Register to logic 1 if either $ERR=1$ or $X=0$ or the transaction is abandoned prematurely (see Section 12.2).

15.5.4 *Delayed Response Bits.* The Delayed Response (DSQ, DSX) bits in the Status Register are also associated with recovery from situations in which an assembled Reply message has been corrupted or lost. They provide a means of access to the Q and X response of the previous command executed by the SCC.

Before completing or abandoning a Command/Reply transaction the SCC sets the DSQ and DSX bits of the Status Register to correspond with the Q and X responses, respectively, from the addressed module or feature of SCC (see Section 12.2).

If the Command/Reply transaction is abandoned before executing the command so that there are no valid Q and X responses related to the transaction, the DSQ and DSX bits should preferably be reset to logic 0.

15.6 Error Recovery Using the Reread Command. The optional Reread feature of SCC (see Section 11.2.2) allows data derived from a previous Read operation, but not successfully transmitted to the SD, to be recovered from the Read-data register of the SCC. The ERR and DERR bits provide the information needed to determine whether the required Read data should be accessed in the Read-data register of SCC or in its original source location.

After sending a command intended to perform a destructive-read operation (such as Read and Clear) at a module or feature of an SCC, the SD may receive:

(1) A valid reply, with $ERR=0$ and $SX=1$ (The destructive Read operation has taken place, and the required data is in the Read-data field of the reply.)

(2) A valid reply, with $ERR=1$ or $SX=0$ (The destructive read did not take place, and the required data is still available at the source. The SD should repeat the destructive-Read command.)

(3) A corrupted reply or no reply (The ERR bit (if any) of the reply is not a reliable indicator whether the destructive-Read operation took place.)

The error-recovery features are provided to deal with this last case, where the SD receives the expected truncated Command message from the SCC but does not receive a valid reply. Under these conditions the SD should send a Reread command to the SCC (see Section 11.2.2).

If the Status field of the reply to the Reread command indicates $DERR=1$, then the destructive-Read operation did not take place. The SD should repeat the destructive-Read operation to access the data from the source.

If the reply to the Reread command indicates $DERR=0$, the destructive-read operation took place, and the required data is in the Read-data field of the reply to the Reread command.

This procedure permits recovery from a single error (an error in the reply to the original destructive-Read command). The required data may be lost if there is an additional error, for example, if there is an error in the reply to the Reread operation.

The error-recovery procedure described above applied primarily to destructive-Read operations, in which the source cannot be relied upon to provide the same data if the operation is repeated. If the SD does not receive a valid reply after sending any other command (nondestructive Read, Write, or Control), a simpler recovery is possible. The SD should send a command to read the Status Register of the SCC. This provides access to the previous Command status and the appropriate decision can be taken to repeat the command if necessary.

INTERFACE SYSTEM (CAMAC) ANSI/IEEE Std 595-1982

16. Summary: Sequence of Actions in Serial Crate Controller

This section summarizes the response of SCC to the received byte stream in terms of the transitions between nine major states, as a result of normal and error conditions. This information is also presented diagrammatically in Fig 16-1 (in full) and Fig 16-2 (in simplified form omitting all error conditions).

The nine major states of SCC are:
Find Header
Receive Command
Execute Command
Send Reply
Find END
Send Demand
Lost Byte Sync
Lost Message Sync
Pass Message

Each of these states is characterised by:
(1) The conditions for entering the state
(2) The bytes normally received by SCC
(3) The bytes normally transmitted by SCC
(4) The normal conditions for exit from the state
(5) The error conditions for exit from the state

**Fig 16-1
Major-State Sequence in SCC**
(Numbered Exits from States Correspond to Sections in Text, For Example, Exit 1.1 is Described in Section 16.1.1)

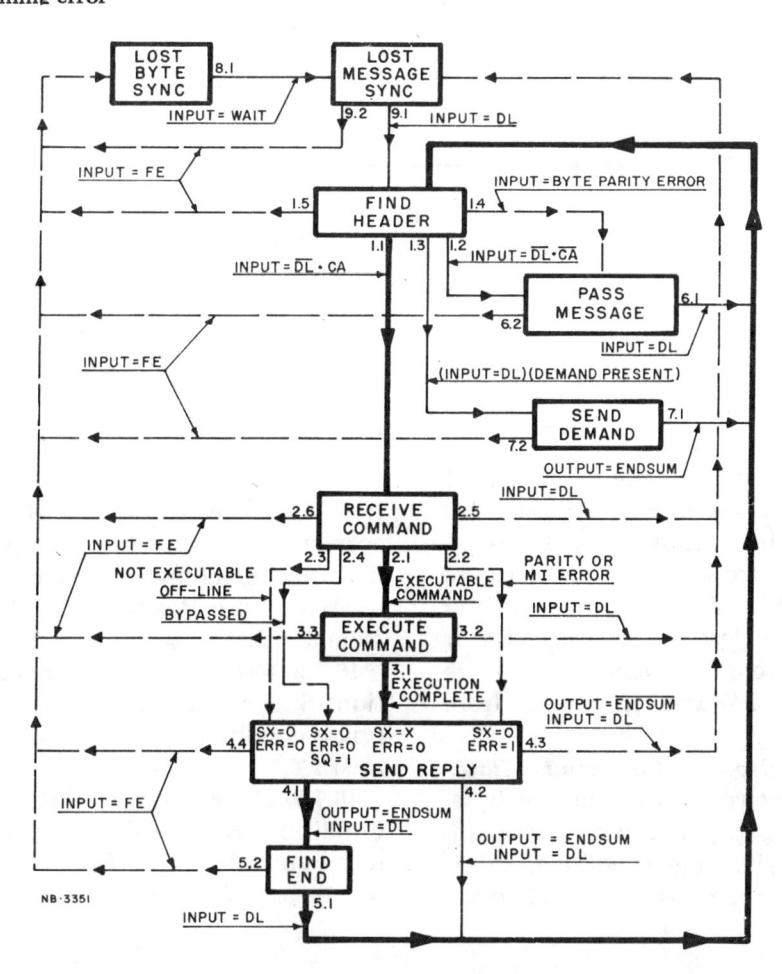

Solid lines — Normal exits
Broken lines — Error Exits
DL — Delimiter byte
CA — Crate Addressed
FE — Framing error

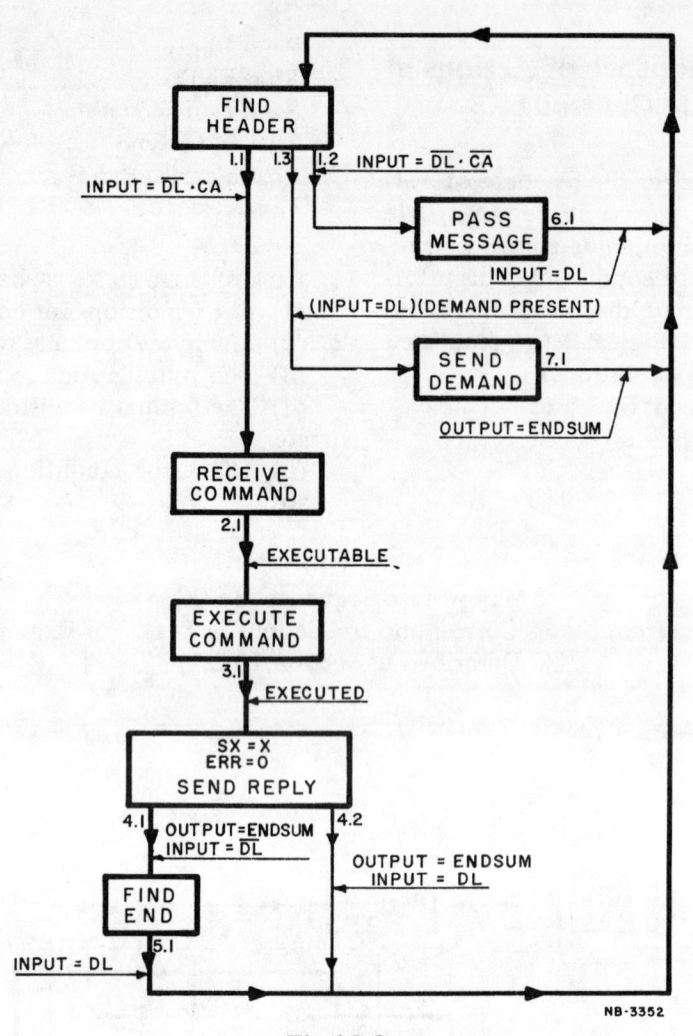

Fig 16-2
Major-State Sequence in SCC: Omitting All Error
Conditions

16.1 Find Header. This is the normal quiescent state of SCC. The SCC receives Delimiter bytes (typically WAIT bytes), followed eventually by a Nondelimiter byte treated as the Header byte of a message. The SCC retransmits all received bytes, including the Header, either directly or with a delay of up to 3 byte periods. There are three normal exits and two error exits from this state.

16.1.1 *Exit to Receive Command State.* Condition: SCC receives a Nondelimiter byte, with correct byte parity and with Crate Address field matching its assigned address. This is the Header byte of a message addressed to the SCC.

16.1.2 *Exit to Pass Message State.* Condition: SCC receives a Nondelimiter byte, with correct byte parity but with Crate Address field not matching its assigned address. This is treated as the Header byte of a message to or from another crate.

16.1.3 *Exit to Send Demand State.* Condition: SCC receives a Delimiter byte, and appropriate conditions exist for initiating a Demand message.

16.1.4 *Error Exit to Pass Message State.* Condition: SCC receives any byte with incorrect byte parity. The Pass Message state is used as a means of waiting for the next message end.

16.1.5 *Error Exit to Lost Byte Sync State.* Condition: SCC detects a byte-framing error while operating in bit-serial mode.

16.2 Receive Command. The Header byte of a Command message has been received and retransmitted. The SCC receives a sequence of bytes constituting the remainder of the Command message up to, and including, the SUM byte. The SCC transmits an END byte followed by WAIT bytes (assuming that the retransmitted Command message is truncated, see Section 3.1.2). When the SCC receives the SUM byte it checks the column parity. There is one normal exit and five error exits.

16.2.1 *Exit to Execute Command State.* Condition: SCC receives an executable command with correct byte and column parity, and correct MI field (if checked).

16.2.2 *Error Exit to Send Reply State (ERR=1).* Condition: SCC receives a command message with wrong byte or column parity, or wrong MI field (if checked). The command is not executed. The reply is a 3-byte error message with ERR=1 and SX-0.

16.2.3 *Error Exit to Send Reply State (Off-Line).* Condition: SCC is Off line and receives a command with correct parity (and MI field if checked) with N field in range N1-N23. The command is not executed. The reply is a message with ERR=0 and SX=0 and length as determined by the F field of the command.

16.2.4 *Error Exit to Send Reply State (Bypassed).* Condition: SCC is bypassed and receives a command with correct parity (and MI field, if checked) that does not reset the Bypass bit in the Status Register.

The command is not executed. The reply is a message with ERR=0, SX=0, and SQ=1 and length as determined by the F field of the command.

16.2.5 *Error Exit to Lost Message Sync State.* Condition: SCC receives any Delimiter byte. No reply is sent. The DERR bit in the Status Register is set to logic 1.

16.2.6 *Error Exit to Lost Byte Sync State.* Condition: SCC detects a byte-framing error while operating in bit-serial mode. No reply is sent. The DERR bit in the Status Register is set to logic 1.

16.3 Execute Command. An executable command has been received. This state continues until the required operation has been performed (including any delay periods associated with Bypass or Loop Collapse devices), or until the operation is abandoned. The SCC normally receives SPACE bytes but will accept any Nondelimiter bytes with correct or incorrect parity and transmits WAIT bytes. There is one normal exit and two error exits from this state.

16.3.1 *Exit to Send Reply State.* Condition: SCC completes the operation cycle (at least to a stage where the Q and X responses and the Read data are available). This can occur before the first SPACE byte has been received.

The reply is a message with ERR=0 and length determined by the F field of the command.

16.3.2 *Error Exit to Lost Message Sync State.* Condition: SCC receives any Delimiter byte. No reply is sent. If stage t_3 of execution has not been reached, the operation is abandoned, and the DERR bit in the Status Register is set to logic 1. If stage t_3 has been reached, then the operation is completed and, because ERR=0, DERR = $\overline{X}$.

16.3.3 *Error Exit to Lost Byte Sync State.* Condition: SCC detects a byte-framing error while operating in bit-serial mode. No reply is sent. If stage t_3 of execution has not been reached, the operation is abandoned, and the DERR bit in the Status Register is set to logic 1. If stage t_3 has been reached, then the operation is completed and, because ERR=0, DERR = $\overline{X}$.

16.4 Send Reply. A command has been received by SCC, and has been executed, or rejected as unexecutable. The SCC normally receives SPACE bytes (but will accept any Nondelimiter byte with correct or incorrect parity), possibly followed by an END byte. It transmits a Reply message concluding with an ENDSUM byte. There are two normal exits and two error exits from this state.

16.4.1 *Exit to Find END State.* Condition: SCC transmits the ENDSUM byte of the reply before it receives the END byte. This is typical of the excess-SPACE mode of operation.

16.4.2 *Exit to Find Header State.* Condition: SCC transmits the ENDSUM byte as the END byte (or any other Delimiter byte) is received.

16.4.3 *Error Exit to Lost Message Sync State.* Condition: SCC receives any Delimiter byte before ENDSUM is transmitted. Transmission of the Reply message is abandoned.

16.4.4 *Error Exit to Lost Byte Sync State.* Condition: SCC detects a byte-framing error while operating in bit-serial mode. Transmission of the Reply message is abandoned.

16.5 Find END. The reply has been sent, but END has not been received. This is typical of the excess-SPACE mode of operation. The SCC receives SPACE bytes, followed eventually by an END byte. It transmits WAIT bytes. There is one normal exit and one error exit from this state.

16.5.1 *Exit to Find Header State.* Condition: SCC receives any Delimiter byte (typically an END byte).

16.5.2 *Error Exit to Lost Byte Sync State.* Condition: SCC detects a byte-framing error while operating in bit-serial mode.

16.6 Pass Message. This state is entered from the Find Header state if the SCC has received a Nondelimiter byte; either a valid Header byte addressed to another crate, or a byte with wrong byte parity. The SCC receives the remaining bytes of the message (not necessarily conforming to the information fields of the standard messages), followed by a Delimiter byte. All received bytes are retransmitted. There is one normal exit and one error exit from this state.

16.6.1 *Exit to Find Header State.* Condition: SCC receives any Delimiter byte

16.6.2 *Error Exit to Lost Byte Sync State.* Condition: SCC detects a byte-framing error while in bit-serial mode.

16.7 Send Demand. The SCC has received a Delimiter byte, and the appropriate conditions exist for initiating a Demand message (see Section 5.1). The SCC receives any sequence of bytes, with correct or incorrect parity. These bytes are delayed in a buffer while the Demand message is transmitted. There is one normal exit and one error exit from this state.

16.7.1 *Exit to Find Header State.* Condition: SCC transmits the ENDSUM byte of the Demand message.

16.7.2 *Error Exit to Lost Byte Sync State.* Condition: SCC detects a framing error while operating in bit-serial mode.

16.8 Lost Byte Sync. The SCC has detected a byte-framing error while operating in bit-serial mode. It removes any delay introduced as a result of sending a previous Demand message and retransmits any arbitrary bit stream that it receives. The byte resynchronising procedure is started. There is only one exit from this state.

16.8.1 *Exit to Lost Message Sync State.* Condition: SCC identifies a received WAIT byte by the method defined in Section 10.2.3.

16.9 Lost Message Sync. The SCC has encountered a Delimiter byte in a prohibited context or is required to reestablish message synchronism as a precaution after having regained byte synchronism. It retransmits any arbitrary byte stream that it receives. The message resynchronizing procedure is started. There is one normal exit and one error exit from this state.

16.9.1 *Exit to Find Header State.* Condition: SCC receives at least one Delimiter byte.

16.9.2 *Error Exit to Lost Byte Sync State.* Condition: SCC detects a byte-framing error while operating in bit-serial mode.

Appendix

[This Appendix is not a part of ANSI/IEEE Std 595-1982, IEEE Standard Highway Interface System (CAMAC).]

A1. Specification of CAMAC Serial Crate Controller Type L2

A1.1 Interpretation. This appendix specifies the physical and operational characteristics of the recommended SCC Type L2.

The aim is to specify SCC Type L2 in such a way that it is possible to use any SCC Type L2 as a direct replacement for any other, without affecting the nominal hardware or software operational performance. There may, however, be differences in secondary characteristics such as internal structure, test facilities, front panel layout, etc.

> In order to conform to the specification of SCC Type L2, a crate controller must conform to all the mandatory requirements of this appendix, and those of the main text of this standard (Sections 3 to 16) and ANSI/IEEE Std 583-1982.
>
> An SCC Type L2 must have no other features, in addition to those required by this appendix, that could affect its full operational interchangeability with other crate controllers conforming to this appendix.

The remaining sections of this appendix define the mandatory requirements for SCC Type L2 that are additional to those in the main text of this standard. Supplementary information on SCC Type L2, including an example of the internal structure, is given in Section A2 of this appendix.

Other SCCs conforming to the main text of this standard do not necessarily have all the mandatory features of SCC Type L2 and may have additional features. However, it is recommended that such crate controllers should be uniform with SCC Type L2 in respect to those features that they have in common.

A1.2 General Features of Serial Crate Controller Type L2

A1.2.1 *Format*

> SCC Type L2 must be a CAMAC multi-width plug-in unit, preferably double-width, conforming to the mechanical standards of ANSI/IEEE Std 583-1982.

A1.2.2 *Dataway Connectors*

> SCC Type L2 must have Dataway connectors for the Control station and at least one Normal station.
>
> Connections to the free bus lines $P1$ and $P2$, and to the Free-use patch contacts $P3$-$P7$ are prohibited.

A1.3 Messages for Serial Crate Controller Type L2

A1.3.1 *Message Formats*

> SCC Type L2 must use Command, Reply, and Demand messages with the standard formats defined in Sections 3.1, 3.2, and 3.3.

A1.3.2 *Message Sequences*

> While receiving a Command message addressed to it, SCC Type L2 must transmit a truncated form of the message consisting only of the Header byte and an END byte, as shown in Figs 3-3 and 3-4.

A1.3.3 *Message Fields*

> SCC Type L2 must ignore the MI field of Command messages that it receives.
>
> When transmitting an Error-Reply message, in which the ERR bit of the Status field is at logic 1, the *SX* bit in the Status field must be at logic 0.

A1.3.4 *Demand Message Generation*

> SCC Type L2 must be able to generate Demand messages as defined in Section 3.3. A 3-byte Delay buffer must be provided (see Section 5.2). All 3 bytes must be switched into the path of the incoming bytes when SCC Type L2 begins to transmit a Demand message. After the Demand message has been transmitted, all 3 bytes of the buffer must be switched out when the contents of the buffer consist of 3 WAIT bytes and the previous byte transmitted at the *D*-output port has been a Delimiter byte.
>
> When SCC Type L2 detects a byte-framing error (see Section 10.2.2), the 3-byte Delay buffer must be switched out of the incoming byte path immediately.

A1.4 Serial Highway *D* Ports on Serial Crate Controller Type L2

> The SCC Type L2 must implement, via its *D* ports, the bit-serial and byte-serial modes of transmission. It must be capable of operating at any clock rate up to 5.0 MHz, in either mode.
>
> The SCC Type L2 must not have other SH ports (*U* ports) in addition to the two *D* ports. It must not use Contact 25 of the *D*-input connector, which is reserved for a third control signal.

A1.5 Internal Structure of Serial Crate Controller Type L2

A1.5.1 *Synchronization*

> When operating in bit-serial mode, SCC Type L2 must identify the Start bit of each byte frame by searching for logic 0 after logic 1 in the data-bit sequence.

> In order to establish byte synchronism when operating in bit-serial mode, SCC Type L2 must search for the WAIT byte pattern with Start and Stop bits $(1,11100000,0_2)$ by inspecting, at each bit clock time, the last 10 bits received (see Section 10.2.3).
>
> In order to establish message synchronism (either when operating in byte-serial mode or after having established byte synchronism in bit-serial mode) SCC Type L2 must successfully receive two consecutive Delimiter bytes if it was previously unaddressed, or one Delimiter byte if it was addressed.

Following this, the next Nondelimiter byte will be interpreted by SCC Type L2 as being the Header byte of a message (see Section 4.1.1).

A1.5.2 *Status Register*

> The Status Register of SCC Type L2 must implement all the bit allocations shown in Table 11-2. It must not use any bits shown in Table 11-2 as Reserved or Free use.
>
> As a result of any Command/Reply transaction in which SCC Type L2 does not perform an effective Dataway operation or corresponding internal operation, the DSQ and DSX bits in the Status Register must be set to logic 0 (see Section 12-2).

A1.5.3 *Commands Executed*

> SCC Type L2 must implement all the commands shown in Table 11-1, and no others. The Read LAM-pattern command must not be executed when SCC Type L2 is in the Dataway Off-line state.

Thus, SCC Type L2 has a Status Register, it saves the data resulting from Read operations in a Reread data register (see Section 11.2.2), and provides access to the pattern of Dataway L signals by a real or virtual LAM-pattern register (see Section 11.2.1).

A1.6 Front Panel Features of Serial Crate Controller Type L2

A1.6.1 *Manual Controls and Indicators*

> SCC Type L2 must have all the manual controls and indicators recommended in Section 13.
> The Initialize and Clear switches must be ineffective when SCC Type L2 is in the Dataway On-line state.
> The Crate Address selection must cover the full range, and the Crate Address indicator must show the decimal address 01_{10} to 62_{10}.

A1.6.2 *Other Front Panel Features*

> Any front panel features in addition to those required by A1.6.1 must not affect operational interchangeability.

This does not preclude the provision of monitoring points, etc, associated with maintenance and other nonoperational activities.

A1.7 SGL-Encoder Connector on Serial Crate Controller Type L2

> SCC Type L2 must have a rear-mounted SGL-Encoder connector of the type defined in Section 14.1 and with the contact allocations defined in Table 14-1. It must not use any contacts shown as Free-use in Table 14-1.
> In order to permit use in conjunction with an auxiliary controller, SCC Type L2 must accept a coded-N input to its N decoder (see Section 14.6.1) and must generate the ACL signal (see Section 14.6.2).

A2. Supplementary Information

The supplementary information in this appendix does not form part of the specification of SCC Type L2. It consists of a state diagram and a flow chart, summarizing a procedure for processing the incoming byte stream and generating the outgoing byte stream, and a block diagram suggesting a possible internal structure for SCC Type L2. These are intended to be compatible with the specification and should not be interpreted as modifying any of its mandatory requirements.

A2.1 Transition Diagram. Figure A2-1 shows the allowed transitions between the principal states depicted on the Flow Chart, Fig A2-2.

A2.2 Flow Chart. A procedure for processing the incoming byte stream and generating an outgoing byte stream is defined (in part explicitly and in part implicitly) by the mandatory features of SCC Type L2 concerned with synchronism, errors, and the structure and sequence of messages.

The actions and decisions involved in this procedure are shown in Fig A2-2 as an implementation-independent flow chart, which is closely related to the state-sequence diagrams in Section 16 of the standard. For clarity, all decision elements (other than those concerned with major states) indicate whether the decision is based on the currently received byte, a previous byte, or an asynchronous condition such as the presence of a demand.

Suggested uses of the flow chart are as a basis for a practical design of the message handling logic, or as a theoretical model against which the performance of actual designs can be compared. However, it is not necessary to follow the nomenclature and detailed structure of the flow chart in order to conform to the specification of SCC Type L2 in Section A1 of the appendix.

A2.3 Block Diagram. The block diagram of SCC Type L2, shown in Fig A2-3, is an example of a specific implementation. Suggested

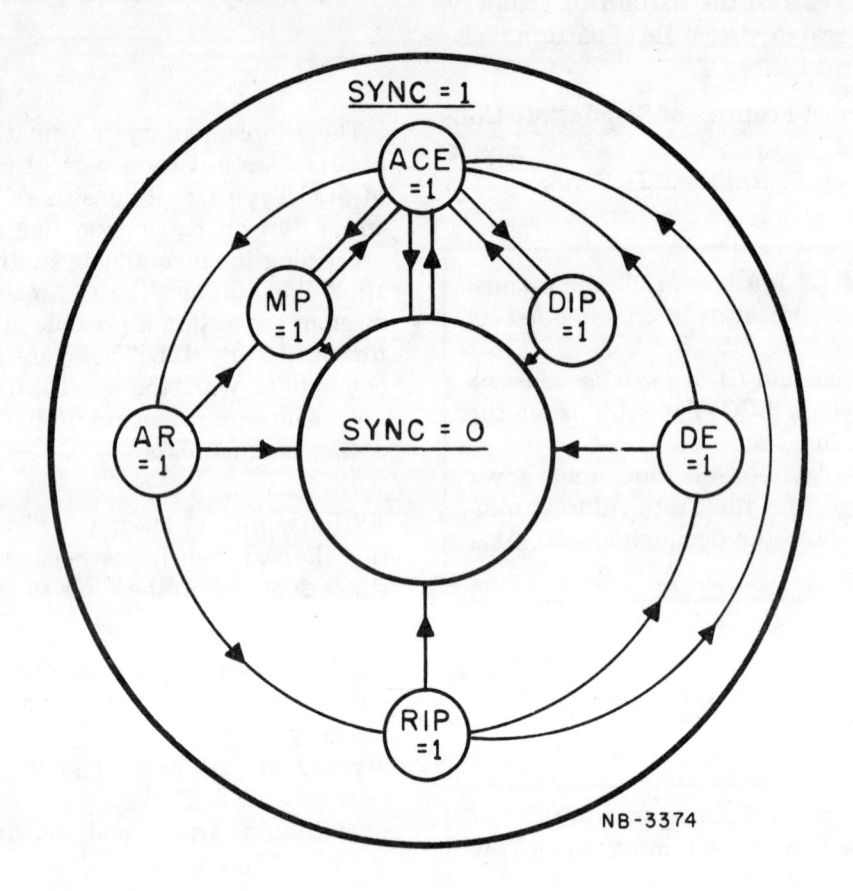

**Fig A2-1
Transition Diagram for Serial Crate Controller Type L2**

This shows the allowed transitions between the principal states depicted on the Flow Chart, Fig A2-2. The normal operating condition of SCC is SYNC = 1, which denotes that message synchronization is achieved. The state SYNC = 0 is entered if message synchronism is lost. (In bit serial mode it is also entered if byte synchronism is lost.) These principal states may be identified with the major states listed in Section 16 of ANSI/IEEE Std 595-1982.

NOTES: ACE — Address Check Enabled (See Find Header in section 16), Normally entered after a Delimiter byte has just been detected; the normal quiescent state of SCC.
AR — Address Received (Receive Command and Execute Command), Entered when SCC receives a Header byte addressed to itself.
DE — Delimiter Expected (Find End), Entered while awaiting Delimiter byte after Reply message has been sent.
DIP — Demand in Progress (Send Demand), Entered while SCC is sending a Demand message.
MP — Message Passing through SCC (Pass Message), Entered if message is for another SCC or if Header byte had incorrect parity.
RIP — Reply in Progress (Send Reply), Entered while SCC is sending a Reply message.

uses of this diagram are as a direct basis for a detailed design or as a model against which the performance and facilities of other designs can be compared. However, it is not necessary to follow the nomenclature and structure of Fig A2-3 in order to conform to the definition of SCC Type L2 in Section A1 of the appendix.

The following sections of this appendix are an explanatory commentary on the main features of this particular block diagram. The terminology and some other details are not necessarily directly applicable to other implementations or to the message handling flow chart in Fig A2-2.

A2.3.1 *Serial Input and Output.* In bit-serial mode the bit clock received at the input D port shifts the incoming bit stream into a serial-to-parallel converter. This generates an internal byte clock derived from the first Stop bit of each byte, and also an output byte clock derived from the Start bit of each byte.

When the crate controller is not Busy (is not processing a Command/Reply transaction), a replica of the received serial bit stream is retransmitted at the output D port, delayed by either 1 bit or 3 bytes, depending on the state of the switchable 3-byte delay. When the crate controller is Busy, the input is routed to the message handling logic, either directly or through the 3-byte delay. The output is synthesized as parallel bytes by the output multiplexer (MUX) and passed through a parallel-to-serial converter to the output D port. Each output byte is initiated by the byte clock derived from the Start bit of an input byte.

In byte-serial mode the byte clock received at the input D port samples the incoming parallel bytes. When the crate controller is not-Busy, a replica of this incoming byte stream is retransmitted at the output D port, delayed by either 1 byte or 4 bytes, depending on the state of the switchable 3-byte delay. When the crate controller is busy, the input is taken to the message handling logic, and the output is synthesized by the output MUX, as described in the preceding text. In this case, the parallel bytes from the MUX are taken to the byte-serial data lines of the output D port.

A2.3.2 *Command/Reply Transaction.* When the crate controller is not-Busy (is not processing a Command/Reply transaction), the crate comparator examines the Crate Address field of each incoming Nondelimiter byte that has correct parity ($\overline{TP\,BAD}$). When the crate comparator detects a match, it generates Crate Addressed. If the controller is Awaiting Header (i.e., has no message passing through it), Controller Busy is set, and the timing generator is enabled in state $T1$. Bits 1 through 6 of the Header byte and all successive bytes, are processed in the longitudinal parity checker.

As successive bytes are received, END/WAIT bytes are transmitted at the output D port, the timing generator steps through states $T2$, $T3$, etc, and the contents of the SA, SF, SN, and SW fields are transferred to the appropriate message handling registers. The $SF16$ and $SF8$ bits determine the sequence of the timing generator. For Write commands states, $T1$ to $T8$ are generated in sequence. For Read and Control commands $T3$ is followed by $T8$.

In state $T8$ the SUM byte is received, and compared to the output of the longitudinal parity checker. If LP Bad is not set, and TP Bad has not been set by a transverse parity error, the Dataway cycle timing generator is enabled.

When the Dataway cycle is complete, state $T9$ is established, and the Reply message synchronizer is enabled. The output MUX selects the SC field and, as the next input byte is received, the first byte of the Reply message is transmitted.

The timing generator enters state $T10$, and the next input byte causes the status field of the reply to be transmitted. A transverse parity bit is generated for each output byte, and Bits 1 through 6 contribute to the checksum in the Longitudinal Parity Generator.

For Write and Control commands, or Command messages that have set Error owing to transverse or longitudinal parity errors, the timing generator goes directly from state $T10$ to state $T15$. For normal Read commands it continues in sequence through states $T11$ to $T14$ to select the 4 Read field bytes.

If message synchronization is lost while the SCC is Busy, the $\overline{\text{MESSAGE SYNC}}$ state is asserted, the outgoing bytes are derived from the incoming bytes, and the Timing Generator continues to be driven by the byte clock to the $T15$ or $>T15$ state. No Dataway operation takes place at $T8$.

If byte synchronization is lost while the SCC is Busy, the $\overline{\text{BYTE SYNC}}$ and $\overline{\text{MESSAGE SYNC}}$ states are asserted, the outgoing bits are derived from the incoming bits, and the byte resynchronization process is initiated. When byte synchronism is reacquired, the derived byte clock continues to drive the timing generator to the $T15$ or $>T15$ state. No Dataway operation takes place at $T8$.

Then, when the SCC receives a Delimiter byte, after reaching state $T15$ or $>T15$, MESSAGE SYNC is reestablished and the Controller Busy state ceases.

In state $T15$ the next input byte causes the ENDSUM byte to be transmitted, and thereafter $T15$ is maintained, together with $>T15$. Any Delimiter byte resets Controller Busy and $>T15$. Any excess SPACE bytes cause WAIT bytes to be transmitted.

A2.3.3 *Onward Transmission of Other Messages.* The previous section deals with commands addressed to the SCC. If, while SCC is in the not-Busy and Awaiting Header condition, it receives any Delimiter byte, its condition is unchanged. If it receives any Nondelimiter byte that does not set Controller Busy, all received bytes are then retransmitted with the appropriate delay, until a Delimiter byte is encountered, and this once again establishes the Awaiting Header condition.

A2.3.4 *Demand Message.* Demand messages are enabled if SCC has received a valid Delimiter byte, is Awaiting Header, and has its demands enabled. Then a DMI signal via the SGL-Encoder connector can assert Demand Present. If the 3-byte delay is switched out, the Demand Message Synchronizer is enabled, and the Delay and Demand Transmit conditions established.

The output MUX synthesizes a 3-byte Demand message, using timing generator states $T9$, $T10$ and $T15$. The input byte stream is taken through the 3-byte delay, but note that the first 3 bytes shifted out of the delay are irrelevant and their contents are ignored.

The Delay condition is maintained until such time as the Delay decoder detects that the 3-byte delay is empty (contains only WAIT bytes), and the SCC is Awaiting Header and not Busy. Then the 3-byte delay is switched out and the undelayed byte stream is once again used.

In the meantime the delayed input byte stream is used and is processed by the Delimiter code checker, parity checkers, and the crate address comparator.

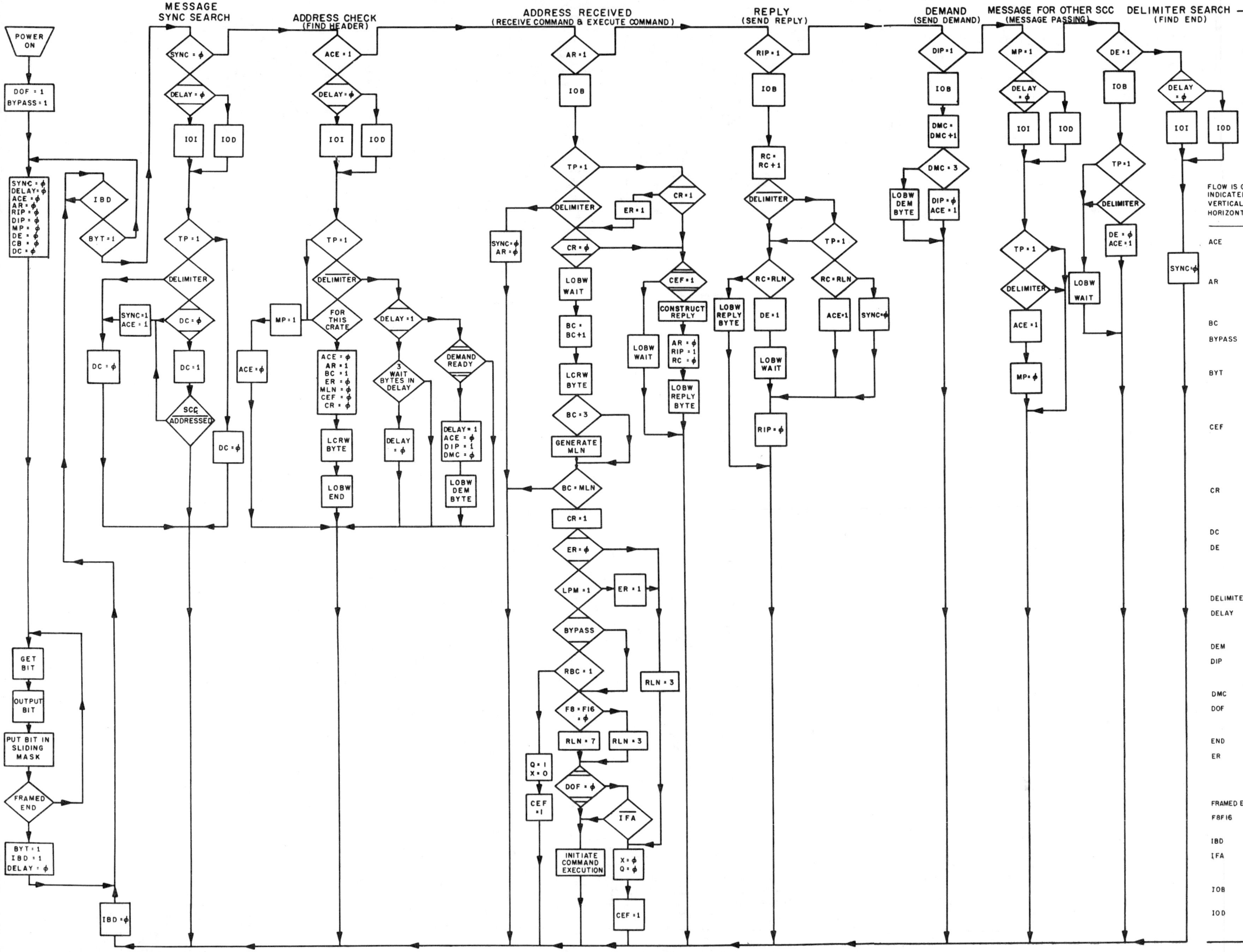

REPRESENTS A DECISION MADE EITHER ON CONTENTS OF INCOMING BYTE OR DECISION MADE DUE TO PRESENCE OF INCOMING BYTE.

REPRESENTS A DECISION MADE EITHER ON CONTENTS OF PREVIOUS BYTE OR DUE TO PRESENCE OF PREVIOUS BYTE.

REPRESENTS A DECISION DUE TO AN ASYNCHRONOUS EVENT.

FLOW IS GENERALLY FROM TOP TO BOTTOM, EXCEPT WHERE INDICATED BY ARROWS.
VERTICAL EXITS FROM DECISIONS ARE YES.
HORIZONTAL EXITS FROM DECISIONS ARE NO.

ACE	ADDRESS CHECK ENABLE (FIND HEADER) = 1 CONTENTS OF NEXT BYTE ENTERING WILL BE TESTED FOR ADDRESS MATCH WITH THIS CRATE. = 0 OTHERWISE
AR	ADDRESS RECEIVED (RECEIVE & EXECUTE COMMAND) = 1 THE HEADER BYTE OF A COMMAND MESSAGE CONTAINED THE ADDRESS FOR THIS CRATE. = 0 OTHERWISE
BC	BYTE COUNTER
BYPASS	CRATE BYPASS = 1 BYPASS ENABLED = 2 BYPASS DISABLED
BYT	BYTE MODE = 1 BYTE-SERIAL MODE OR BIT-SERIAL MODE IF, AND ONLY IF, BYTE CONTAINED CORRECT FRAMING START AND STOP BITS. = 0 OTHERWISE
CEF	CONSTRUCT ENABLE = 1 CONSTRUCT REPLY MESSAGE AS RESULT OF: (a) DATAWAY CYCLE (b) ERROR IN TRANSMISSION (c) DOF = 1 AND $\overline{IFA}$ = 0 OTHERWISE
CR	COMMAND RECEIVED = 1 ALL COMMAND MESSAGE (THRU Σ BYTE) RECEIVED BY SCC = 0 OTHERWISE
DC	DELIMITER-BYTE COUNTER
DE	DELIMITER SEARCH ENABLE (FIND END) = 1 SEARCH FOR DELIMITER WHEN EXCESS SPACE-BYTES RECEIVED BY SCC AFTER REPLY MESSAGE SENT. = 0 OTHERWISE
DELIMITER	ANY BYTE WITH TP = 1 AND BIT # 7 = 1
DELAY	THREE-BYTE DELAY ENABLE = 1 THREE-BYTE DELAY ACTIVE = 0 OTHERWISE
DEM	DEMAND
DIP	DEMAND IN PROGRESS (SEND DEMAND) = 1 A DEMAND MESSAGE BEING SENT. = 0 OTHERWISE
DMC	DEMAND MESSAGE BYTE COUNTER
DOF	DATAWAY OFF LINE STATUS = 1 DATAWAY IS OFF-LINE = 0 DATAWAY IS ON-LINE
END	END BYTE
ER	MESSAGE PARITY ERROR STATUS = 1 TRANSVERSE PARITY (TP) OF A BYTE OR LONGITUDINAL PARITY (LP) OF A MESSAGE WAS INCORRECT. = 0 OTHERWISE
FRAMED END	END BYTE WITH BYT = 1
F8F16	FUNCTION BITS 8 AND 16 - MSB OF FUNCTION CODE BITS IN BYTE # 3 OF A COMMAND MESSAGE.
IBD	INPUT BYTE DETECTED
IFA	INTERNAL FEATURE ADDRESSED $N(30)\cdot\{A(0)\cdot[F(1)+F(17)+F(19)+F(23)]+A(1)\cdot F(0)\}$.
IOB	INITIATE OUTPUT BYTE - SEND BYTE CONTAINED IN OUTPUT BUFFER
IOD	INITIATE OUTPUT DELAY - SEND BYTE OBTAINED FROM OUTPUT OF THREE-BYTE DELAY BUFFER
IOI	INITIATE OUTPUT OF INPUT - SEND BYTE RECEIVED AT SCC INPUT
LCRW	LOAD COMMAND REGISTERS WITH
LOBW	LOAD OUTPUT BUFFER WITH
LPM	LONGITUDINAL PARITY OF MESSAGE = 1 CORRECT LP = 0 INCORRECT
MLN	MESSAGE LENGTH - EXPECTED LENGTH FROM F CODE BITS AND < M1, M2 > IDENTIFIER BITS (IF USED). BITS CONTAINED IN COMMAND MESSAGE BYTES.
MP	MESSAGE PASSING = 1 WHEN (a) HEADER BYTE TESTED WITH ACE = 1 FAILED TO MATCH ADDRESS. = 1 WHEN (b) CRATE BYPASSED AND COMMAND WAS NOT TO REMOVE BYPASS. = 0 OTHERWISE (OR ON DETECTION OF DELIMITER AFTER HAVING MP = 1).
N	STATION NUMBER - SPECIFIED BY COMMAND MESSAGE TO THIS SCC
RBC	REMOVE BYPASS COMMAND = 1 COMMAND MESSAGE TO SWITCH BYPASS OUT TO THIS CRATE. = 0 OTHERWISE
RC	REPLY BYTE COUNTER
RIP	REPLY IN PROGRESS = 1 REPLY MESSAGE BEING SENT = 0 OTHERWISE
RLN	REPLY MESSAGE LENGTH
SYNC	SYNCHRONISM (MESSAGE) = 1 SYNCHRONISM OBTAINED = 0 OTHERWISE
TP	TRANSVERSE PARITY = 1 TP CORRECT = 0 TP INCORRECT
WAIT	WAIT BYTE

ND-610-J

CAMAC SERIAL SYSTEM

IMPLEMENTATION INDEPENDENT FLOW CHART OF SCC - L2

ND-610-J

Fig A2-2
Implementation Independent Flow Chart of SCC Type L2

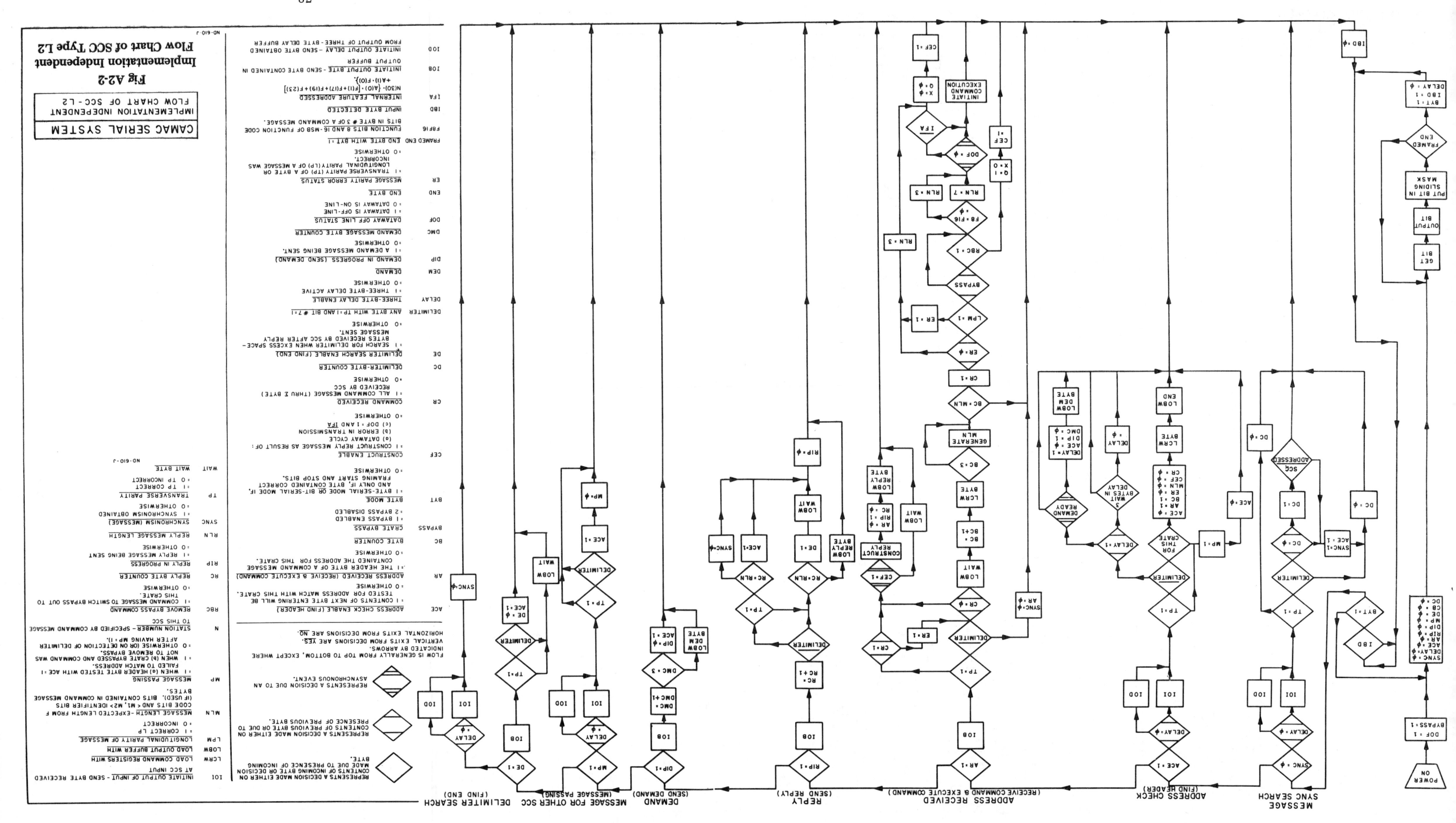

Fig A2-3 SCC Type L2 Block Diagram

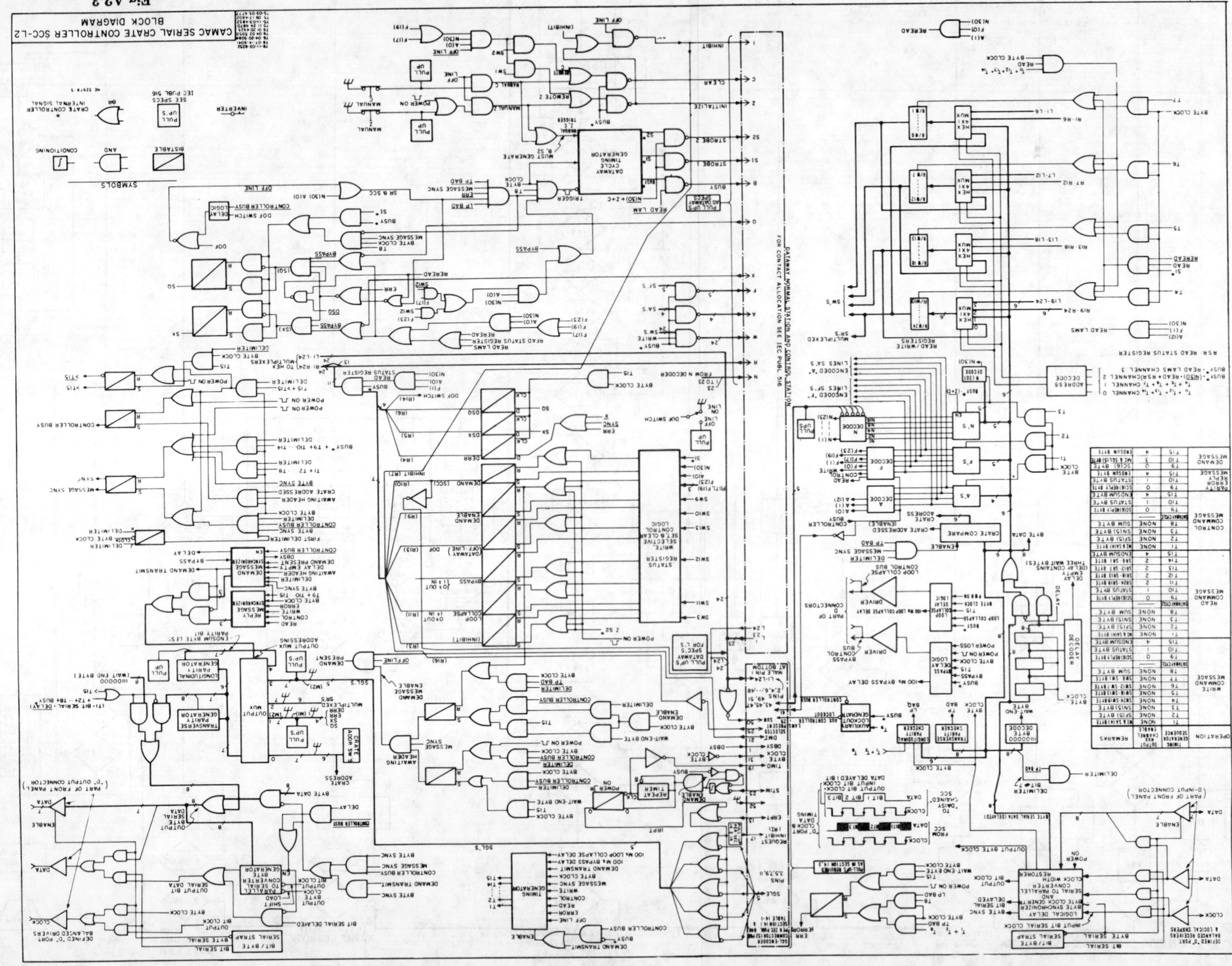

A3. Bibliography

ANSI/IEEE Std 583-1982, Modular Instrumentation and Digital Interface System (CAMAC).

ANSI/IEEE Std 596-1982, Parallel Highway Interface System (CAMAC).

COSTRELL, L. Highways for CAMAC Systems — A Brief Introduction. *IEEE Transactions on Nuclear Science*, vol NS-21, Feb 1974, pp 870-875.

COSTRELL, L. Standardized Instrumentation System for Computer Automated Measurement and Control. *IEEE Transactions on Industry Applications*, vol IA-11, May/Jun 1975, pp 319-323.

ESONE Report ESONE/SH/01, CAMAC Serial System Organization, A Description, November 1974.

FASSBENDER, P., and BEARDEN, F. Demonstrating Process Control Standards; An Exercise in Success. *IEEE Conference Record of the 1974 Ninth Annual Meeting of the IEEE Industry Applications Society*, Oct 7-10, Pittsburgh, PA, pp 457-462.

HORELICK, D., and LARSEN, R. S. CAMAC: A Modular Standard. *IEEE Spectrum*, vol 13, Apr 1976, pp 50-55.

JOERGER, F.A., and KLAISNER, L.A. Functional Instrumentation Modules, *IEEE Transactions on Industry Applications*, vol IA-11, Nov/Dec 1975.

KIRSTEN, F. A. Some Characteristics of Interfaces Between CAMAC and Small Computers, *IEEE Transactions on Nuclear Science*, Vol NS-20, Apr 1973, pp 42-49.

KLAISNER, L.A. A Serial Data Highway for Remote Digital Control. *IEEE Conference Record of the 1974 Ninth Annual Meeting of the IEEE Industry Applications Society*, Oct 7-10, Pittsburgh, PA, pp 453-456.

LYON, W. T., and ZOBRIST, D. W. System Design Considerations When Using Computer-Independent Hardware. *IEEE Transactions on Industry Applications*, vol IA-11, May/Jun 1975, pp 324-327.

MACHEN, D. R. The CAMAC Serial Systems Description for Long Line, Multicrate Applications, *IEEE Transactions on Nuclear Science*, vol NS-21, Feb 1974, pp 876-880.

RADWAY, T.D. A Standardized Approach to Interfacing Stepping Motors to Computers. *IEEE Conference Record of the 1974 Ninth Annual Meeting of the IEEE Industry Applications Society*, Oct 7-10, Pittsburgh, PA, pp 257-259.

US Energy Research and Development Administration Report TID-26488, CAMAC Serial System Organization, A Description, Dec 1973.

US Energy Research and Development Administration Report TID-26614, CAMAC Specification of Amplitude Analogue Signals Within a 50 Ohm System, Oct 1974.

US Energy Research and Development Administration Report TID-26616, Block Transfers in CAMAC Systems, Feb 1976. (To be issued also by ESONE Committee as Supplement to EUR 4100e).

US Energy Research and Development Administration Report TID-26618, CAMAC Tutorial Articles, Oct 1976.

WILLARD, F. G. Interfacing Standardization in the Large Control System. *IEEE Transactions on Industry Applications*, vol IA-11, Jul/Aug 1975, pp 362-364.

Index

All references are to section numbers unless Fig or Table is stated. The most important references are given first, followed by other references in parenthesis. To assist identification, index terms are usually capitalized in the text of this standard.

The following abbreviations are used in the index:

SH	Serial Highway
SD	Serial Driver
SCC	Serial Crate Controller

A, Dataway Signal	See Subaddress
Abbreviations and symbols	following Contents
ACL signal	See Auxiliary Controller Lockout
Auxiliary Controller	14.6
— access to N decoder	14.6.1, (14.2.12, A1.7)
— interlock with SCC	14.6.1, 14.6.3
Auxiliary Controller Lockout signal (ACL)	14.2.13, 14.6.2, (12.4.1, 14.6.3, A1.7, Table 14-1)
B, Dataway signal	See Busy
BCK signal	See Byte Clock
Bibliography	A3
Bit-clock signal	2.3
Bit-serial mode	9.2, 7, (2.3)
Block diagram of SSC Type L2	A2, Fig. A2-3
Branch Highway, Parallel	1.1, 14
Bus 1 and Bus 2 contacts	7.1.2, Fig. 7-1
Busy, Dataway signal B	11.2.1, 12.4.1
Bypass	12.4.2, (4.1.3)
— switching	15.1.1, Fig 15-1
— indicator on SCC	13.2.7
Byte clock	2.3
— extraction from byte frames	9.2.3
— at SGL-Encoder connector (BCK)	14.2.8, Fig 14-2, (Table 14-1)
Byte frame	2.3, Fig 2-3, (9.2)
Byte parity	15.3.2, (4.1.2, 4.1.3)
Byte serial mode	9.1, 7, (2.3)
Byte stream	8.2
Byte synchronization	10.2
C, Dataway signal	See Clear
CBY signal	See Controller Busy
Checksum	See Column parity
Clear	
— Dataway signal C	12.1.1
— manual control on SCC	13.1.3
Clock (system clock)	2.4, 8.3.1
— frequency	8.1
— period (T, T_{min})	8.3, 8.3.1, Fig 8-1
Cluster, of crate assemblies	2.5
Column parity	15.3, (3.5.2, 4.1.2, 4.1.3)
Command accepted bit SX	15.5.2, Fig 3-5, (3.4.8, Table 15-1)
Command messages	3.1, Figs 3-1, 3-2, (3.6.1)
Command/Reply transaction	4, (3)
— length of	Table 3-1
Commands	
— implemented by SCC	Table 11-1
— implemented by SCC Type L2	A1.5.3
Compatibility, with SH	1.2
— compatible devices	2.7, Fig 2-6
Configuration, of SH	2.1, Fig 2-1
— reconfiguration options	12.4
Conform, with specification	1.2, A1.1
Connectors	
— D port	7.1, 13.3, Table 7-1
— SGL Encoder	14.1, Table 14-1, (A1.7)
— Dataway	2.7, A1.2.2
— others on SCC	13.4, A1.4

INTERFACE SYSTEM (CAMAC)

Contact assignments	
— *D* port	7.1.2, Table 7-1
— SGL Encoder	14.1, Table 14-1
Control operation	4.4, Figs 4-5, 4-6
Control signals, *D* port	7.3
— signal standards	7.3.2, Table 7-4
— sources and receivers	7.3.1, Fig 7-4
— reserved	7.1.2, 15.1.2, A1.4
Controller Busy Signal (CBY)	14.2.11, (Table 14-1)
Crate Address	
— field (SC)	3.4.1
— switch on SCC	13.1.1
— indicator on SCC	13.2.1, A1.6.1
Crate Addressed indicator	13.2.5
Data extraction interval	8.3, 8.3.3, Fig 8-1
Dataway Off-line	
— state	12.4.1, Table 12-1
— manual control on SCC	13.1.2
Dataway On-line	
— state	12.4.1
— indicator on SCC	13.2.2
DBSY signal	See Demand Busy
Delay	
— buffer in SCC	5.2, (10.3)
— buffer in SCC Type L2	A1.3.4
— propagation	8.4
Delayed Command Accepted bit (DSX)	12.2, 15.5.4
Delayed-Error bit (DERR)	12.2, 15.5.3, Fig 3-5, (3.4.8)
Delayed *Q*-response bit (DSQ)	12.2, 15.5.4
Delayed Reply message	12.4.2, 12.4.3
Delimiter	
— bit	2.2
— byte	2.2, 3.5.1
Demands, identification of	5.3
Demand *L*24	12.3.2
Demand Busy signal (DBSY)	14.2.6, (Table 14.1)
Demand message	
— format	3.3, Figs 3-7, 3-8, (6.4)
— enable	5.1, 12.3.1
— generation by SCC	5, Figs 5-1, 5-2
— generation by SCC Type L2	A1.3.4
— indicator on SCC	13.2.6
Demand Message Initiate signal (DMI)	14.2.4, Fig 14-3, (5.1, Table 14-1)
DERR bit	See Delayed Error
DMI signal	See Demand Message Initiate
Downstream	2.1
D ports	2.5, 7, (13.3, Fig 2-4, Table 7-1)
— on SCC Type L2	A1.4
DSQ bit	See Delayed *Q* response
DSX bit	See Delayed Command Accepted
Encoded Serial Graded-*L* signals (SGLE)	14.2.3, (Table 14-1)
END byte	3.5.3, Figs 3-1 to 3-4
ENDSUM byte	3.5.2, 3.5.5, Figs 3-5 to 3-8
ERPT signal	See External Repeat
ERR bit	See Error bit
Error bit (ERR)	15.5.1, Fig 3-5, Table 15-1, (3.4.8)
Error Detection	
— by Geometric code	2.2, 15.3
— by context	15.3.4
by module	15.3.5
Error Reply message	15.4, 4.1.4
Execute Command state	4.1.3, 16.3, Figs 16-1, 16-2
External Repeat signal (ERPT)	14.2.7, 14.4.2, (Table 14-1)

Fields, of messages	3.4, Figs 3-2, 3-4, 3-6, 3-8, (A1.3.3)
Find END State	16.5, Figs 16-1, 16-2
Find Header state	4.1.1, 16-1, Figs 16-1, 16-2
Flow Chart of SCC Type L2	A2, Fig A2-2
Free-use, Dataway Contact *P*	A1.2.2
Front panel	
— of SCC	13
— of SCC Type L2	A1.6
Function field *SF*	3.4.4
Geometric code, error detection	
— principle	15.3.1, Fig 15-3, (2.2)
— implementation on SH	15.3.2, Fig 15-4
— performance	15.3.3
Header byte	2.2, 4.1.1
Hung Demand	14.4
— SGL bit pattern	3.4.9
— false	14.4, 14.5.4
I, Dataway signal	See Inhibit
Indicators on SCC	13.2, A1.6
Inhibit	
— Dataway signal *I*	12.1.2
— indicator on SCC	13.2.3
Initialize	
— Dataway signal *Z*	12.1.1
— manual control on SCC	13.1.3
L, Dataway signal	See Look-at-Me, and Demand *L*24
LAM pattern	11.2.1, Table 11.1, (5.3)
— reading while Off-line	12.4.1
Look-at-Me signal *L*	14.2.1, 12.3.2
Loop Collapse	12.4.3
— switching	15.1.2, Fig 15-2
Lost Byte Sync state	10.2.2, 16.8, Figs 16-1, 16-2, (10.3, A1.3.4)
Lost Message Sync state	10.1.2, 16.9, Figs 16-1, 16-2, (4.1.4, 10.3)
Lost Sync	15.2, 10.3
L-sum	14.2.2 (Table 14-1)
*M*1, *M*2 bits	See Message Identification
Manual controls, on SCC	13.1, A1.6
May, permitted practice	1.2
Messages	
— formats	2.2, 3, Fig 2-2, (A1.3)
— sequences	4, Figs 4-1 to 4-6, Fig 5-1
Message identification	
— field (MI)	3.4.7, 4.1.2, Table 3-2, (4.1.3, 6)
— ignored by SCC Type L2	A1.3.3
— by SCC and SD	6
Message synchronization	10.1
MI	See Message Identification
Must, mandatory requirement	1.2
N decoder	See Station Number
N, Dataway signal	See Station Number
Nondelimiter bytes	3.5.1
Non-Return-to-Zero-Level	7
NRZL	See Non-Return-to-Zero-Level
Off-line state	See Dataway Off-line
On-line state	See Dataway On-line

Parity	See byte-parity, column parity
Parallel Highway	1.1
Pass Message state	16.6, Figs 16-1, 16-2
Pause bits, noncontinguous bytes	9.2.1, Fig 2-3
PH	See Parallel Highway
Port	2.5
Q, Dataway signal	See Response, Q
Read Data field *SR*	3.4.6
Read operation	4.2, Figs 4-1, 4-2
Receive Command state	4.1.2, 16.2, Figs 16-1, 16-2
Receiver, balanced signals	7.2.4, Table 7-3, Fig 7-3
References	See Bibliography
Registers, in SCC	11, Table 11-1
	See also Status, LAM pattern, Reread Data
Reply message	3.2, Figs 3-5, 3-6, (6.3)
Reply space	4.6, Fig 3-2, (3.5.7)
— length of	4.6.3, 12.4.2, 12.4.3
Request Inhibit	14.2.14
Reread Data, register in SCC	11.2.2, Table 11-1
— use in error recovery	4.6.2, 15.6, (4.6.2)
Response, Q response *SQ*	3.4.8, Fig 3-5, Table 15-1
RS-422, Electronic Industries Association	
Signal Standard	7.2
S, prefix to SH fields and bits	3
S1, S2 Dataway signals	See Strobe
SA field	See Subaddress
SC field	See Crate Address
SCC, SCC Type L2	See Serial Crate Controller
SD	See Serial Driver
Selected-LAM Present	
— bit in Status Register	12.3.3
— signal (SLP)	14.2.5, (Table 14-1)
Send Demand state	16.7, Figs 16-1, 16-2
Send Reply state	4.1.4, 16.4, Figs 16-1, 16-2
Serial Crate Controller (SCC)	2.8
— Type L2	A1, A2
Serial Driver (SD)	2.1, 2.6
Serial Graded L Field	3.4.9, 14.5.3
Serial Graded L Signals	See Encoded Serial Graded L
Serial Highway (SH)	1.1, 2
SF field	See Function
SGL field	See Serial Graded L Field
SGLE signals	See Encoded Serial Graded *L*
SGL Encoder	14, 14.5, Fig 14-1
SH	See Serial Highway
Should, recommended practice	1.2
Signals, *D* port	
— data	7.2, 8.3.2, 8.3.3, 8.3.4
— clock	7.2
— control	See Control Signals
— timing	8.3, Fig 8-1
— standards, balanced signals	7.2
Signals, SGL Encoder	
— definitions	14.2
— standards	14.3, Table 14-2
SLP signal	See Selected LAM present
SN field	See Station Number
SPACE byte	3.5.7, Figs 3-1, 3-2
— excess	4.6.2
SQ bit	See Response
SR field	See Read data
Start bit	9.2, Fig 2-3, A1.5.1
Start Timer signal (STIM)	14.2.9, Table 14.1
States, of SCC	4.1, 16, Figs 16-1, 16-2, A2
Station Address	See Station Number
Station Number	
— field *SN*	3.4.2, Fig 3-2
— use of *N*(30) in SCC	11
— signals *N*1 to *N*16	14.2.12, 14.6.1 (Table 14-1)

Status field	3.4.8, Fig 3-6
Status register	11.1, 12, Tables 11-1, 11-2
— initial state	Table 11-3
— in SCC Type L2	A1.5.2
STIM signal	See Start Timer
Stop bit	2.3, Fig 2-3, 9.2
Strobe signals, Dataway $S1$, $S2$	12.4.1
Subaddress A	11.1, Table 11-1
Subaddress field SA	3.4.3, Fig 3-2
SUM	
— byte	3.5.2, 3.5.6, Figs 3-1, 3-2
SW field	See Write Data
SX bit	See Command Accepted
Synchronization	10, A1.5.1
System clock	See Clock
Symbols	See Abbreviations and Symbols
T, T_{min}	See Clock period
Three-byte delay	5.2, A1.3.4
Time out, Hung demands	14.4, 14.5.4
Time-out signal (TIMO)	14.2.10, 14.4.1, Fig. 14-3, (Table 14-1)
TIMO signal	See Time-out
Timer, Hung demands	14.4
— in SCC	14.4.1
— in SGL Encoder	14.4.2
Timing, of D-port Signal	8
Transaction	See Command/Reply Transaction
Transition Diagram, of SCC Type L2	A2, Fig A2-1
Transmission lines (D port)	7.2.1
Transmitter, balanced signals	7.2.3, Table 7-2, Fig 7-2
Truncated Command message	3.1.2, 4.5, Figs 3-3, 3-4, (6.2)
— generation by SCC Type L2	A1.3.2
Unsynchronized state	See Lost Sync
— indicator on SCC	13.2.4
U-ports	2.5, Fig 2-5
— prohibited on SCC Type L2	A1.4
Upstream	2.1
WAIT byte	3.5.4, Figs 4-1 to 4-6, Fig 5-1
Write data field SW	3.4.5
Write operation	4.3, Figs 4-3, 4-4
X, Dataway signal	15.3.5, 15.5.2, 3.4.8
Z, Dataway signal	See Initialize

ANSI/IEEE
Std 596-1982
(Revision of ANSI/IEEE
Std 596-1976)

An American National Standard

IEEE Standard Parallel Highway Interface System (CAMAC*)

Sponsor

Instruments and Detectors Committee of
the IEEE Nuclear and Plasma Sciences Society

Approved September 17, 1981
IEEE Standards Board

Approved December 15, 1981
American National Standards Institute

*Computer Automated Measurement and Control

© Copyright 1982 by

The Institute of Electrical and Electronics Engineers, Inc
345 East 47th Street, New York, NY 10017

*No part of this publication may be reproduced in any form,
in an electronic retrieval system or otherwise,
without the prior written permission of the publisher.*

IEEE Standards documents are developed within the Technical Committees of the IEEE Societies and the Standards Coordinating Committees of the IEEE Standards Board. Members of the committees serve voluntarily and without compensation. They are not necessarily members of the Institute. The standards developed within IEEE represent a consensus of the broad expertise on the subject within the Institute as well as those activities outside of IEEE which have expressed an interest in participating in the development of the standard.

Use of an IEEE Standard is wholly voluntary. The existence of an IEEE Standard does not imply that there are no other ways to produce, test, measure, purchase, market, or provide other goods and services related to the scope of the IEEE Standard. Furthermore, the viewpoint expressed at the time a standard is approved and issued is subject to change brought about through developments in the state of the art and comments received from users of the standard. Every IEEE Standard is subjected to review at least once every five years for revision or reaffirmation. When a document is more than five years old, and has not been reaffirmed, it is reasonable to conclude that its contents, although still of some value, do not wholly reflect the present state of the art. Users are cautioned to check to determine that they have the latest edition of any IEEE Standard.

Comments for revision of IEEE Standards are welcome from any interested party, regardless of membership affiliation with IEEE. Suggestions for changes in documents should be in the form of a proposed change of text, together with appropriate supporting comments.

Interpretations: Occasionally questions may arise regarding the meaning of portions of standards as they relate to specific applications. When the need for interpretations is brought to the attention of IEEE, the Institute will initiate action to prepare appropriate responses. Since IEEE Standards represent a consensus of all concerned interests, it is important to ensure that any interpretation has also received the concurrence of a balance of interests. For this reason IEEE and the members of its technical committees are not able to provide an instant response to interpretation requests except in those cases where the matter has previously received formal consideration.

Comments on standards and requests for interpretations should be addressed to:

    Secretary, IEEE Standards Board
    345 East 47th Street
    New York, NY 10017
    USA

Foreword

[This Foreword is not a part of ANSI/IEEE Std 596-1982, IEEE Standard Parallel Highway Interface System (CAMAC).]

The interface system on which this standard is based was developed by the ESONE Committee of European Laboratories** with the collaboration of the NIM Committee of the US Department of Energy.* This standard is supplementary to ANSI/IEEE Std 583-1982, Modular Instrumentation and Digital Interface System (CAMAC), and is based on ERDA† Reports TID-25876, March 1972 (corresponding to ESONE Report EUR 4600e) and TID-25877. The mandatory features in this standard are identical to those in the preceeding reports and in Publication 552 of the International Electrotechnical Commission (IEC). A serial highway interface system, also intended for use with ANSI/IEEE Std 583-1982, as defined in ANSI/IEEE Std 595-1982, Serial Highway Interface System.

This standard was reviewed and balloted by the Nuclear Instruments and Detectors Committee of the IEEE Nuclear and Plasma Sciences Society. Because of the broad applicability of this standard, coordination was established with numerous IEEE Societies, Groups, and Committees, including the Communications Society, Computer Society, Industry Applications Society, Industrial Electronics and Control Instrumentation Group, Instrumentation and Measurement Group, and the Power Generation and Nuclear Power Engineering Committees of the Power Engineering Society.

The revision of this standard was in conjunction with the 1981 review (1982 issue) of the entire family of IEEE CAMAC standards undertaken to incorporate existing addenda and corrections into the standards.

At the time of approval of this standard, the membership of the Nuclear Instruments and Detectors Committee of the IEEE Nuclear and Plasma Sciences Society was as follows:

D. C. Cook, *Chairman*   **Louis Costrell,** *Secretary*

J. A. Coleman	T. R. Kohler	P. L. Phelps
J. F. Detko	H. W. Kraner	J. H. Trainor
F. S. Goulding	W. W. Managan	S. Wagner
F. A. Kirsten	G. L. Miller	F. J. Walter
	D. E. Persyk	

At the time it approved this standard, the American National Standards Committee on Nuclear Instruments, N42, had the following representatives:

Louis Costrell, *Chairman*   **D. C. Cook,** *Secretary*

Organization Represented	*Name of Representative*
American Conference of Governmental Industrial Hygienists	Jesse Lieberman
American Nuclear Society	Frank W. Manning
Health Physics Society	J. B. Horner Kuper
	J. M. Selby *(Alt)*
Institute of Electrical and Electronics Engineers	Louis Costrell
	Julian Forster *(Alt)*
	David C. Cook *(Alt)*
	A. J. Spurgin *(Alt)*
Instrument Society of America	M. T. Slind
	J. E. Kaveckis *(Alt)*
Lawrence Berkeley National Laboratory	Lee J. Wagner
Oak Ridge National Laboratory	Frank W. Manning
	D. J. Knowles *(Alt)*
US Department of the Army, Materiel Command	Abraham E. Cohen
US Department of Commerce, National Bureau of Standards	Louis Costrell
US Department of Energy	Gerald Goldstein
US Federal Emergency Management Agency	Carl R. Siebentritt, Jr
US Nuclear Regulatory Commission	Robert E. Alexander
US Naval Research Laboratory	David C. Cook
Members-at-Large	J. G. Bellian
	O. W. Bilharz, Jr
	John M. Gallagher, Jr
	Voss A. Moore
	R. F. Shea
	E. J. Vallario

---

†Energy Research and Development Administration (now part of the Department of Energy).

*NIM Committee[1]
L. Costrell, Chairman

C. Akerlof	N. W. Hill	L. B. Mortara
E. J. Barsotti	D. Horelick	V. C. Negro
B. Bertolucci	M. E. Johnson	L. Paffrath
J. A. Biggerstaff	C. Kerns	D. G. Perry
A. E. Brenner	F. A. Kirsten	I. Pizer
R. M. Brown	P. F. Kunz	E. Platner
E. Davey	R. S. Larsen	S. Rankowitz
W. K. Dawson	A. E. Larsh, Jr	S. J. Rudnick
S. R. Deiss	R. A. LaSalle	G. K. Schulze
S. Dhawan	N. Latner	W. P. Sims
R. Downing	F. R. Lenkszus	D. E. Stilwell
T. F. Droege	R. Leong	J. H. Trainor
C. D. Ethridge	C. Logg	K. J. Turner
C. E. L. Gingell	S. C. Loken	H. Verweij
A. Gjovig	D. R. Machen	B. F. Wadsworth
B. Gobbi	D. A. Mack	L. J. Wagner
D. B. Gustavson	J. L. McAlpine	H. V. Walz
D. R. Heywood		D. H. White

NIM Executive Committee for CAMAC
L. Costrell, Chairman

E. J. Barsotti	F. A. Kirsten
J. A. Biggerstaff	R. S. Larsen
S. Dhawan	D. A. Mack
J. H. Trainor	

NIM Dataway Working Group
F. A. Kirsten, Chairman

E. J. Barsotti	C. Kerns
J. A. Biggerstaff	P. F. Kunz
L. Costrell	R. S. Larsen
S. Dhawan	D. R. Machen
A. Gjovig	L. Paffrath
D. R. Heywood	S. Rankowitz
D. Horelick	S. J. Rudnick

NIM Software Working Group
D. B. Gustavson, Jr, Chairman
W. K. Dawson, Secretary

A. E. Brenner	P. A. LaSalle
R. M. Brown	F. B. Lenkszus
S. R. Deiss	C. A. Logg
S. Dhawan	J. McAlpine
C. D. Ethridge	L. B. Mortara
M. E. Johnson	D. G. Perry

NIM Mechanical and Power Supplies Working Group
L. J. Wagner, Chairman

L. Costrell
C. Kerns
S. J. Rudnick
W. P. Sims

NIM Analog Signals Working Group
D. I. Porat, Chairman

L. Costrell
C. E. L. Gingell
N. W. Hill
S. Rankowitz

NIM Serial System Subgroup
D. R. Machen, Chairman

E. J. Barsotti
D. Horelick
F. A. Kirsten
R. G. Martin
L. Paffrath
S. J. Rudnick

NIM Block Transfer Subgroup
E. J. Barsotti, Chairman

W. K. Dawson
F. A. Kirsten
R. A. LaSalle
F. R. Lenkszus
R. G. Martin
R. F. Thomas, Jr

NIM Multiple Controllers Subgroup
P. F. Kunz, Chairman

E. J. Barsotti	D. R. Machen
F. A. Kirsten	R. G. Martin

---

[1] National Instrumentation Methods Committee of the US Department of Energy.

**ESONE Committee[2]

W. Attwenger, Austria, *Chairman* 1980–81; H. Meyer, Belgium, *Secretary*

Representatives of ESONE Member Laboratories

W. Attwenger, Austria	E. Kwakkel, Netherlands	A. C. Peatfield, England
R. Biancastelli, Italy	J. L. Lecomte, France	I. C. Pyle, England
L. Binard, Belgium	J. Lingertat, D. R. Germany	B. Rispoli, Italy
J. Biri, Hungary	M. Lombardi, Italy	M. Sarquiz, France
B. Bjarland, Finland	M. Maccioni, Italy	W. Schoeps, Switzerland
D. A. Boyce, England	P. Maranesi, Italy	R. Schule, F. R. Germany
B. A. Brandt, F. R. Germany	C. H. Mantakas, Greece	P. G. Sjolin, Sweden
F. Cesaroni, Italy	D. Marino, Italy	L. Stanchi, Italy
P. Christensen, Denmark	H. Meyer, Belgium	R. Trechcinski, Poland
W. K. Dawson, Canada	K. D. Muller, F. R. Germany	M. Truong, France
M. DeMarsico, Italy	J. G. Ottes, F. R. Germany	P. Uuspaa, Finland
C. A. DeVries, Netherlands	A. D. Overtoom, Netherlands	H. Verweij, Switzerland
H. Dilcher, F. R. Germany	E. C. G. Owen, England	A. J. Vickers, England
B. V. Fefilov, USSR	L. Panaccione, Italy	S. Vitale, Italy
R. A. Hunt, England	M. Patrutescu, Roumania	M. Vojinovic, Yugoslavia
W. Kessel, F. R. Germany	R. Patzelt, Austria	K. Zander, F. R. Germany
R. Klesse, France		D. Zimmermann, F. R. Germany

ESONE Executive Group (XG)

W. Attwenger, Austria, XG, *Chairman* 1980–81
H. Meyer, Belgium, *Secretary*

P. Christensen, Denmark	M. Sarquiz, France
M. Dilcher, F. R. Germany	R. Trechcinski, Poland
B. Rispoli, Italy	A. Vickers, England
H. Verweij, Switzerland	

ESONE Technical Coordination Committee (TCC)

A. C. Peatfield, England, *TCC Chairman*
P. Christensen, Denmark, *TCC Secretary*

W. Attwenger, Austria	W. Kessel, F. R. Germany
R. Biancastelli, Italy	J. Lukacs, Hungary
G. Bianchi, France	R. Patzelt, Austria
H. Dilcher, F. R. Germany	P. J. Ponting, Switzerland
P. Gallice, France	W. Schoeps, Switzerland
S. Vitale, Italy	

ECA/ESONE CAMAC Document Maintenance Study Group (DMSG)

P. Gallice, France, *Chairman*

R. C. M. Barnes, England	F. Iselin, Switzerland
L. Besse, Switzerland	H. Meyer, Belgium
J. Davis, England	H. J. Trebst, F. R. Germany

When the IEEE Standards Board approved this standard on September 17, 1981, it had the following membership:

Irvin N. Howell, Jr, *Chairman* **Irving Kolodny,** *Vice Chairman*

Sava I. Sherr, *Secretary*

G. Y. R. Allen	Jay Forster	F. Rosa
J. J. Archambault	Kurt Greene	Robert W. Seelbach
James H. Beall	Loering M. Johnson	Jay A. Stewart
John T. Boettger	Joseph L. Koepfinger	W. E. Vannah
Edward Chelotti	J. E. May	Virginius N. Vaughan, Jr
Edward J. Cohen	Donald T. Michael*	Art Wall
Len S. Corey	J. P. Riganati	Robert E. Weiler

*Member emeritus.

---

[2]European Standards on Nuclear Electronics Committee.

Contents

SECTION	PAGE
CAMAC and NIM Standards and Reports	8
1. Introduction	9
2. Interpretation	9
3. The Branch	10
4. Use of Lines at a Branch Highway Port	11
4.1 Command	11
4.1.1 Crate Address $BCR1 - BCR7$	11
4.1.2 Station Number $BN1, 2, 4, 8, 16$	13
4.1.3 Subaddress $BA\,1, 2, 4, 8$	13
4.1.4 Function $BF1, 2, 4, 8, 16$	13
4.2 Data and Status	13
4.2.1 Read and Write $BRW1 - BRW24$	13
4.2.2 Response BQ	13
4.2.3 Command Accepted BX	13
4.2.4 Command Accepted BX Response to Graded-L Request BG	13
4.2.5 Command Accepted X and BX Disabling (Supplementary Information)	14
4.3 Timing (BTA, $BTB1 - BTB7$)	14
4.4 Demand Handling	14
4.4.1 Branch Demand BD	14
4.4.2 Graded-L Request BG	14
4.5 Common Controls	15
4.5.1 Branch Initialize BZ	15
4.5.2 Dataway Initialize Z, Clear C, and Inhibit I	15
4.6 Reserved Lines $BV6$ and $BV7$ and Free Lines $BV1 - BV5$	15
5. Branch Operations	15
5.1 Command Mode Operations	19
5.1.1 Read Operations: Phase 1	19
5.1.2 Read Operations: Phase 2	19
5.1.3 Read Operations: Phase 3	20
5.1.4 Read Operations: Phase 4	20
5.1.5 Write Operations	20
5.1.6 Other Command Operations	20
5.2 Graded-L Operations	21
5.3 Differential Delays (Skew)	21
5.4 Identification of On-Line Crate Controllers	21
6. Connectors	22
6.1 Connection to Shield of Branch Highway Cable	25
7. Signal Standards at Branch Highway Ports	25
7.1 Inputs	28
7.2 Outputs	28
7.3 Terminations	28
7.4 Off-Line Power-Off Conditions	29

TABLES

Table 1	Signal Lines at Branch Highway Ports	12
Table 2	Station Number Codes Used in Crate Controllers	12
Table 3	Sequence of Command Mode Operation	16
Table 4	Sequence of Graded-L Operation	17
Table 5	Standard Connector for Branch Highway Ports	22

TABLES

		PAGE
Table 6	Contact Assignments at Branch Highway Ports: By Signal	25
Table 7	Contact Assignments at Branch Highway Ports: By Contact Numbers	26
Table 8	Signal Standards at Branch Highway Ports	27
Table 9	Commands Implemented by CAMAC Crate Controller Type A-1	33
Table 10	Contact Assignments for Rear Connector of Crate Controller Type A-1	35

FIGURES

Fig 1	CAMAC Branch: Chain Configuration	10
Fig 2	CAMAC Branch: Example of an Alternative Configuration	10
Fig 3	Timing of Branch Read Operation	18
Fig 4	Timing of Branch Write Operation	18
Fig 5	Branch Highway Ports: Arrangement of Connectors on Crate Controllers	23
Fig 6	Branch Highway Ports: Contact Layout	24
Fig 7	CAMAC Crate Controller Type A-1 (Appendix)	38

APPENDIX

A1	Specification of CAMAC Crate Controller Type A-1	29
	A1.1 CAMAC Crate Controller Type A-1	29
	A1.2 Other Crate Controllers	29
	A1.3 General Features	29
	A1.4 Front Panel	30
	A1.5 Dataway Signals	30
	A1.5.1 Data Signals	30
	A1.5.2 Command Signals	30
	A1.5.3 Common Control Signals	31
	A1.5.4 Patch Connections	31
	A1.6 Demand Handling	31
	A1.6.1 Branch Demand	31
	A1.6.2 Graded L	31
	A1.6.3 Pull Up for GL and L Lines	32
	A1.7 Timing Requirements	32
	A1.7.1 Command Mode Operations with Dataway $S1, S2,$ and B	32
	A1.7.2 Graded-L Operations	32
	A1.7.3 Command Mode Operations Without Dataway $S1, S2,$ or B	32
	A1.8 Commands Implemented by Crate Controller Type A-1	33
	A1.9 LAM-Grader Connector	33
	A1.9.1 Signal Standards	34
	A1.9.2 Timing — Branch Demand	34
	A1.9.3 Timing — Graded-L Operation	35
	A1.9.4 Timing — Command Mode Operations	35
	A1.10 Off-Line State	35
	A1.11 Dataway Inhibit I in Off-Line State	35
A2	Branch Highway Length Limitations	38
	A2.1 Parallel Branch Highway	38
	A2.2 Balanced Highway	39
A3	Bibliography	40

APPENDIX FIGURES

Fig 7	CAMAC Crate Controller Type A-1	37
Fig A2.1	Effect of Branch Highway Length on Noise Margin	38
Fig A2.2	Balanced Transmission System with Single-Ended-to-Balanced Converters	38
Fig A2.3	Differential Receiver Input Voltages	39

INDEX 41

CAMAC and NIM Standards and Reports

Title	IEEE, ANSI Std No	IEC No	DOE No	EURATOM (EUR) or ESONE No
CAMAC Instrumentation and Interface Standards*	SH08482* (Library of Congress No 8185060)	—	—	—
Modular Instrumentation and Digital Interface System (CAMAC)	ANSI/IEEE Std 583-1982	516	TID-25875† and TID-25877†	EUR 4100e
Serial Highway Interface System (CAMAC)	ANSI/IEEE Std 595-1982	640	TID-26488†	EUR 6100e
Parallel Highway Interface System (CAMAC)	ANSI/IEEE Std 596-1982	552	TID-25876† and TID-25877†	EUR 4600e
Multiple Controllers in a CAMAC Crate	ANSI/IEEE Std 675-1982	729	DOE/EV-0007	EUR 6500e
Block Transfers in CAMAC Systems	ANSI/IEEE Std 683-1976 (Reaff 1981)	677	TID-26616†	EUR 4100 suppl
Amplitude Analog Signals within a 50 Ω System	—	—	TID-26614	EUR 5100e
The Definition of IML A Language for Use in CAMAC Systems	—	—	TID-26615	ESONE/IML/01
CAMAC Tutorial Articles	—	—	TID-26618	—
Real-Time BASIC for CAMAC	ANSI/IEEE Std 726-1982	§	TID-26619†	ESONE/RTB/03
Subroutines for CAMAC	ANSI/IEEE Std 758-1979 (Reaff 1981)	713	DOE/EV-0016†	ESONE/SR/01
Recommendations for CAMAC Serial Highway Drivers and LAM Graders for the SCC-L2	—	—	DOE/EV-0006	ESONE/SD/02
Definitions of CAMAC Terms	Included in SH08482	678	DOE/ER-0104	ESONE/GEN/01
Standard Nuclear Instrument Modules NIM	—	547**	TID-20893 (Rev 4)	—

†Superseded by corresponding IEEE Standard listed.

*This is a hard cover book that contains ANSI/IEEE Std 583-1982, ANSI/IEEE Std 595-1982, ANSI/IEEE Std 596-1982, ANSI/IEEE Std 675-1982, ANSI/IEEE Std 683-1976 (Reaff 1981), ANSI/IEEE Std 726-1982 and IEEE Std 758-1979 (Reaff 1981), plus introductory material and a glossary of CAMAC terms.

**Covers only mechanical features and connector pin assignments.

§ In preparation.

NOTE: *Availability of Documents*
ANSI Sales Department, American National Standards Institute, 1430 Broadway, New York, NY 10018.
IEEE IEEE Service Center, 445 Hoes Lane, Piscataway, New Jersey 08854, USA.
IEC International Electrotechnical Commission, 1, rue de Varembé, CH-1211 Geneva 20, Switzerland.
DOE and TID Reports National Bureau of Standards, Washington, D.C. 20234, USA, Attn: L. Costrell.
EURATOM Office of Official Publications of the European Communitities, P.O. Box 1003, Luxembourg.
ESONE Commission of the European Communities, CGR-BCMN, B-2440 GEEL, Belgium, Attn: ESONE Secretariat, H. Meyer.

An American National Standard

IEEE Standard Parallel Highway Interface System (CAMAC)

1. Introduction

The Dataway defined in ANSI/IEEE Std 583-1982, Modular Instrumentation and Digital Interface System, is the basis for all CAMAC* systems. It provides the means of interconnection between modules and a crate controller within one crate. This standard defines the CAMAC Parallel Highway Interface System for interconnecting up to seven CAMAC crates (or other devices) and a system controller. The Parallel Highway, so called because the data bits are transmitted along the Highway in parallel, is capable of handling higher data rates than nonparallel highways. However, in some applications, for example, where there are long distances between crates or where simplicity of interconnections is particularly important, the Serial Highway of ANSI/IEEE Std 595-1982, Serial Highway Interface System (CAMAC), may be more suitable.

This standard defines the signals, timing, and logical organization of the connections from crate controllers and Parallel Highway Drivers to the Parallel Highway through a standard connector. The internal structures of crate controllers and Parallel Highway Drivers, and the physical construction of the Parallel Highway System are defined only where they affect compatibility between parts of the system.

An appendix defines in more detail those features of the crate controller that affect hardware and software compatibility. The appendix can be used as a formal specification of a standard crate controller, Crate Controller Type A-1 (CCA-1), or as a general recommendation to promote uniformity between crate controllers.

Since a number of Parallel Highways can be used to assemble a large system (as discussed in Section 3), a CAMAC Parallel Highway, together with its connected equipment and its Driver, is designated a CAMAC Branch. The CAMAC Parallel Highway Driver is thus commonly referred to as the CAMAC Branch Driver. Associated terms are Branch Demand, Branch Address, Branch Initialize, Branch operation, etc. The use of these designations is continued in this standard. Where the meaning is clear, shortened terms, usually with the prefix CAMAC omitted, are also used.

2. Interpretation

This standard is a reference text describing and defining the CAMAC Parallel Highway Interface System (CAMAC Parallel Highway System). It should be read in conjunction with, and is supplementary to, ANSI/IEEE Std 583-1982. No part of this standard is intended to supersede or modify ANSI/IEEE Std 583-1982.

In this standard there are mandatory requirements, recommendations, and examples of permitted practice.

> Mandatory clauses of the specification are enclosed in a box, as here, and usually include the word *must*.

Definitions of recommended or preferred practice (to be followed unless there are sound reasons to the contrary) include the word *should*.

Examples of permitted practice generally include the word *may*, and leave freedom of choice to the designer or user.

> In order to *conform* with the specification of the CAMAC Parallel Highway System, an equipment or system must satisfy all the mandatory requirements in this standard, excluding the appendix. If constructed as a CAMAC plug-in unit the equipment must also satisfy the mandatory requirements of ANSI/IEEE Std 583-1982.

Section A1 of this standard's appendix defines Parallel Crate Controller Type A-1 in such a way that operationally interchangeable units can be produced by different manufacturers. See Section A1.1 of the Appendix regarding conforming with the specification of the CAMAC Crate Controller Type A-1.

> The single letter designations *A*, *B*, *C*, *D*, and *E* (prefixed in some cases by "CC") are reserved for future use in connection with this standard and are not to be used until assigned.

> In order to be *compatible* with the CAMAC Parallel Highway System, equipment need not satisfy all the mandatory requirements but must not interfere with the full operation of all the features of the Parallel Highway System and crate controllers (including Type A-1) as defined in this standard.

No part of this document is intended to exclude the use of equipment that is compatible in the above sense, even if it does not conform fully to the specification or is not constructed as CAMAC plug-in units.

3. The Branch

A multicrate CAMAC system consists of one or more *Branches*, each having a *Branch Highway* which is the means of interconnection between a *Branch Driver* and *crate controllers*. During each *Branch operation* the Branch Driver can communicate with a maximum of seven crate controllers. All Branch Drivers and crate controllers have standard *Branch Highway ports*[3] by which they are connected to the Highway. Each port consists of a 132-contact connector (for 65 signals and their individual return lines, plus cable shield) with defined contact allocations and signal conventions. Each crate controller has two identical internally linked ports in order to allow the Branch to have the chain configuration shown in Fig 1. Other configurations are possible, such as that shown in Fig 2, where the Branch Driver is not at the end of the Branch and some crates are connected by only one port.

In addition to their normal *on-line* state, crate controllers have an *off-line* state which allows them to remain physically connected to the Branch while ignoring (and not impeding) all Branch operations. If required, the Branch Driver can recognize which crate addresses are associated with on-line crate controllers.

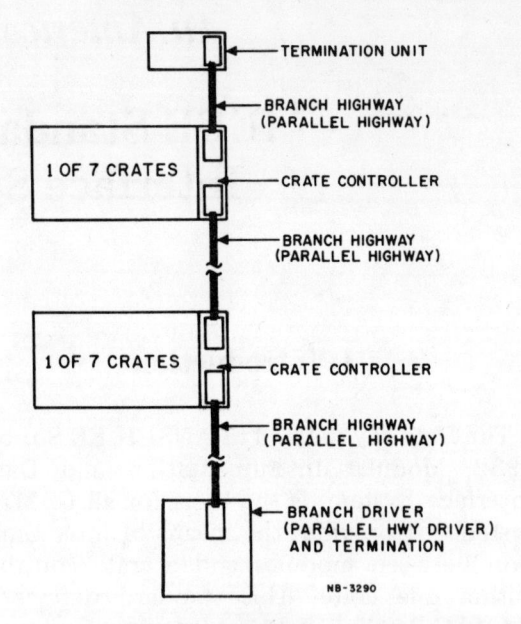

**Fig 1
CAMAC Branch: Chain Configuration**

**Fig 2
CAMAC Branch: Example of an Alternative Configuration**

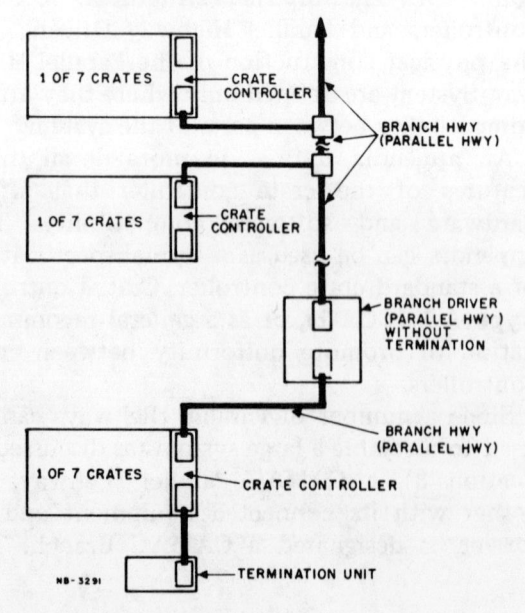

---

[3] In the sense, that a *port* is 'an entrance or exit of a network, etc.'

The basic mode of operation of the Branch is the *command mode*. The Branch Driver, which is typically associated with a system controller or computer, issues a command during each Branch operation. This command includes crate address information to select one or more crate controllers. Each addressed crate controller accepts the command from the Branch Highway and generates the corresponding Dataway command (station number, subaddress, and function). During *read* operations data signals are generated by a module on the Dataway Read lines, transferred to the data lines of the Branch Highway by the crate controller, and accepted by the Branch Driver. During *write* operations the Branch Driver generates data signals on the Branch Highway, and these are transferred to the Dataway Write lines by the crate controller, and accepted by a module. During other command operations there is no transfer of read or write data via the Branch Highway.

The Branch has two *demand handling* features which allow the Branch Driver to respond to Look-at-Me (LAM) signals from modules. For single-level demand handling, which merely indicates the presence of demands without identifying them, the crate controllers combine the LAM signals to form a common *Branch Demand* signal. For multilevel demand handling, which allows the Branch Driver to identify twenty-four different demands, there is the *Graded-L mode* of Branch operation. The Branch Driver issues a Graded-*L* Request (typically as the result of receiving the Branch Demand signal) and each on-line crate controller responds by selecting or rearranging its LAM signals to form a 24 bit Graded-*L* word. The Graded-*L* words from all crates are combined on the Branch Highway and presented to the Branch Driver.

At a Branch Highway port the *Data* lines are used in the command mode for information transfers in either direction between crate controllers and the Branch Driver. These lines are also used to convey the pattern of demands in the Graded-*L* mode.

Transfers in either mode through a Branch Highway port are controlled by interlocking *timing signals*, which automatically adjust the timing of each Branch operation to suit the actual transmission delays and controller performance that are encountered.

Initialize is the only *common control* signal that is transmitted through the Branch Highway port to the Dataway.

Highway Length Limitations (Supplementary Information). The standard Branch Highway uses a single-ended (unbalanced) mode of signal transmission. This mode gives reliable transmission in systems where the total length of the Highway is not excessive. The Branch Highway length can be extended considerably by the use of single-ended to balanced converters. This subject is discussed in Section A2 of the appendix.

4. Use of Lines at a Branch Highway Port

Each line at a Branch Highway port must be used in accordance with the mandatory requirements detailed in the following sections. Table 1 shows the titles, the standard designations, and the sources of the signals. Lines at a port are distinguished from corresponding lines in the Dataway by the prefix B, for example, the function code is carried by F lines in the Dataway and BF lines at a Branch Highway port.

4.1 Command

The command signals are used to control operations in the command mode, at which time the signal on the BG line (see Section 4.4.2) must be in the **0** state. They are transmitted by the Branch Driver on the BCR, BN, BA, and BF lines at the Branch Highway port (see following text).

4.1.1 *Crate Address BCR1—BCR7*

The seven crate controllers that can be addressed during any Branch operation must each be associated with a different BCR line (although all Branch Highway ports have provision for all seven BCR lines).

Each crate controller must therefore include means, such as a switch or patch connection, for selecting the appropriate BCR line (referred to as BCR_i). The assignment of BCR lines to crates is not necessarily related to the physical arrangement of

Table 1
Signal Lines at Branch Highway Ports

Title		Designation	Generated by	Signal Lines	Use
Command	Crate Address	$BCR1 - BCR7$	Branch Driver	7	each line addresses one crate in the branch
	Station Number	$BN1, 2, 4, 8, 16.$	Branch Driver	5	binary coded station number
	Subaddress	$BA1, 2, 4, 8.$	Branch Driver	4	as on Dataway A lines
	Function	$BF1, 2, 4, 8, 16.$	Branch Driver	5	as on Dataway F lines
Data	Read/Write	$BRW1 - BRW24$	Branch Driver W or Crate Controller R, GL	24	for Read data, Write data, and Graded-L
Status	Response	BQ	Crate Controller	1	as on Dataway Q line
	Command Accepted	BX	Crate Controller	1	as on Dataway X line
Timing	Timing A	BTA	Branch Driver	1	indicates presence of command, etc
	Timing B	$BTB1 - BTB7$	Crate Controller	7	each line indicates presence of data, etc, from one crate controller
Demand Handling	Branch Demand	BD	Crate Controller	1	indicates presence of demand
	Graded-L Request	BG	Branch Driver	1	requests Graded-L operation
Common Control	Initialize	BZ	Branch Driver	1	as on Dataway Z line
Reserved		$BV6$ and $BV7$		2	for future requirements
Free		$BV1$-$BV5$		5	for non-standard user requirements

NOTE: An individual return line is provided for each signal line. Two lines are provided for a connection to the shield, if any, of the Branch Highway cable.

Table 2
Station Number Codes Used in Crate Controllers

N Code	Use	B, $S1$, and $S2$	Remarks
$N(0)$	reserved		
$N(1) - (23)$	address the corresponding normal station	yes	
$N(24)$	address preselected normal stations	yes	normal stations occupied by the controller need not be addressed
$N(26)$	address all normal stations	yes	
$N(28)$	address crate controller only	yes	
$N(30)$	address crate controller only	no	no Dataway operation
$N(25, 27, 29, 31)$	reserved		

crates within the Branch. The Branch Driver is permitted to generate signals simultaneously on more than one BCR line in order to select several crates for the same operation.

It is recommended that the crate controller should include a means of protection against spurious signals on the selected BCR line. For example, the incoming BCR signal or an internal signal derived from it may be conditioned by integration.

It will be seen later, in Section 4.3, that each crate controller is associated with not only one of the BCR lines but also the corresponding one of seven BTB lines.

The Branch is not in a valid operating condition if more than one on-line crate controller is connected to the same BCR line. A means of reducing the risk of this occurring is suggested in Section 5.4.

INTERFACE SYSTEM (CAMAC)

4.1.2 Station Number BN1, 2, 4, 8, 16

> Signals on these five lines indicate the binary coded station number to be used within the selected crate or crates, and are decoded in the crate controller. In a crate controller the 32 codes are allocated as shown in Table 2.

At least one normal station is occupied by the crate controller, and there are station number codes to address the remaining 23 normal stations individually. In addition there are codes to multiaddress all normal stations or those stations indicated by the contents of a *Station Number Register* (SNR). Two further station number codes address the controller and its extensions irrespective of their location in the crate.

4.1.3 Subaddress BA1, 2, 4, 8

> Signals on these four lines must be retransmitted on the Dataway Subaddress lines (*A*1, 2, 4, 8) by an addressed crate controller whenever it is on line during a command mode operation.

4.1.4 Function BF1, 2, 4, 8, 16

> Signals on these five lines must be retransmitted on the Dataway Function lines (*F*1, 2, 4, 8, 16) by an addressed crate controller whenever it is on line during a command mode operation.

4.2 Data and Status

4.2.1 Read and Write BRW1 – BRW24

> These twenty-four lines are used in command mode read operations to transmit data from the addressed crate controllers to the Branch Driver, with *BRW*1 corresponding to the Dataway *R*1, etc. They are also used in command mode write operations to transmit data from the Branch Driver to the crate controllers, with *BRW*1 corresponding to the Dataway *W*1, etc. In the Graded-*L* mode they are used to transmit the pattern of demands from all on-line crate controllers in the Branch to the Branch Driver. The generation of 1 state outputs to these lines is restricted to Branch Drivers during command mode write operations and to addressed on-line crate controllers during Graded-*L* operations and command mode read operations.

4.2.2 Response BQ

> During a command mode operation with an associated Dataway operation each crate controller that is on line and addressed must generate *BQ* corresponding to Dataway *Q* (*BQ* = *Q*). During a command mode operation that tests the status of a feature of the crate controller, without a Dataway operation, the crate controller must generate the appropriate *BQ* response. At all other times crate controllers must generate *BQ* = 0. The signal on the *BQ* line at the Branch Driver is the OR combination of the signals from all crate controllers.

4.2.3 Command Accepted BX

> During a command mode operation with an associated Dataway operation each crate controller that is on line and addressed must generate *BX* corresponding to Dataway *X* (*BX* = *X*). During all other command operations the crate controller must generation *BX* = 1, if it accepts the command and *BX* = 0 if it does not accept the command. The signal on the *BX* line at the Branch Driver is the OR combination of the signals from all crate controllers.

4.2.4 Command Accepted BX Response to Graded-L Request BG.
The generation of *BX* by Crate Controller Type A-1 is fully defined for command mode operations (Sections 4.2.3 and A1.8). Graded-*L* operations, however, are generally multiaddressed, in which case the *BX* signal at the Branch Driver is an unreliable indication that all crates have responded to the operation. Therefore, the Command Accepted *BX* response to a Graded-*L* Request *BG* is not defined. For the guidance of designers, it is recommended that:

(1) When Crate Controller Type A-1 is addressed in a Graded-*L* operation, it should generate *BX* = 0

(2) During a Graded-*L* operation the Branch Driver should not respond to the state of the *BX* line.

4.2.5 Command Accepted X and BX Disabling (Supplementary Information). CAMAC computer interfaces, whether crate controllers or Branch Drivers, that include provisions for monitoring the Command Accepted X or BX response should also include a mode of operation in which an X (or BX) = 0 response does not result in an automatic system alarm. This mode is necessary to permit "normal" operation of a system that includes early plug-in units that do not have provision for generating or transmitting the Command Accepted signals. Such plug-in units always "respond" with $X = 0$. When performing an Address Scan block transfer the combination $Q = 0$, $X = 0$ should not result in an automatic system alarm.

4.3 Timing BTA, $BTB1 - BTB7$

> The timing of all command mode and Graded-L Branch operations is controlled by Branch timing signals. The Branch Driver initiates operations by signals on the common BTA line, and each addressed crate controller responds with a signal on its individual BTB line. All seven BTB lines are provided at each Branch Highway port, but each crate controller uses the line BTB_i, corresponding to the line BCR_i, by which it is addressed.
>
> Each on-line crate controller must generate $BTB_i = 1$ when it is not addressed. The Branch Driver (and other crate controllers) can thus distinguish between BTB lines associated with on-line crates ($BTB_i = 1$) and off-line or absent crates ($BTB_i = 0$) (see Section 5.4).
>
> The Branch Driver generates $BTA = 1$ to indicate that it is presenting a command or Graded-L request at its port, and maintains the signal until it has accepted the resulting BRW or BQ information. Each crate controller generates $BTB_i = 0$ when it has established data or BQ information during Branch operations.
>
> The timing signals must be generated through intrinsic OR outputs and must have 10-90 percent signal transition times in the range 100 ± 50 ns.

It is recommended that the crate controller should include a means of protection against spurious signals on the BTA line. For example, the incoming BTA signal or an internal signal derived from it may be conditioned by integration.

The full timing sequence is described in Section 5.

4.4 Demand Handling. LAM L signals from units in any part of the Branch typically demand that an appropriate command or sequence of commands be generated. The Branch therefore has two demand handling features, one associated with the Branch Demand signal and the other with the Graded-L Request signal.

4.4.1 Branch Demand BD

> Each crate controller can generate a demand signal, as any logical function of the L signals on the Dataway, through an intrinsic OR connection to the common Branch Demand line BD. No restriction is placed on the time at which the BD signal may change, and therefore its 10-90 percent transition time must be in the range 100 ± 50 ns. The delay between the time when an L signal at the control station of the crate controller reaches a maintained **1** or **0** state and the time when the BD signal at the Branch Highway port of the same crate controller reaches a corresponding maintained **1** or **0** state must not exceed 400 ns.

This maximum delay may be due partly to the crate controller and partly to some other unit involved in processing the L signals (for example, the LAM Grader associated with Crate Controller Type A-1. The maximum delay due to Crate Controller Type A-1 is defined in Section A1.9.2 of the appendix.

4.4.2 Graded-L Request BG

> The Branch Driver initiates Graded-L mode operations by generating the Graded-L Request signal BG, accompanied by BCR signals to all on-line crates. Each addressed crate controller generates a 24-bit Graded-L word on the BRW lines, and the Branch Driver reads the OR combination of these words. The Dataway L signals in each crate are graded, to select the relevant signals and assign them to the appropriate bits of the Graded-L word.

The grading process may, for example, be organized so that the Branch Driver reads a word indicating which crates require attention, or which actions (such as program interrupts or autonomous transfers) are required. If the Graded-L requests from the Branch are arranged in priority order in the word, it is recommended, for uniformity, that a request on line $BRW(n+1)$ should have priority over a request on $BRW(n)$.

Crate Controller Type A-1 provides an additional means of access to the LAM information (see Section A1.9.4 and Table 9).

4.5 Common Controls
4.5.1 *Branch Initialize BZ*

> The Branch Initialize signal BZ is generated by the Branch Driver and has absolute priority over all other signals in the Branch. The normal Branch timing signals are not used with BZ. In order to allow crate controllers to discriminate against spurious signals of short duration the Branch Driver must maintain $BZ = 1$ for a minimum of $10\ \mu s$. It must not generate a Graded-L or command mode operation during the following $5\ \mu s$ period.

4.5.2 *Dataway Initialize Z, Clear C, and Inhibit I*

> A crate controller receiving a Branch Initialize signal whose duration exceeds a minimum value (specified as $3\pm 1\ \mu s$) must initiate the generation of Dataway Initialize Z together with Busy B and Strobe $S2$ as required by ANSI/IEEE Std 583-1982. The generation of $S1$, in addition to the mandatory B and $S2$, is optional and cannot be relied upon by other units connected to the Dataway.
>
> All crate controllers must include some means of generating Dataway Clear C and Inhibit I signals.

There are no Branch Highway lines for the Dataway common control signals Clear and Inhibit. A crate controller should generate Dataway Z and C signals, and generate and remove Dataway I, in response to command mode operations as defined in Table 9.

A crate controller may also generate Dataway common control signals in response to front panel signals, unless this is specifically prohibited (as in the case of Crate Controller Type A-1).

4.6 Reserved Lines $BV6$ and $BV7$ and Free Lines $BV1$-$BV5$

> Signal and return lines $BV1$-$BV7$ are provided at all Branch Highway ports. Where more than one port is provided, as in crate controllers, these lines must be linked across between corresponding contacts (see also Section A1.1). $BV6$ and $BV7$ signal and return lines are reserved for future requirements and must not be used until assignments have been made. (Any future allocation will be in the order $BV7$, $BV6$.)
>
> Signal and return lines $BV1$-$BV5$ are designated as Free Lines for any use. However their use must conform to the requirements of this standard. Hence, signals on these lines must conform to Section 7 (for example, these lines cannot be used for other types of signals or for power supplies).

Any signal that is asynchronous with respect to the Branch operation should be generated from a source that defines the transition time in accordance with Sections 4.3 and 4.4.1. It should be noted that the BV lines are terminated at one end of the Highway (and preferably at both ends) as specified in Section 7.3 and Table 8.

No standard uses are defined for $BV1$-$BV5$, and there may be conflicts between items of equipment using these lines in different ways. Where there are well-established conventions for the use of these lines, conforming to such conventions may limit conflicts.

5. Branch Operations

> All transfers of information (read data and write data, Q, X, and Graded-L) through Branch Highway ports are organized as Branch operations. The timing of each operation is controlled by the Branch timing signals BTA and $BTB1$-$BTB7$ and can be divided into four phases as defined in Tables 3 and 4 and Figs 3 and 4.

Table 3
Sequence of Command Mode Operation

Phase	Action in Branch Driver	Timing Signal Change and Direction	Action in Crate Controller i
1	(1) Establishes branch command (and write data) (2) Compensates for skew	$BTA \;\; 0 \rightarrow 1$	
2	(1) Waits for $BTB_i = 0$ from all addressed crate controllers (2) Compensates for skew (3) Accepts BQ and BX (and read data)	$BTB_i \;\; 0 \leftarrow 1$	(1) Initiates Dataway Operation (2) Establishes BQ and BX (and read data on Branch Highway or write data on Dataway)
3		$BTA \;\; 0 \leftarrow 1$	
4	(1) Waits for $BTB_i = 1$ from all addressed crate controllers (2) Removes command (and write data) or Begins Phase 1 of next operation.	$BTB_i \;\; 0 \rightarrow 1$	(1) Completes Dataway Operation (2) Removes BQ and BX (and read data) from Branch Highway

Branch Operation ←→ Dataway Operation

NOTES:
(1) Throughout the operation $BG = 0$
(2) Actions shown in brackets apply when required by the command.

INTERFACE SYSTEM (CAMAC)

Table 4
Sequence of Graded-L Operation

Phase	Action in Branch Driver	Timing Signal Change and Direction	Action in Crate Controller i
1	(1) Establishes BG and BCR for on-line crates (2) Compensates for skew	BTA: 0→1 ↑	
2	(1) Waits for $BTB_i = 0$ from all addressed crate controllers (2) Compensates for skew (3) Accepts GL information	↓ BTB_i: 0←1	Establishes GL information on Branch Highway
3		BTA: 0←1 ↑	
4	(1) Waits for $BTB_i = 1$ from all addressed crate controllers (2) Removes BG and BCR or Begins Phase 1 of next operation	↓ BTB_i: 0→1	Removes GL information

Branch Operation

NOTES:
(1) Throughout the operation $BG = 1$
(2) Command signals BN, BA, and BF are ignored.

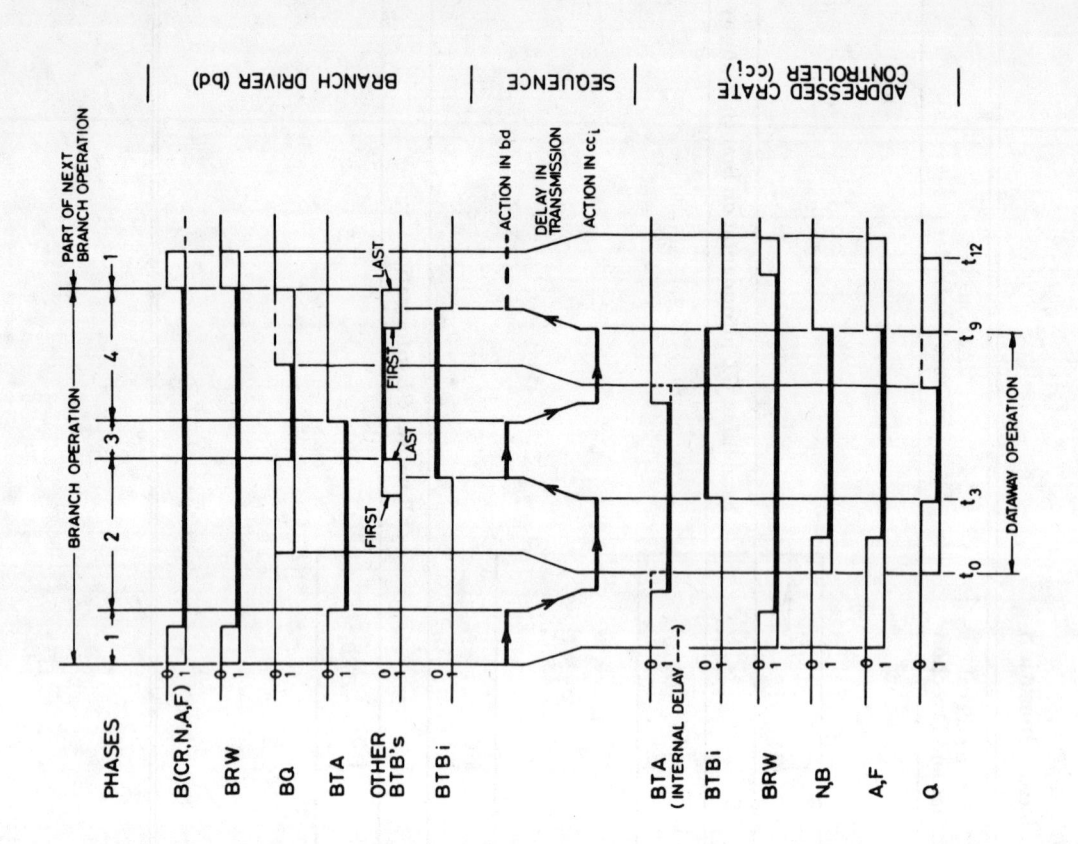

Fig 4
Timing of Branch Write Operation

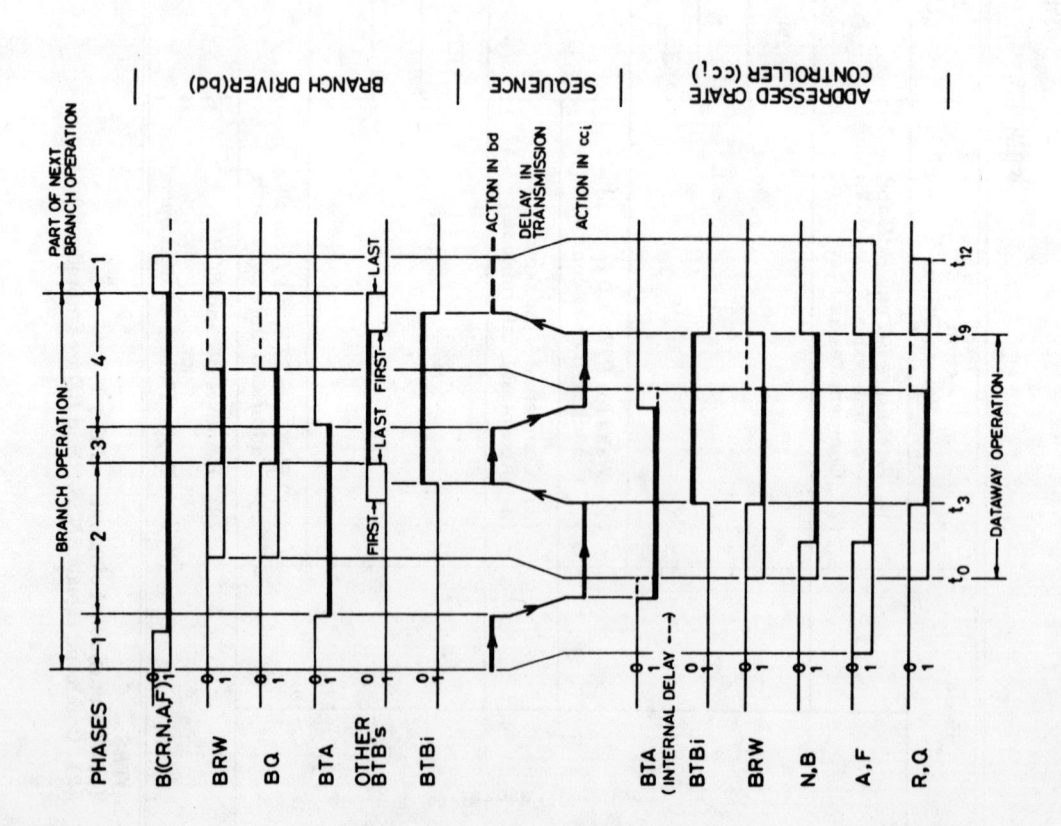

Fig 3
Timing of Branch Read Operation

During *Phase 1* the Branch Driver presents at its port one or more crate addresses either included in a command (together with write data if required by the command), or accompanying a Graded-*L* Request. After a delay which compensates for signal skew it generates $BTA = 1$ to start the next phase.

During *Phase 2* each addressed crate controller responds to $BTA = 1$ either by initiating the Dataway operation required by the command and presenting Q, X, and any read data at its port, or by presenting Graded-*L* information. It then generates $BTB_i = 0$ on its individual BTB line. The Branch Driver starts the next phase when it has received $BTB_i = 0$ from all addressed crate controllers.

During *Phase 3* the Branch Driver introduces a delay to compensate for signal skew and then accepts Q, X, and read data, or the Graded-*L* information. It generates $BTA = 0$ to start the next phase.

During *Phase 4* each addressed crate controller responds to $BTA = 0$ either by completing the Dataway operation and removing Q, X, and read data presented at its port, or by removing the Graded-*L* information. It then generates $BTB_i = 1$ on its individual BTB line.

The Branch Driver ends Phase 4 when it has received $BTB_i = 1$ from all addressed crate controllers, and is then free to begin another Branch operation, either immediately (in which case new command, write data, or Graded-*L* Request signals are set up) or later (in which case the existing signals are removed).

The BTB lines corresponding to off-line or absent crate controllers remain in the 0 state throughout the operation, and those corresponding to unaddressed on-line crates remain in the 1 state.

The timing of the four phases is automatically adjusted by the sequence of timing signals to suit the actual signal delays occurring in the Highway and the response times of crate controllers, etc.

The timing sequence for command mode operations is described in detail in Section 5.1. Graded-*L* operations are described in Section 5.2.

In practice the various Branch, command, and data signals are unlikely to have precisely the same transmission delay, and this problem of signal skew is discussed in Section 5.3.

Branch operations will not be completed if the Branch Driver or addressed crate controllers fail to respond to the timing signals in the correct sequence. Branch Drivers should therefore include some form of time-out feature to detect when an operation has not been completed within a reasonable time, so that appropriate action can be taken. Precautions against operations that would otherwise fail due to addressing absent or off-line crate controllers may be based on the means of recognizing these crates described in Section 5.4.

> The relationship between the Branch Highway operation and the Dataway operation in addressed crates must satisfy the requirements of Table 3 and ANSI/IEEE Std 583-1982.

The relative timing of the Dataway strobes $S1$ and $S2$ and the Branch timing signals BTA and BTB is specified in detail for Crate Controller Type A-1 (see Section A1.7). In other crate controllers the relative timing will depend, for example, on whether or not there are registers for data and command.

5.1 Command Mode Operations. The sequence during a command mode operation is shown in Table 3.

The following sections detail the four phases of a read operation and then outline the minor respects in which other operations differ. One or more crates may be addressed in any operation.

5.1.1 *Read Operations: Phase 1.* The sequence during a read operation (Function codes $0 - 7$) is illustrated in Fig 3.

Phase 1 involves actions in the Branch Driver, which presents the complete command [BCR, BN, BA, $BF(0$ to $7)$] at its port and then, after a delay to compensate for skew (see Section 5.3), generates $BTA = 1$ to initiate Phase 2.

5.1.2 *Read Operations: Phase 2.* After the transmission delay and signal transition time of the Branch Highway, each crate controller receives the command signals and then, when they are stable, the timing signal $BTA = 1$. Phase 2 involves actions in all addressed crate controllers.

Each addressed crate controller ($BCR_i = 1$) responds to $BTA = 1$ by beginning the timing sequence for a Dataway operation. In Fig 3 the

Dataway operation is initiated by *BTA* after conditioning by integration, as recommended in Section 4.3. At time t_0 of this operation (see Fig 9 of ANSI/IEEE Std 583-1982) the Dataway Busy *B* and command signals must be generated. It is recommended that *B* and the *N* signals (derived by decoding the *BN* signals) should be generated when the crate controller has received *BTA* = 1, although the *A* and *F* signals (reproduced from the corresponding *BA* and *BF* signals) may be generated earlier (see Figs 3 and 4).

The addressed module responds to the command by transmitting *Q*, *X*, and read data which are established on the Dataway at time t_3 (see Fig 9 of ANSI/IEEE Std 583-1982). These signals are reproduced by the crate controller on the *BRW*, *BQ*, and *BX* lines at its Branch Highway port, and are maintained during Phase 3. (If the command addresses a register in the crate controller, the read data and *Q* information need not be transferred via the Dataway.) When the controller has presented these *BRW*, *BQ*, and *BX* signals, it generates BTB_i = 0.

The Branch Driver initiates Phase 3 at some later time when it has received *BTB* = 0 from all addressed crates. Fig 3 shows BTB_i = 0 from a particular crate, and also earlier and later *BTB* signals from other addressed crates. The Branch Driver waits for the last *BTB* signal. For example, it may detect the condition:

$$(\overline{BCR1} + \overline{BTB1}) \cdot (\overline{BCR2} + \overline{BTB2}) \cdots$$
$$(\overline{BCR7} + \overline{BTB7}) = 1$$

For each unaddressed crate $\overline{BCR_i}$ = 1, and therefore the state of $\overline{BTB_i}$ is ignored. For each addressed crate $\overline{BCR_i}$ = 0, and therefore the condition is satisfied only when $\overline{BTB_i}$ = 1.

5.1.3 *Read Operations: Phase 3.* During Phase 3 the Branch Driver introduces a delay to allow for signal skew and then takes whatever action is necessary to accept the information from the *BRW*, *BQ*, and *BX* lines. When it has accepted this information it generates *BTA* = 0 to initiate Phase 4.

5.1.4 *Read Operations: Phase 4.* Each addressed crate controller receives *BTA* = 0 at some later time and is then free to change its signal outputs to the *BRW*, *BQ*, and *BX* lines. During Phase 4 the crate controller takes any further action necessary to complete its Dataway operation.

This may result in the read data and *Q* signals changing (shown by broken lines in Fig 3), due to actions in addressed modules in response to Strobe *S*2.

At the end of the Dataway operation t_9 the crate controller removes the Dataway *B* and *N* signals. It also removes any 1 state outputs to the *BRW*, *BQ*, and *BX* lines. It may do this immediately after the end of the Dataway operation at t_9 (as shown in Fig 3) if it has gates between the Dataway and Branch Highway lines. This is mandatory for Crate Controller Type A-1. Alternatively, it may remove the *BRW* and *BQ* signals within 400 ns of the end of the Dataway operation at t_{12} by relying on the addressed modules removing their outputs to the *R* and *Q* lines when they receive *N* = 0.

In either case the crate controller generates BTB_i = 1 when it has removed all 1 state outputs to the branch *BRW*, *BQ*, and *BX* lines, and the Dataway *B* and *N* lines.

The Branch Driver ends Phase 4 at some later time when it has received BTB_i = 1 from all addressed crates. For example it may detect the condition:

$$(\overline{BCR1} + BTB1) \cdot (\overline{BCR2} + BTB2) \cdots$$
$$(\overline{BCR7} + BTB7) = 1$$

For each unaddressed crate $\overline{BCR_i}$ = 1, and therefore the state of BTB_i is ignored. For each addressed crate $\overline{BCR_i}$ = 0, and therefore the condition is satisfied only when BTB_i = 1. The Branch Driver is then free to remove the command signals and to begin another command mode or Graded-*L* operation. The extreme case shown in Fig 3, is when Phase 1 of the next operation follows immediately, so that the Branch Driver removes the command signals of one operation while setting up the command or Graded-*L* Request signals for the next.

5.1.5 *Write Operations.* The sequence during a write operation (Function codes 16 − 23) is shown in Fig 4. The sequence is similar to that for a read operation (described previously), except that write data signals are generated by the Branch Driver for the same period as command signals. The signal BTB_i = 1 from the crate controller during Phase 4 has the additional significance that the write data has been accepted.

5.1.6 *Other Command Operations.* Operations with Function codes 8 − 15 and 24 − 31,

INTERFACE SYSTEM (CAMAC)

which do not use the Read or Write lines of the Dataway, nevertheless use the Dataway Q and Branch BQ lines. Their timing is therefore similar to that of read operations as described previously. The Dataway Q signal is allowed to change during these operations (see ANSI/IEEE Std 583-1982, Section 5.4.3), hence the BQ signal may also change at any time.

5.2 Graded-L Operations.
The Graded-L operation is equivalent to a multicrate read operation in which the normal command is replaced by the Graded-L Request signal ($BG = 1$) and crate address signals to all on-line crate controllers. The station number, subaddress, and function signals are not used, and are ignored by crate controllers during this operation. It is typical, but not essential, that Graded-L operations are initiated by the Branch Demand signal, $BD = 1$.

The sequence during a Graded-L operation is shown in Table 4.

> During the Graded-L operation the Branch Driver generates a set of BCR signals which must have $BCR_i = 0$ on all lines corresponding to absent or off-line crate controllers and $BCR_i = 1$ on all lines corresponding to on-line crates. The BCR signals are accompanied by $BG = 1$.

It is recommended that the Branch Driver should derive the necessary information about the state of the crate controllers from the BTB lines (see Section 5.4).

When the Branch Driver has presented the BCR signals and the Graded-L Request signal it generates $BTA = 1$. In response to BG, BCR_i, and BTA each on-line crate controller generates its Graded-L word through intrinsic OR outputs to the BRW lines at its Branch Highway port, without generating Dataway signals B, $S1$, or $S2$.

The grading process which forms the Graded-L word need not take place in the crate controller, but may involve another unit, such as the LAM Grader associated with Crate Controller Type A-1 (see Section A1.9). The Dataway L signals are free to change at any time, and hence the BRW signals may also change.

Each addressed crate controller generates $BTB_i = 0$ when it has presented its Graded-L information to the BRW lines. The process of establishing the Graded-L information involves two special causes of delay. First, if the L signal from a module has been gated off the Dataway by a preceding command mode operation, there will be a delay of up to 400 ns before L is re-established at the crate controller. Second, the Crate Controller Type A-1 which is specified in Section A1 of the appendix requires a separate *LAM-Grader* unit for processing L signals. This can involve additional delays in establishing L signals at the LAM-Grader unit, and in establishing the Graded-L signals at the crate controller.

When the Branch Driver has received $BTB_i = 0$ from all addressed crates, it introduces a delay to allow for signal skew and then accepts the Graded-L word from the BRW lines. Having done this it generates $BTA = 0$.

When the crate controller receives $BTA = 0$, it removes the Graded-L information from the BRW lines and generates $BTB_i = 1$. The operation is completed when the Branch Driver receives $BTB_i = 1$ from all addressed crate controllers and is free to remove the Graded-L Request and crate address signals.

5.3 Differential Delays (Skew)

> The delays encountered by the BTA and BTB signals are used to adjust the timing of the Branch operation. However, there may be skew, or differential delays, between BTA and the individual bits of the command and write data signals received at the crate controller, and between BTB and the individual bits of the BRW, BQ, and BX signals received at the Branch Driver. The Branch Driver must introduce an appropriate delay before generating $BTA = 1$, in order to ensure that all command signals are established at crate controllers before they receive BTA. It must also delay its internal action in response to $BTB = 0$, in order to ensure that all data, BQ and BX signals have become established.

This correction for skew may be either fixed, to cover a stated maximum skew, or adjustable to suit the specific application. Additional compensation for skew is permitted elsewhere in the Branch.

5.4 Identification of On-Line Crate Controllers.
During the period between the end of Phase 4 of one Branch operation and the beginning of Phase 2 of the next operation, the

Branch Driver receives $BTB_i = 1$ from on-line crate controllers and $BTB_i = 0$ from off-line or absent crate controllers. The state of the BTB lines may therefore be sampled by the Branch Driver immediately before any operation in order to identify the on-line crate controllers.

It is strongly recommended that the Branch Driver should identify the on-line crate controllers in this way immediately before each Graded-L operation in order to fulfill the mandatory requirement of Section 5.2 that all on-line crate controllers are addressed. Hence, the Branch Driver should generate $BCR_i = 1$ if $BTB_i = 1$.

The Branch Driver may also identify the on-line crate controllers before all command mode operations and compare them with the crate addresses specified in the command. This allows prompt detection of operations that would otherwise fail through addressing off-line or absent crate controllers, and avoids the much slower process of relying on a time-out feature which operates after the operation has failed (see Section 5).

A further application of this method of identifying on-line crate controllers would be to ensure that a crate controller cannot come on-line if there is already an on-line controller with the same address (see Section 4.1.1). Each crate controller could check that the condition $(BTB_i + BCR_i) = 0$ is satisfied before switching to the on-line state. It would remain off-line if there is already an on-line controller with the same address, either in the unaddressed state ($BTB_i = 1$) or in the addressed state ($BCR_i = 1$).

6. Connectors

The Branch Highway ports use the 132-contact connectors defined in Table 5, or equivalent types fully mateable with those in Table 5. The fixed member used on the Branch Driver, crate controller, and termination unit has 132 sockets. The free member used on cables has 132 pins.

The contact layout and outline dimensions of the fixed and free members are given (for information only) in Figs 5 and 6.

The assignment of the signal and return lines is defined in Table 6, arranged by signals, and in Table 7, arranged by contact numbers.

At least two fixed connectors must be mounted at the front of each crate controller, with all corresponding contacts joined to provide a continuous path through the controller. The correct orientation of these connectors is important. Contact 1 must be uppermost on the top connector and lowermost on the bottom connector (see Fig 5).

Branch Drivers must have at least one fixed connector. If they do not contain the terminations of the signal lines (see Section 7.3) they must have at least two fixed connectors.

Table 5
Standard Connector for Branch Highway Ports*

Original Manufacturer	Hughes Aircraft Company
Connector Type	WSS Subminiature Rectangular Connector
Number of Contacts	132
Polarizing Code	BN
Catalog Numbers	
Fixed member (socket moulding)	WSS 0132 SOO BN 000
Free member (pin moulding)	WSS 0132 Pxx BN yyy
	where Pxx yyy denotes type of jackscrew
Hood for free member	WAC 0132 H005 (for example)

*Equivalent connectors that are fully mateable with those listed here may be used.

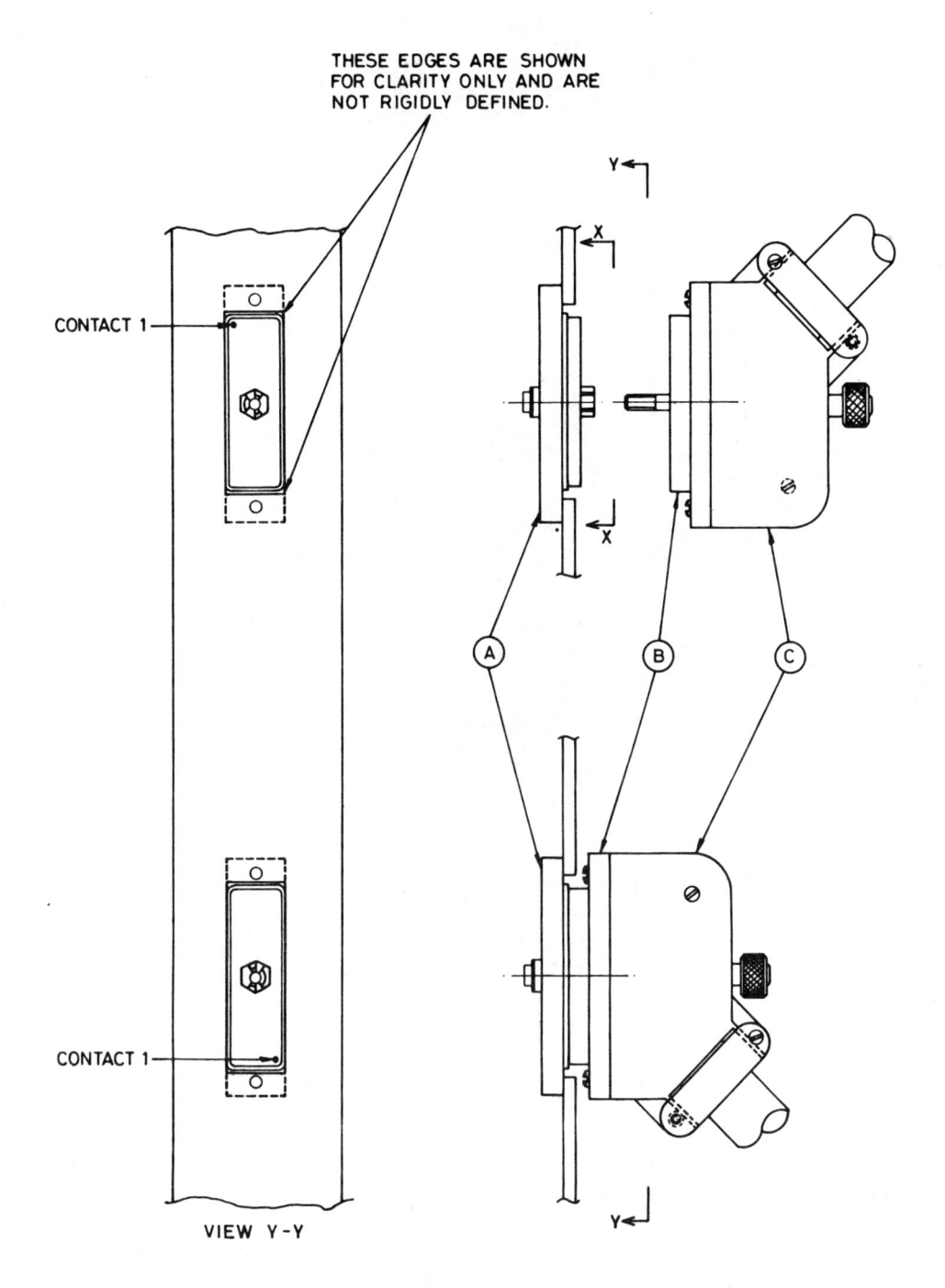

INTERFACE SYSTEM (CAMAC)

THESE EDGES ARE SHOWN FOR CLARITY ONLY AND ARE NOT RIGIDLY DEFINED.

VIEW Y-Y

NOTES:
(1) View *X-X* shown in Fig 6
(2) For information only (see Section 6).

A Fixed Connector—Socket body without jackscrew
B Free Connector—Pin body with jackscrew
C Hood—Recommended assembly with cable entry adjacent to Contact 1

**Fig 5
Branch Highway Ports: Arrangement of Connectors on Crate Controllers**

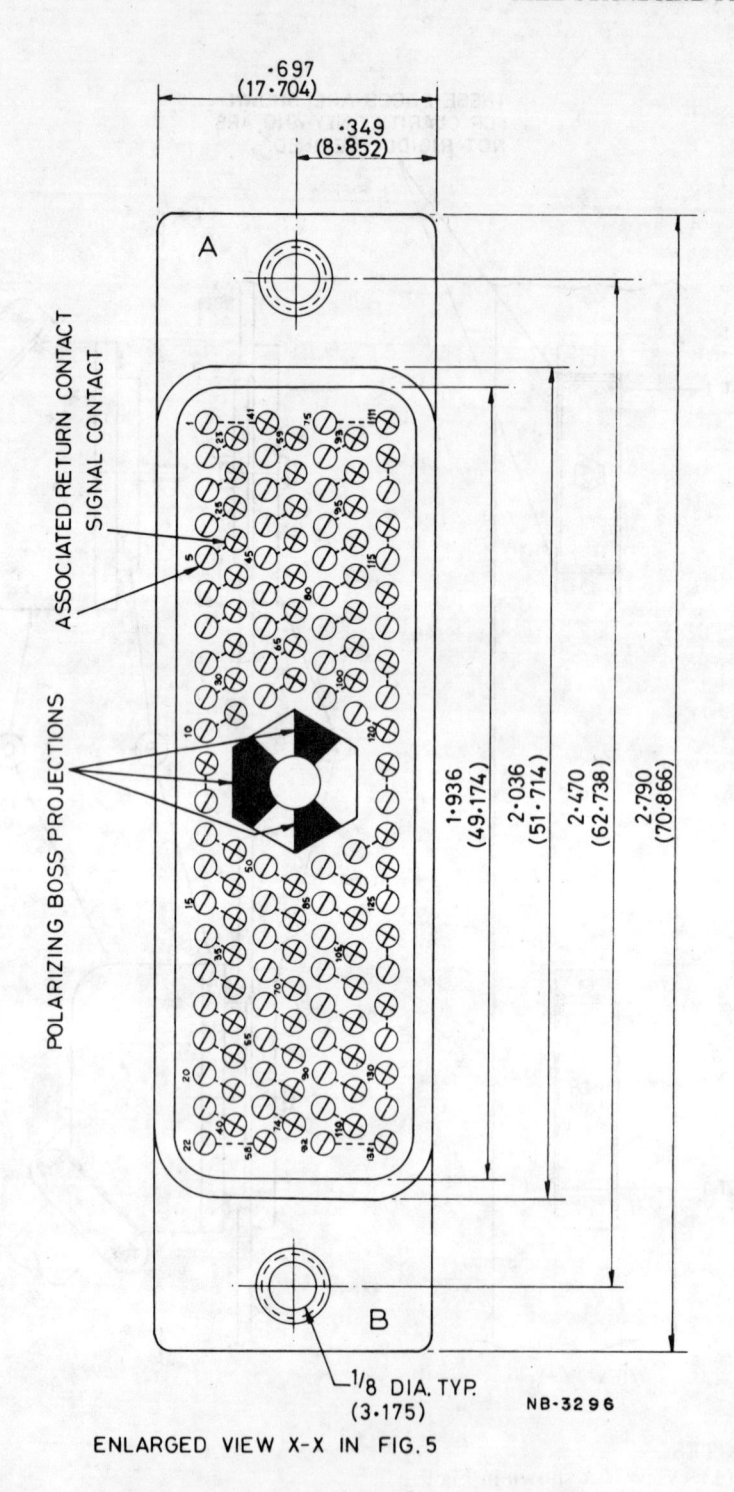

**Fig 6
Branch Highway Ports: Contact Layout
(Front View of Fixed Connector)**

NOTES:
(1) Dimensions in inches. Approximate millimeter equivalents in brackets
(2) For information only (see Section 6).

Table 6
Contact Assignments at Branch Highway Ports: By Signals

Signal Contact	Return Contact	Signal		Signal Contact	Return Contact	Signal	
32	13	BCR1	Crate Address	93	76	BRW1	Read/Write Lines
33	14	BCR2	Crate Address	94	77	BRW2	Read/Write Lines
34	15	BCR3	Crate Address	95	78	BRW3	Read/Write Lines
35	16	BCR4	Crate Address	96	79	BRW4	Read/Write Lines
67	50	BCR5	Crate Address	97	80	BRW5	Read/Write Lines
68	51	BCR6	Crate Address	98	81	BRW6	Read/Write Lines
69	52	BCR7	Crate Address	99	82	BRW7	Read/Write Lines
36	17	BN1	Station Address	100	83	BRW8	Read/Write Lines
37	18	BN2	Station Address	103	84	BRW9	Read/Write Lines
38	19	BN4	Station Address	104	85	BRW10	Read/Write Lines
39	20	BN8	Station Address	105	86	BRW11	Read/Write Lines
40	21	BN16	Station Address	106	87	BRW12	Read/Write Lines
41	1	BA1	Subaddress	107	88	BRW13	Read/Write Lines
23	2	BA2	Subaddress	108	89	BRW14	Read/Write Lines
24	3	BA4	Subaddress	109	90	BRW15	Read/Write Lines
25	4	BA8	Subaddress	110	91	BRW16	Read/Write Lines
				112	113	BRW17	Read/Write Lines
70	53	BF1	Function Code	114	115	BRW18	Read/Write Lines
71	54	BF2	Function Code	116	117	BRW19	Read/Write Lines
72	55	BF4	Function Code	118	119	BRW20	Read/Write Lines
73	56	BF8	Function Code	124	125	BRW21	Read/Write Lines
74	57	BF16	Function Code	126	127	BRW22	Read/Write Lines
				128	129	BRW23	Read/Write Lines
61	44	BQ	Response	130	131	BRW24	Read/Write Lines
63	46	BTA	Timing	26	5	BV1	Free Lines
31	10	BTB1	Timing	27	6	BV2	Free Lines
11	12	BTB2	Timing	28	7	BV3	Free Lines
58	22	BTB3	Timing	29	8	BV4	Free Lines
132	92	BTB4	Timing	30	9	BV5	Free Lines
123	102	BTB5	Timing	64	47	BV6	Reserved Lines
120	101	BTB6	Timing	65	48	BV7	Reserved Lines
121	122	BTB7	Timing	66	49	BX	Command Accepted
60	43	BD	Demand	111	75	BSC	Cable Shield
59	42	BG	Graded-L Request				
62	45	BZ	Initialize				

Extra connectors may be provided on Branch Drivers and crate controllers, unless this is specifically prohibited (as in the case of Crate Controller Type A-1).

6.1 Connection to Shield of Branch Highway Cable. The contacts designated BSC and BSC(R) are available for making a connection through the Branch Highway port to the shield, if any, of the Branch Highway cable. These two contacts are normally used in parallel, and do not carry Branch Highway signals.

Units that terminate the Branch Highway signal lines (see Section 7.3) must connect BSC and BSC(R) to ground. All other units must provide the option of connecting these contacts to ground.

7. Signal Standards at Branch Highway Ports

All units connected to the Branch Highway must conform to the absolute limits of the signal standards at the Branch Highway ports, as specified in Table 8.

ANSI/IEEE
Std 596-1982

IEEE STANDARD PARALLEL HIGHWAY

Table 7
Contact Assignments at Branch Highway Ports: By Contact Numbers

Contact	Signal	Contact	Signal	Contact	Signal
1	$BA1(R)$	34	$BCR3$	67	$BCR5$
2	$BA2(R)$	35	$BCR4$	68	$BCR6$
3	$BA4(R)$	36	$BN1$	69	$BCR7$
4	$BA8(R)$	37	$BN2$	70	$BF1$
5	$BV1(R)$	38	$BN4$	71	$BF2$
6	$BV2(R)$	39	$BN8$	72	$BF4$
7	$BV3(R)$	40	$BN16$	73	$BF8$
8	$BV4(R)$	41	$BA1$	74	$BF16$
9	$BV5(R)$	42	$BG(R)$	75	$BSC(R)$
10	$BTB1(R)$	43	$BD(R)$	76	$BRW1(R)$
11	$BTB2$	44	$BQ(R)$	77	$BRW2(R)$
12	$BTB2(R)$	45	$BZ(R)$	78	$BRW3(R)$
13	$BCR1(R)$	46	$BTA(R)$	79	$BRW4(R)$
14	$BCR2(R)$	47	$BV6(R)$	80	$BRW5(R)$
15	$BCR3(R)$	48	$BV7(R)$	81	$BRW6(R)$
16	$BCR4(R)$	49	$BX(R)$	82	$BRW7(R)$
17	$BN1(R)$	50	$BCR5(R)$	83	$BRW8(R)$
18	$BN2(R)$	51	$BCR6(R)$	84	$BRW9(R)$
19	$BN4(R)$	52	$BCR7(R)$	85	$BRW10(R)$
20	$BN8(R)$	53	$BF1(R)$	86	$BRW11(R)$
21	$BN16(R)$	54	$BF2(R)$	87	$BRW12(R)$
22	$BTB3(R)$	55	$BF4(R)$	88	$BRW13(R)$
23	$BA2$	56	$BF8(R)$	89	$BRW14(R)$
24	$BA4$	57	$BF16(R)$	90	$BRW15(R)$
25	$BA8$	58	$BTB3$	91	$BRW16(R)$
26	$BV1$	59	BG	92	$BTB4(R)$
27	$BV2$	60	BD	93	$BRW1$
28	$BV3$	61	BQ	94	$BRW2$
29	$BV4$	62	BZ	95	$BRW3$
30	$BV5$	63	BTA	96	$BRW4$
31	$BTB1$	64	$BV6$	97	$BRW5$
32	$BCR1$	65	$BV7$	98	$BRW6$
33	$BCR2$	66	BX	99	$BRW7$
				100	$BRW8$
				101	$BTB6(R)$
				102	$BTB5(R)$
				103	$BRW9$
				104	$BRW10$
				105	$BRW11$
				106	$BRW12$
				107	$BRW13$
				108	$BRW14$
				109	$BRW15$
				110	$BRW16$
				111	BSC
				112	$BRW17$
				113	$BRW17(R)$
				114	$BRW18$
				115	$BRW18(R)$
				116	$BRW19$
				117	$BRW19(R)$
				118	$BRW20$
				119	$BRW20(R)$
				120	$BTB6$
				121	$BTB7$
				122	$BTB7(R)$
				123	$BTB5$
				124	$BRW21$
				125	$BRW21(R)$
				126	$BRW22$
				127	$BRW22(R)$
				128	$BRW23$
				129	$BRW23(R)$
				130	$BRW24$
				131	$BRW24(R)$
				132	$BTB4$

NOTE: $BRW1(R)$ is the return line corresponding to $BRW1$, etc.

26

Table 8
Signal Standards at Branch Highway Ports

Condition At Branch Highway ports	Logic State	Absolute Limits	Recommended Values
Inputs			
(1) Voltage range accepted by unit	0 1	+2.4 V to +5.5 V 0 V to +1.2 V*	
(2) Maximum current supplied by unit (see Section 7.1)	0 1	±0.3 mA +1.6 mA (±0.3 mA for Crate Controller Type A—1)	±0.3 mA†,††
Outputs			
(3) Voltage range generated by unit	1	0 V to +0.5 V	0 V to +0.3 V
(4) Minimum current sinking capability**	1	127 mA	133 mA
Termination			
(5) Open-circuit voltage	0	+4.5 V maximum	+4.1 V preferred††
(6) Short-circuit current	1	50 mA maximum	
(7) Terminating impedance			100 Ω preferred††
Branch Highway			
(8) Characteristic impedance		70 Ω minimum	100 Ω maximum††

*Higher than Transistor Transistor Logic voltage (TTL) levels provide an increased noise margin taking into account cable losses and reflections due to mismatches.

†Low input currents result in smaller reflections. Receivers with high input impedance may feed current into the line or draw current from the line.

**The current sinking capability is given by

$$\frac{V_0 - V_{\text{out low}}}{Z/2} + 8 \cdot I_{\text{in low}} = \begin{cases} 127 \text{ mA absolute minimum} \\ 133 \text{ mA recommended minimum} \end{cases}$$

where V_0 = 4.5 V maximum open-circuit voltage

$$V_{\text{out low}} = \begin{cases} 0.5 \text{ V absolute} \\ 0.3 \text{ V recommended} \end{cases} \text{maximum low state output voltage}$$

Z = 70 Ω minimum characteristic impedance

$I_{\text{in low}}$ = 1.6 mA maximum low state input current.

††Recommended values refer to a set of design values for a preferred terminating circuit.

In addition, Table 8 gives recommended values for certain characteristics. The recommended value for input current, to closer limits than the absolute value, leads to a set of design values for a preferred terminating circuit.

The signal standards assume that the Branch Highway presents, at all ports, conditions equivalent to a twisted-pair cable with a characteristic impedance of at least 70 Ω (see Table 8, Part 8).

A unit behaves with respect to a particular line as either an input (accepting signals from the Highway) or an output (generating signals on the Highway) or a termination (biasing the signal lines to the 0 state and terminating them with approximately the characteristic impedance). In some cases a unit may perform several of these roles. For example, the *BRW* lines have inputs and outputs at crate controllers and Branch Drivers, and may also have terminations in Branch Drivers. Such units must satisfy the parts of Table 8 that are appropriate to each particular role.

> Any capacitive load imposed on the signal lines by shaped outputs and integrated inputs (see Sections 4.3, 4.4.1, and A1.7) must be small compared with the characteristic impedance of the Highway, taking into account the transition time of the signals.

7.1 Inputs

> All inputs that receive signals from the Branch Highway ports must accept the voltage ranges specified in Table 8 at Part (1) and must not impose current loadings greater than those specified in Table 8 at Part (2). The specified input current loading refers to the total current supplied to any signal line at a Branch Highway port by a unit that is receiving signals from the line, including the effect of any output circuits connected to the same line. A maximum of eight units is allowed on each signal line.

The absolute value for current loading corresponds to typical Transistor Transistor Logic (TTL) devices, but a lower value is recommended for all units, and is mandatory for Crate Controller Type A-1 (see Section A1.3).

7.2 Outputs

> All outputs that transmit signals through Branch Highway ports must be sources that allow wired-OR connections. In the 1 state the sources must produce signals within the voltage range specified in Table 8 at Part (3) and have the current sinking capability specified in Table 8 at Part (4), in order to drive eight inputs (see Section 7.1) and two terminations (see Section 7.3) under dynamic conditions. The *BD*, *BTA*, and *BTB* signals must be generated from sources that define the transition times (see Sections 4.3 and 4.4.1). The generation of other signals with defined transition times is permitted.

If the Branch Driver includes one termination for the Highway, its current sinking capability at the appropriate lines of the port may be reduced accordingly.

7.3 Terminations

> All 65 signal lines must be terminated at one end of the Branch Highway with a circuit providing the appropriate "pull-up" current to bias the line to the 0 state, and the appropriate terminating impedance to limit signal reflections. All return lines and the connections to the cable shield must be connected to ground at this point. The current from each termination circuit into the Branch Highway line in the logic 1 state must not exceed the short-circuit current specified in Table 8 at Part (6).

It is strongly recommended that all 65 signal lines be terminated at both ends of the Branch Highway. It is suggested that there be a termination unit that can be used at either end of the Highway by connecting it to the second Branch Highway port of the last crate controller or to the second port of the Branch Driver if this does not have internal terminations and is at the end of the Branch.

> If such a termination unit is provided, it must terminate all 65 signal lines and ground the return lines and connections to the cable shield.

If all inputs connected to the Branch Highway impose the lower current loading recommended in Table 8 at Part (2), and the Highway has a characteristic impedance between 70 and 100 Ω, the terminating circuits should be designed to have the target values for impedance and open circuit voltage given in Table 8 at Parts (7) and (5) in order to achieve optimum speed and noise margin. If inputs have the higher absolute value for current loading, it will be necessary to design the terminating circuits for the appropriate compromise between speed and noise margin to suit the particular application.

INTERFACE SYSTEM (CAMAC)

7.4 Off-Line and Power-Off Conditions

> A crate controller must not generate 1 state outputs at its Branch Highway ports when in the off-line state and receiving normal power supplies.

It is recommended that a crate controller should not interfere with the operation of the Branch when in the off-line state without power supplies. This, which applies to all inputs and outputs through the ports, is in order to allow power to be switched off (for example, for maintenance and changing modules) without disturbing the remainder of a system.

Appendix

[This Appendix is not a part of ANSI/IEEE Std 596-1982, Parallel Highway Interface System (CAMAC).]

A1. Specification of CAMAC Crate Controller Type A-1

A1.1 CAMAC Crate Controller Type A-1

> In order to conform with the specification for CAMAC Crate Controller Type A-1, a crate controller must have all the mandatory features defined in this appendix. It must have no other features that would affect its interchangeability with any other Crate Controller Type A-1, taking into account the effect of such interchange on both hardware and software. It must be fully interchangeable with one conforming to Fig 7, although it need not have identical structure, internal signals (shown without the prefix *B* in Fig 7) or logical expressions.

A1.2 Other Crate Controllers.
It is recommended that other crate controllers should be interchangeable with Crate Controller Type A-1 in respect to those features that they have in common, although they need not have all the mandatory features of Crate Controller Type A-1 and may have additional features.

A1.3 General Features

> The crate controller must conform fully with the mandatory requirements of ANSI/IEEE Std 583-1982, and of ANSI/IEEE Std 596-1982 (Sections 1 – 7 of this standard). It is mandatory that all signal inputs at the Branch Highway ports of Crate Controller Type A-1 must satisfy the lower input current standard (± 0.3 mA) shown in Table 8.
>
> Crate Controller Type A-1 must not occupy more than three stations. It should preferably be a double-width unit which engages with the Dataway at the control station and the adjacent normal station.
>
> In addition to the two front panel connectors for Branch Highway ports (see Section A1.4), the crate controller must have a rear-mounted connector for a link to an optional separate LAM-Grader unit (see Section A1.9).

A1.4 Front Panel

The crate controller must have all the following front panel features, and no others that would affect interchangeability (for example, the addition of indicators for test purposes is permitted).

(1) There must be two connectors for Branch Highway ports, as defined in Section 6 of ANSI/IEEE Std 596-1982, mounted with the correct orientation and with all corresponding contacts joined.

(2) There must be a means of indicating the selected crate address (1 — 7). There must not be easy access at or through the front panel to the means of changing the crate address.

(3) There must be a means of selecting off-line status of the crate controller (see Section A1.10).

(4) There must be a coaxial connector for the Inhibit signal input. The type of connector and the signal standard are specified in ANSI/IEEE Std 583-1982, Sections 4.2.5 and 7.2.1, respectively.

(5) There must be two push buttons, or equivalent manual controls, for Initialize and Clear. These are only effective in the off-line state, and the front panel layout or markings should indicate this.

A1.5 Dataway Signals

A1.5.1 *Data Signals*

When the crate controller is on line and addressed during a read command operation with a station number other than $N(30)$, it must retransmit the signals from the twenty-four Dataway Read lines through intrinsic OR outputs to the BRW lines. Crate Controller Type A-1 must have gates between the R and BRW lines so that this transfer of read data occurs only when the crate controller is addressed and on line, for example when $BCR_i \cdot (BTA + \overline{BTB_i}) = 1$. During write operations with station number other than $N(30)$ it must retransmit the signals from the twenty-four BRW lines to the Dataway Write lines.

It is recommended that all crate controllers should include gates between the R and BRW lines, and between the BRW and W lines, so that these transfers of data are only effective when the crate controller is addressed and on line. These gates may further limit the transfers to read operations $\overline{BF16} \cdot \overline{BF8} = 1$ and write operations $BF16 \cdot \overline{BF8} = 1$, respectively. However, the crate controller is permitted to generate signals on the Dataway Write lines during any operation, but other units connected to the Dataway can only rely on the presence of such signals during Dataway write operations.

A1.5.2 *Command Signals.* The Branch Highway command signals BN, BA, and BF should be conditioned in the crate controller, for example, by integration or by staticizing at a time related to BTA $0 \rightarrow 1$, in order to protect the Dataway command lines from the effects of crosstalk into Branch Highway command lines.

The subaddress and function signals from the BA and BF lines must be retransmitted by the crate controller on the Dataway A and F lines during all command mode operations when the controller is on line and addressed.

In a double-width crate controller each of the Station Number codes $N(1)$ through $N(23)$ must be decoded in the crate controller to produce a signal on the corresponding Dataway line $N1$ to $N23$.

Command operations with $N(26)$ must generate Dataway signals on all the lines $N1$ through $N23$. Command operations with $N(24)$ generate Dataway signals on $N1$ through $N23$ as determined by the contents of a 23 bit Station Number Register (SNR). This register is loaded from $BRW1$-$BRW23$ by the command $N(30) \cdot A(8) \cdot F(16)$. The bit of the Station Number Register that is loaded from $BRW1$ controls the state of $N1$, etc. The register is not reset by the Dataway Initialize signal Z.

A triple-width controller may alternatively have a 22 bit Station Number Register, decode $N(1)$ through $N(22)$, and generate signals on Dataway lines $N1$ through $N22$.

INTERFACE SYSTEM (CAMAC)

A1.5.3 *Common Control Signals*

The Dataway Initialize signal Z must be generated in response to the command $N(28) \cdot A(8) \cdot F(26)$ and to the Branch Initialize signal (see Section 4.5.1). It must also be generated in response to the manual Initialize control, but only when the crate controller is in the off-line state.

The Dataway Clear signal C must be generated in response to the command $N(28) \cdot A(9) \cdot F(26)$. It must also be generated in response to the manual Clear control, but only when the crate controller is in the off-line state.

The Dataway Initialize Z and Clear C signals must be generated with the timing specified for unaddressed operations in ANSI/IEEE Std 583-1982, Fig 10. They must be associated with a sequence including B and $S2$ signals, also with the timing specified in ANSI/IEEE Std 583-1982, Fig 10. The sequence is permitted to include $S1$, but this is not mandatory, and other units connected to the Dataway must not rely on the generation of $S1$ with Z and C.

The Dataway Inhibit signal I must be initiated when an on-line crate controller generates Dataway Initialize Z, and must reach a maintained 1 state not later than time t_3 (see ANSI/IEEE Std 583-1982, Fig 10). When some other unit generates Initialize (accompanied by Inhibit), an on-line crate controller must generate Inhibit in response to Dataway Z gated by $S2$. The inhibit signal must also be generated in response to the command $N(30) \cdot A(9) \cdot F(26)$. In all these cases the Inhibit signal must be maintained by the crate controller until reset by the command $N(30) \cdot A(9) \cdot F(24)$. It must also be generated while the front panel Inhibit signal is present.

The command $N(30) \cdot A(9) \cdot F(27)$ must produce a $Q = 1$ response if there is a 1 state signal on the Dataway Inhibit line.

A1.5.4 *Patch Connections*

Crate Controller Type A-1 must not use the patch pins of the Dataway stations that it occupies.

A1.6 Demand Handling

A1.6.1 *Branch Demand*

The Branch Demand signal BD must be derived, subject to the following conditions, from the OR combination of an External Demand signal from contact 48 of the LAM-Grader connector and an Internal Demand signal which is the OR of the 24 GL signals received via the LAM-Grader connector.

The output of the Branch Demand signal to the BD line must be disabled by the command $N(30) \cdot A(10) \cdot F(24)$ or by the Dataway Initialize signal Z with $S2$. It must be enabled by the command $N(30) \cdot A(10) \cdot F(26)$. The command $N(30) \cdot A(10) \cdot F(27)$ must give a $BQ = 1$ response if the output of BD is enabled. The command $N(30) \cdot A(11) \cdot F(27)$ must give a $BQ = 1$ response if the OR of the Internal and External Demands is in the 1 state, even if the output of BD is disabled.

The Internal Demand signal must be inhibited by the 1 state of the Inhibit Internal D signal from contact 51 of the LAM-Grader connector.

A1.6.2 *Graded L*

In response to a Graded-L Request signal $BG = 1$, accompanied by $BCR_i = 1$, the crate controller must generate the Graded-L operation signal on contact 1 of the LAM-Grader connector. It must accept the Graded-L signals $GL1$–$GL24$ from the LAM-Grader connector and transmit them to the BRW lines ($GL1$ to $BRW1$, etc).

The crate controller must also accept the Graded-L signals from the LAM-Grader connector and transmit them to the BRW lines in response to command mode operations with $N(30) \cdot A(0-7) \cdot F(0)$ (see Section A1.9.4).

In both cases the GL information must be transferred from the LAM-Grader connector to the BRW lines with minimum delay, and the signals must not appear on the Dataway Read lines.

A1.6.3 Pull Up for GL and L Lines

> Pull-up current sources in accordance with ANSI/IEEE Std 583-1982, Table 6, must be provided on all *GL* lines in the crate controller, and must not be provided on the *L* lines, thus allowing a simple LAM Grader to form wired-OR combinations of *L* signals.

A1.7 Timing Requirements.

In command mode operations with Station Number codes other than $N(30)$, the crate controller generates Dataway Strobe signals $S1$ and $S2$, with timing related to that of the Branch timing signals *BTA* and *BTB* as defined in Section, A1.7.1.

Command operations with Station Number code $N(30)$ do not generate $S1$, $S2$, or B signals on the Dataway lines (see Section A1.7.3).

In Graded-*L* operations there are no Dataway Strobe or B signals, and the timing must take into account the signal delays in any non-Dataway connections to a LAM-Grader unit. These timing requirements are defined in Sections A1.7.2 and A1.9.3.

> The internal timing generator of the crate controller must be protected against spurious signals on the *BTA* and *BCR* lines.

One method of protection, shown in Fig 7, is to condition the incoming signals from the *BTA* line and the selected *BCR* line by integration with a time constant of 100 ± 50 ns. Another method is to condition the internal signal *TA* which controls the timing generator. Transitions of the *BTA* and *BCR* signals are detected by the crate controller after a delay (shown in Fig 3 and 4) due to this protection.

A1.7.1 Command Mode Operations with Dataway S1, S2, and B.

> The following timing conditions must be satisfied when the crate controller responds to a command mode branch operation which requires a Dataway operation with signals $S1$, $S2$, and B. In this section the times t_0, t_3, t_5, etc, refer to the corresponding key points on Fig 9 of ANSI/IEEE Std 583-1982.

> In Phase 2 of the operation, after actions by the Branch Driver during Phase 1, the crate controller detects $BTA = 1$, accompanied by $BG = 0$, $BCR_i = 1$, and the appropriate command signals. It must then initiate the required Dataway N signals and B, thus starting the Dataway operation at t_0.
>
> At t_3, which is 400^{+200}_{-0} ns after t_0, the crate controller must initiate the $0 \rightarrow 1$ transition of Dataway Strobe $S1$, and the Branch timing signal transition BTB_i $1 \rightarrow 0$. At t_5, which is 200^{+100}_{-0} ns after t_3, the $1 \rightarrow 0$ transition of Strobe $S1$ must be initiated.
>
> In Phase 4, the crate controller initiates the $0 \rightarrow 1$ transition of Strobe $S2$ at t_6, which is either when it detects $BTA = 0$ or when the interval t_5-t_6 is 100^{+100}_{-0} ns, if this occurs later.
>
> At t_8, which is 200^{+100}_{-0} ns after t_6, the $1 \rightarrow 0$ transition of $S2$ must be initiated.
>
> At t_9, which is 100^{+100}_{-0} ns after t_8, the crate controller must initiate the $1 \rightarrow 0$ transitions of Dataway signals N and B, and must isolate the Dataway Q and R lines from the Branch Highway BQ and BRW lines respectively. It must then initiate the Branch timing signal transition BTB_i $0 \rightarrow 1$.

A1.7.2 Graded-L Operations

> The crate controller must satisfy the following timing conditions during Graded-*L* operations with $BG = 1$ and $BCR_i = 1$. In Phase 2 it must initiate Branch timing signal transition BTB_i $1 \rightarrow 0$ within 400^{+200}_{-0} ns after detecting $BTA = 1$. At the same time it must be presenting to its BRW outputs the GL information received via the LAM-Grader connector (see Section A1.9.3). In Phase 4 it must remove the GL information from its BRW outputs with minimum delay after detecting $BTA = 0$, and initiate the signal transition BTB_i $0 \rightarrow 1$.

A1.7.3 Command Mode Operations Without Dataway S1, S2, or B

> Command mode operations addressed to $N(30)$ are concerned with internal features of the crate controller and with reading Graded-*L* information via the LAM-Grader

Table 9
Commands Implemented by CAMAC Crate Controller Type A—1

Action	N	A	F	Response
Generate Dataway Z	28	8	26	$BQ = 0$
Generate Dataway C	28	9	26	$BQ = 0$
Read GL	30	0 — 7	0	$BQ = 1$
Load SNR	30	8	16	$BQ = 1$
Remove Dataway I	30	9	24	$BQ = 0$
Set Dataway I	30	9	26	$BQ = 0$
Test Dataway I	30	9	27	$BQ = 1$ if $I = 1$
Disable BD output	30	10	24	$BQ = 0$
Enable BD output	30	10	26	$BQ = 0$
Test BD output enabled	30	10	27	$BQ = 1$ if BD enabled
Test demands present	30	11	27	$BQ = 1$ if demands present

connector. The crate controller must not generate signals on the Dataway S1, S2, B, or R lines.

The timing of these operations must follow the requirements for command mode operations (see Section A1.7.1) with the exception that the S1, S2, and B signals are not generated on the Dataway lines, although there may be equivalent internal signals.

A1.8 Commands Implemented by Crate Controller Type A-1

Crate Controller Type A-1 must recognize and implement the commands summarized in Table 9, and must not use any other commands. When addressed with any of these commands it must generate $BX = 1$. The five Function codes $F(0, 16, 24, 26, 27)$ must be fully decoded in the crate controller.

The crate controller must generate $BQ = 1$ in response to all commands that read from or write to its registers, or the LAM-Grader connector. In Crate Controller Type A-1 the two commands to which this applies are $N(30) \cdot A(0 - 7) \cdot F(0)$ and $N(30) \cdot A(8) \cdot F(16)$.

A1.9 LAM-Grader Connector

The rear-mounted connector for a link to an optional separate LAM-Grader unit must be a fifty-two contact Cannon Double-Density fixed member with pins (Type 2DB52P), or equivalent type fully mateable with the free member with sockets, Cannon Type 2DB52S. It must be mounted at the rear of the crate controller above the Dataway connectors within the area for free access (see ANSI/IEEE Std 583-1982, Fig 3), with Contact 1 lowermost. The fifty-two contacts are assigned as shown in Table 10.

The LAM-Grader accepts L signals from the crate controller via the LAM-Grader connector. It generates Graded-L GL signals and, optionally, the External Demand signal. It may include gates, mask registers, etc, for processing the L signals, or may merely consist of passive interconnections between the contacts of the LAM-Grader connector. It may interact with the crate controller in the following ways:

(1) *Branch Demand.* Crate Controller Type A-1 derives the Branch Demand BD signal from the Graded-L signals (and, optionally, the External D signal) which it receives via the LAM-Grader connector.

(2) *Graded-L Operations.* The crate controller generates the *Graded-L operation* signal on Contact 1 to indicate that it requires Graded-L signals.

If the LAM Grader responds to this signal it must satisfy the timing requirements of Section A1.9.3.

(3) *Command Mode Operations.* In response to commands with $N(28)$ or $N(30)$ the crate controller generates the Controller Addressed signal on Contact 50. This allows the LAM Grader to be treated as a detached part of the crate controller that can be addressed inde-

Table 10
Contact Assignments for Rear Connector of Crate Controller Type A-1

Contact	Usage	Contact	Usage
1	Graded-L operation	2	$L1$
3	$GL1$	4	$L2$
5	$GL2$	6	$L3$
7	$GL3$	8	$L4$
9	$GL4$	10	$L5$
11	$GL5$	12	$L6$
13	$GL6$	14	$L7$
15	$GL7$	16	$L8$
17	$GL8$	18	$L9$
19	$GL9$	20	$L10$
21	$GL10$	22	$L11$
23	$GL11$	24	$L12$
25	$GL12$	26	$L13$
27	$GL13$	28	$L14$
29	$GL14$	30	$L15$
31	$GL15$	32	$L16$
33	$GL16$	34	$L17$
35	$GL17$	36	$L18$
37	$GL18$	38	$L19$
39	$GL19$	40	$L20$
41	$GL20$	42	$L21$
43	$GL21$	44	$L22$
45	$GL22$	46	$L23$
47	$GL23$	48	External D
49	$GL24$	50	Controller Addressed
51	Inhibit Internal D	52	0 V

pendently of its location in the crate. The presence of Dataway Busy B distinguishes operations with $N(28)$ from those with $N(30)$. The Controller Addressed signal with Dataway $A(0-7)$, but without B, indicates that the crate controller requires Graded-L signals. In conjunction with a Dataway operation and B the Controller Addressed signal may be used, for example, to access registers in the LAM Grader.

> If the LAM Grader responds to the Controller Addressed signal it must satisfy the timing requirements of Section A1.9.4.
> The *Graded-L operation* signal on Contact 1 must be in the logic 1 state when the crate controller is on line and $(BTA + \overline{BTB_i}) \cdot BG \cdot BCR_i = 1$.
> The *Controller Addressed* signal on Contact 50 must be in the 1 state during command mode operations to $N(28)$ or $N(30)$ when the crate controller is on line and $[N(28) + N(30)] (BTA + \overline{BTB_i}) \cdot \overline{BG} \cdot BCR_i = 1$.

Equivalent conditions for the generation of these two signals, Controller Addressed and Graded-L operation, are shown in Fig 7 in terms of the internal (nonmandatory) signals of a particular implementation of Crate Controller Type A-1.

> All mandatory timing requirements refer to signal conditions at the LAM-Grader connector on the crate controller. The interval between the initiation of a signal by the crate controller and the receipt of an established response from the external unit thus includes delays due to both the external unit and its linking cable.

A1.9.1 *Signal Standards*

> All signals via the LAM-Grader connector must satisfy Section 7.1 of ANSI/IEEE Std 583-1982. The signal standard for N signals applies to the Graded-L operation signal on Contact 1, and also to the Controller Addressed signal on Contact 50. All other signals including External D on Contact 48 and Inhibit Internal D on Contact 51, follow the standard for L signals.

A1.9.2 *Timing — Branch Demand*

> The maximum overall delay between the time when an L signal at the control station of the crate controller reaches a maintained

INTERFACE SYSTEM (CAMAC)

> 1 or 0 state and the time when the *BD* signal at the Branch Highway port of the same crate controller reaches a corresponding maintained 1 or 0 state is defined in Section 4.4.1. When the crate controller is used in conjunction with an external LAM Grader the component of this delay due to the crate controller must not exceed 250 ns.

A1.9.3 *Timing — Graded-L Operations*

> The interval between the generation of the Graded-*L* operation signal, accompanied by *L* signals, and the establishment of corresponding Graded-*L* signals must not exceed 350 ns.

A1.9.4 *Timing — Command Mode Operations*

> The interval between the generation of the Controller Addressed signal (accompanied by *L* signals, and in conjunction with Dataway signals *F*(0), *A*(0 - 7), *B* = 0) and the establishment of corresponding Graded-*L* signals must not exceed 350 ns.
>
> The external unit must present identical *GL* information in response to the Graded-*L* operation signal and to the Controller Addressed signal with *A*(0), *F*(0), and *B* = 0. Subaddresses *A*(1-7) may be used to access different selections of *GL* information.
>
> If the external unit responds to command mode operations with *N*(28)·*A*(0 — 15), *B* = 1, and an *F* code, it must satisfy the normal timing requirements of ANSI/IEEE Std 583-1982 for a CAMAC module and is permitted to make data transfers via the Dataway *R* and *W* lines.

A1.10 Off-Line State

> The off-line state is selected by means of the off-line manual control on the front panel of the crate controller. In this state the controller does not respond to command or Graded-*L* Request signals on the Branch Highway, and does not generate Branch timing or demand signals on the Highway.
>
> The following minimum conditions must be observed in the off-line state:
>
> (1) The front panel manual controls for Initialize and Clear must be effective. (They must be ineffective in the on-line state).
>
> (2) The front panel Inhibit signal input must continue to be effective. Dataway Inhibit must only be generated in response to the front panel Inhibit input.
>
> (3) The crate controller must not respond to *BTA* = 1. It must not generate Dataway *B*, *N*, *S*1, or *S*2 signals in response to *BTA* = 1 with *BG* = 0, or access the Graded-*L* information in response to *BTA* = 1 with *BG* = 1.
>
> (4) The crate controller must not generate 1 state outputs to the *BTB*, *BD*, *BRW*, *BQ*, or *BX* lines. An off-line crate is thus prevented from interfering with Branch operations.
>
> (5) The crate controller must not respond to *BZ* = 1.

The following additional conditions are recommended in the off-line state:

(6) If there are gates between the Branch Highway and Dataway lines *W*, *N*, *A*, and *F*, these should be used to isolate the crate controller. Branch operations are thus prevented from interfering with an off-line crate.

(7) In the absence of power supplies to the crate controller, all inputs and outputs via the Branch Highway ports should be free to assume either the 0 state or the 1 state, as required by other units connected to the Branch, and should not impose abnormal current loadings.

A1.11 Dataway Inhibit *I* in Off-Line State. It is mandatory in ANSI/IEEE Std 583-1982 (Section 5.5.2) that units generating Initialize *Z* must also generate *I*. Units that can generate and maintain *I* must maintain *I* = 1 until specifically reset. See Section 5.5.2 of ANSI/IEEE Std 583-1982. Both these requirements are met by Crate Controller Type A-1 in the on-line state. However, Section A1.10(2) of this standard specifically prohibits the generation of *I* = 1 in the off-line state other than in response to the front panel Inhibit input. The off-line state has been defined in such a way that a manual or test controller can be used to test or set up equipment while the crate is off line. Section A1.10(2) is primarily intended to prevent the generation of maintained $I = 1$ by Crate Controller Type A-1, since this has no manual means of resetting I and would obstruct any such off-line activities.

To be consistent with this aim:

(1) Crate Controller Type A-1 should generate I as defined in Sections A1.5.3 and A1.10 (and as shown in Fig 7).

(2) Any auxiliary means of generating commands in an off-line crate should conform to Section 5.5.2 of ANSI/IEEE Std 583-1982 by generating $I = 1$ in response to $Z \cdot S2$. It should preferably maintain $I = 1$ and provide a means of resetting it.

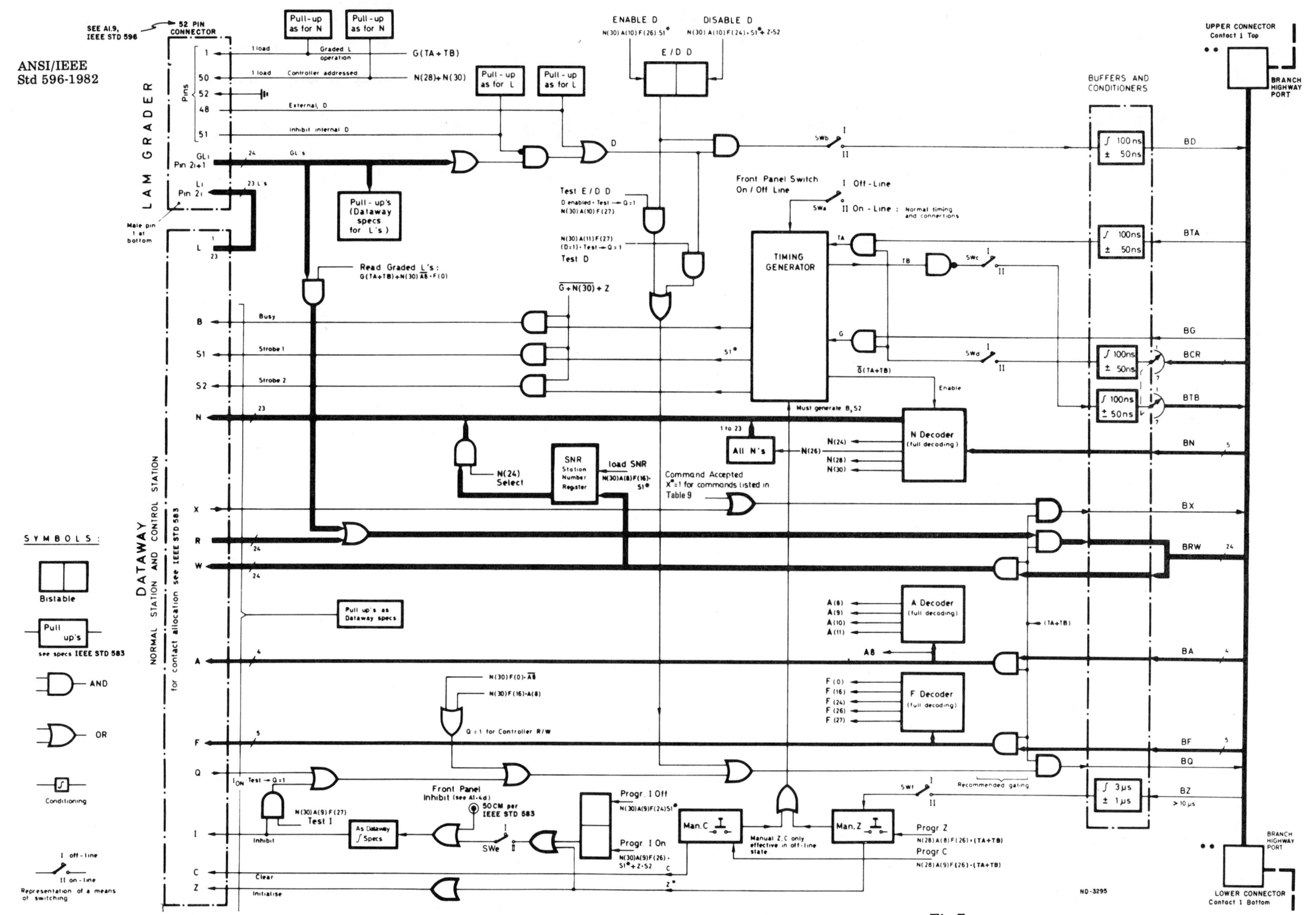

Fig 7
CAMAC Crate Controller Type A-1
(Double Width Unit)

37

A2. Branch Highway Length Limitations

A2.1 Parallel Branch Highway. The Parallel Branch Highway described in ANSI/IEEE Std 596-1982 uses a single-ended (unbalanced) mode of signal transmission. This mode gives reliable transmission in systems where the total length of the Highway is not excessive. The signal levels at the receivers, particularly in the logic 1 state, are affected by voltage drops in the conductors. As Highway length is increased, the noise margin at the receivers is thereby reduced.

Fig A2-1 shows the calculated relation among Highway length, wire size, driver output voltage (1 state), and noise margin. Only voltage drops owing to resistance of the Highway conductors were considered in calculating the curves. They are based on a "worst case" system of seven crates and a termination unit, separated from the Branch Driver by a Branch Highway cable of length indicated on the ordinate, and in which all signal circuits are simultaneously in the 1 state. The recommended values given in Table 8 of ANSI/IEEE Std 596-1982 were used in calculating the two solid curves. For the two dashed curves, the absolute limit for output driver voltage $E_0 = 0.5$ V was used. For $E_0 = 0.3$ V, the noise margin for zero Highway length is $NM = 0.9$ V. Thus the curve for $E_0 = 0.3$V, $NM = 0.45$V gives the Highway lengths at which the noise margin is reduced by half. The Highway lengths corresponding to this condition are 30, 45, 75, and 120 m for #28, #26, #24, and #22 AWG wire sizes, respectively. Available Branch Highway cables commonly use #26 AWG wire.

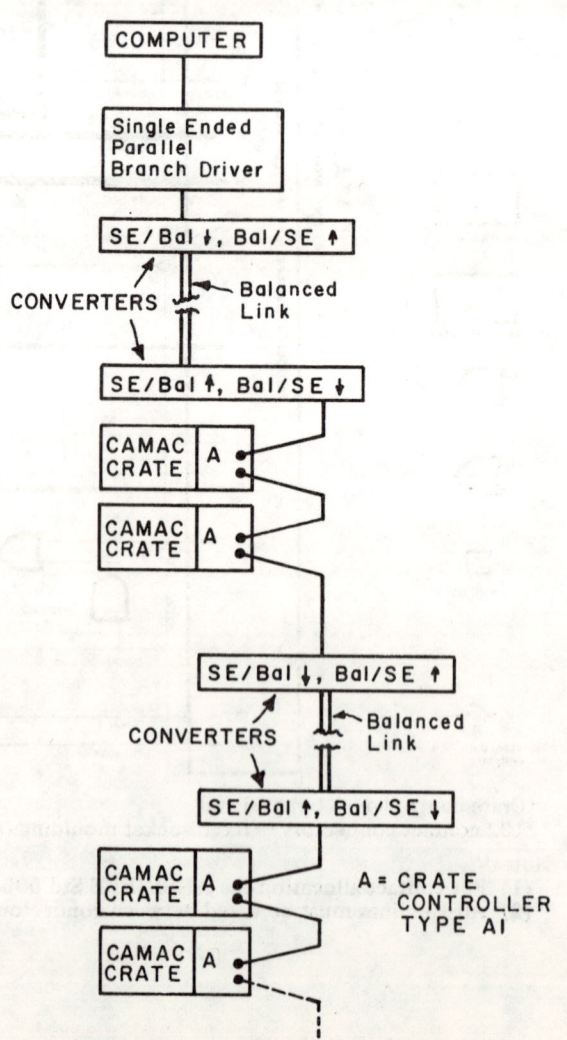

Fig A2.1
Effect of Branch Highway Length on Noise Margin

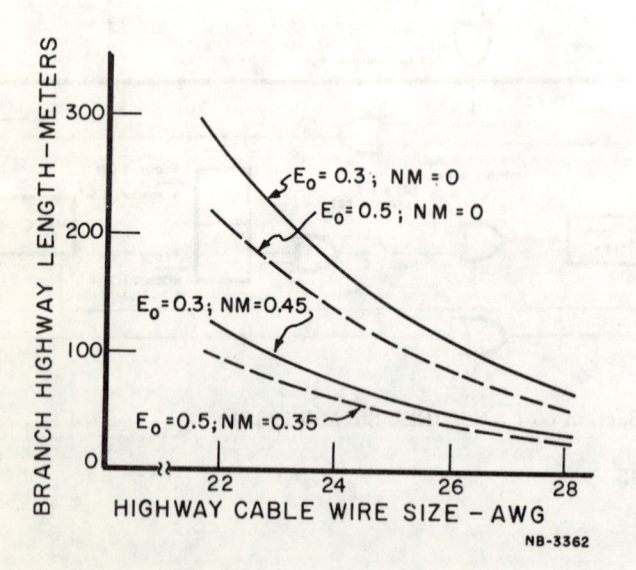

E_0 Driver output voltage at logic 1, in volts
NM Noise margin at receiver, in volts

Fig A2.2
Balanced Transmission System With Single-Ended-to-Balanced Converters

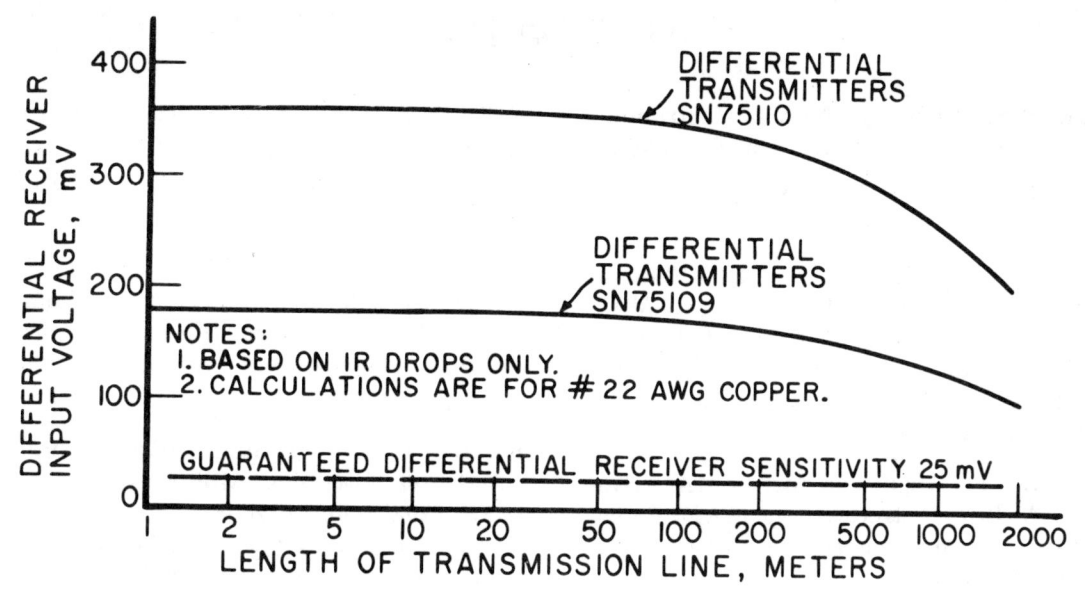

**Fig A2.3
Differential Receiver Input Voltages**

NOTES:
(1) Based on IR drops only
(2) Calculations are for No 22 AWG Copper

A2.2 Balanced Highway. A Branch can be conveniently extended beyond the limits discussed in the preceding text by the use of single-ended-to-balanced converters. In this scheme, signals are transmitted over a long span of distance using a balanced transmission mode. At each end of the span the balanced signals are converted to the standard Branch Highway mode, with appropriate termination of the single-ended signals. Aside from the converters and the balanced span, other components (for example, all Branch Drivers and crate controllers) are completely CAMAC standard. Fig A2-2 shows the configuration of such a system. It has two long distance links each of which utilizes the balanced signal transmission mode. "A" indicates Crate Controller Type A-1 and "SE/Bal ↓, Bal/SA ↑" indicates a single-ended-to-balanced, balanced-to-single-ended converter. Transient protection may be necessary on long lines.

Fig A2-3 shows curves of received signal voltage as a function of transmission line length for a differential (balanced) signal transmission system using current-switching transmitters. The curves are based on *IR* drops in a cable using #22 AWG Conductors.

A3. Bibliography

ANSI/IEEE Std 583-1982, Modular Instrumentation and Digital Interface System (CAMAC).

ANSI/IEEE Std 595-1982, Serial Highway Interface System (CAMAC).

BRILL, A. B; PARKER, J. F.; ERICKSON, J. J.; PRICE, R. R.; and PATTON, J. A., CAMAC Applications in Nuclear Medicine at Vanderbilt: Present Status and Future Plans, *IEEE Transactions on Nuclear Science*, vol NS–21, Feb 1974, pp 892–897.

COSTRELL, L., Highways for CAMAC Systems — A Brief Introduction. *IEEE Transactions on Nuclear Science*, vol NS–21, Feb 1974, pp 870–875.

COSTRELL, L., Standardized Instrumentation System for Computer Automated Measurement and Control. *IEEE Transactions on Industry Applications*, vol IA-11, May/Jun 1975, pp 319–323

DHAWAN, S., CAMAC Crate Controller Type A-1, *IEEE Transactions on Nuclear Science*, vol NS–20, Apr 1973, pp. 35–41.

EURATOM Report EUR 4600e, CAMAC — Organization of Multi-Crate Systems, 1972.

EURATOM Report EUR 6100, CAMAC Serial Highway Interface System, 1976.

FASSBENDER, P., and BEARDEN, F., Demonstrating Process Control Standards; An Exercise in Success. *IEEE Conference Record of the 1974* Ninth *Annual Meeting of the IEEE Industry Applications Society*, Oct 7–10, Pittsburgh, PA, pp 457–462.

HORELICK, D., and LARSEN, R. S., CAMAC: A Modular Standard. *IEEE Spectrum*, vol 13, Apr 1976, pp 50–55.

JOERGER, F. A., and KLAISNER, L. A., Functional Instrumentation Modules, *IEEE Transactions on Industry Applications*, vol IA-11, Nov/Dec 1975.

KIRSTEN, F. A., A Short Description of the CAMAC Branch Highway, *IEEE Transactions on Nuclear Science*, vol NS–20, Apr 1973, pp 21–27.

LYON, W. T., and ZOBRIST, D. W., System Design Considerations When Using Computer-Independent Hardware. *IEEE Transactions on Industry Applications*, vol IA-11, May/Jun 1975, pp 324–327.

RADWAY, T. D., A Standardized Approach to Interfacing Stepping Motors to Computers. *IEEE Conference Record of the 1974 Ninth Annual Meeting of the IEEE Industry Applications Society*, Oct 7–10, Pittsburgh, PA, pp 257–259.

US Energy Research and Development Administration Report TID-25876, CAMAC — Organization of Multi-Crate Systems, Mar 1972.

US Energy Research and Development Administration Report TID-25877, Supplementary Information on CAMAC Instrumentation System, Dec 1972.

US Energy Research and Development Administration Report TID-26616, Block Transfers in CAMAC Systems (to be issued also by ESONE as supplement to EUR 4100e), Feb 1976.

US Energy Research and Development Administration Report TID-26614, CAMAC Specification of Amplitude Analogue Signals Within a 50 Ohm System, Oct 1974.

US Energy Research and Development Column Report TID-26618, CAMAC Tutorial Articles, Oct. 1976.

WILLARD, F. G., Interfacing Standardization in the Large Control System. *IEEE Transactions on Industry Applications*, vol IA-11, Jul/Aug 1975, pp 362–364.

Index

The principal references are given first. Any subsidiary references follow, after a semi-colon. Information specific to Crate Controller Type A-1 is indicated by references to the Appendix, for example, A1.9.2.

The abbreviation *BH* is used for branch highway, and CCA-1 for Crate Controller Type A-1.

Absent crates — identification of	5.4; 4.3
Bibliography	A3
Branch	3; 1, Fig 1, Fig 2
Branch Driver	3; 1, 6, Fig 1, Fig 2
Branch Highway	3; 1, Fig 1, Fig 2
Branch Highway Length Limitations	A2, A3

Branch Highway lines
*BA*1, 2, 4, 8	see Subaddress
*BCR*1—*BCR*7	see Crate Address
BD	see Demand, Branch
*BF*1, 2, 4, 8, 16	see Function
BG	see Graded-*L* Request
*BN*1, 2, 4, 8, 16	see Station Number
BQ	see Response
*BRW*1—*BRW*24	see Read and Write
BSC	see Cable Shield
BTA, *BTB*1—*BTB*7	see Timing, Branch
BV	see Reserved Lines
BX	see Command Accepted
BZ	see Initialize

Branch Highway port	4; 3
— connector for	6; Table 5, Fig 5, Fig 6
— lines at	Table 1; Tables 6, 7
Branch operations	5; 3
Busy — Dataway line *B*	Table 2; 5.2, A1.5.3, A1.7
Cable Shield — at BH port *BSC*	6
Capacitative loading — on BH lines	7
Characteristic impedance — of BH lines	Table 8; 7
Clear — Dataway line *C*	A1.5.3; 4.5.2, Table 9
— manual control	A1.4(5); A1.5.3
Command Accepted — BH line *BX*	4.2.3; A1.8
Command Accepted *X* and *BX* Disabling	4.2.5
Command Accepted *BX* Response to Graded-*L* Request *BG*	4.2.4
Command mode operations	5.1, Table 3; 3, Fig 3, Fig 4
— in CCA-1	A1.7.1, A1.7.3; A1.9.4, Table 10
Command Signals — BH	4.1, A1.5.2
Common control signals	4.5, A1.5.3; 3
Compatibility — definition	2
Conditioning — of input signals at BH ports	7.2, A1.7; 4.1.1, 4.3, 4.5.2, 7, Fig 7
Connector — see BH port connector	
— see LAM-Grader connector	
Controller Addressed Signal — in CCA-1	A1.9
Crate Address — BH lines (*BCR*1—*BCR*7)	4.1.1; A1.4(2)
Crate controller	3, A1.2; 1
Crate Controller Type A-1	A1; 1, 4.4.1
— commands implemented	Table 9; A1.8
— Crate Address selection	A1.4(2)
— delay of *BD*	A1.9.2
— disable/enable *BD*	A1.6.1; Table 9
— front panel features	A1.4; 6, A1.10, Fig 5, Fig 7
— pull-up for *L* and *GL*	A1.6.3; Fig 7
Dataway	1
Dataway Inhibit *I* in Off-Line State	A1.11
Dataway lines	
— see: Busy *B*	
Strobes *S*1, *S*2	
Inhibit *I*	
Initialize *Z*	
Data lines — at BH port, See Read and Write	
Demand, Branch *BD*	4.4.1, A1.6.1; 3, 5.2, A1.9.2
Demand handling	3; 4.4, A1.6
Differential delays — see skew	

Free Lines — Branch Highway ($BV1-BV5$)	4.6; Table 6, Table 7
Function — BH lines $BF1, 2, 4, 8, 16$	4.1.4; A1.5.2
GL — See Graded-L Signals	
Graded-L mode operations	5.2, Table 4; 3, A1.6.2
— timing in CCA-1	A1.7.2
— timing in LAM Grader	A1.9.3
Graded-L Operation Signal — in CCA-1	A1.9
Graded-L Request — BH line BG	4.4.2
Graded-L Signals ($GL1-GL24$)	3; A1.6.2, A1.7.2, Table 10
— grading process	4.4.2; 5.2, A1.9, A1.9.3
— priority order of	4.4.2
Highway	1, also see Branch Highway
Highway Driver	1, also see Branch Driver
Highway length limitations	See Branch Highway length limitations
Inhibit — Dataway line I	A1.5.3; 4.5.2, Table 9
— front panel signal	A1.5.3; A1.4(4), A1.10(2)
Initialize — BH line BZ	4.5.1; A1.10(5)
— Dataway line Z	A1.5.3; 4.5.2, Table 9
— manual control	A1.4(5); A1.5.3
Inputs — signal standards at BH port	7.1, Table 8; A1.10(6)
— isolation in off-line state	A1.10(7)
Integration — of inputs, see Conditioning	
— of outputs, see Shaping	
LAM Grader	A1.9; 4.4.1, 5.2, A1.3
LAM-Grader connector — on CCA-1	A1.9, Table 10; A1.3, Fig 7
— signal standards at	A1.9.1
— timing at	A1.9.2, A1.9.3, A1.9.4
Length limitations, Branch Highway	See Branch Highway length limitations
Manual controls of CCA — see Initialize	
— see Clear	
May — permitted practice, definition	2
Must — mandatory practice, definition	2
Off-line state	A1.10, A1.11, 3, 7.4, A1.4(3)
On-line state — identification of	5.4; 3
Outputs — signal standards at BH port	7.2, Table 8; A1.10(4), A1.10(7)
Parallel Highway	1, also see Branch Highway
Parallel Highway Driver	1, also see Branch Driver
Patch connections — Dataway, at CCA-1	A1.5.4
Phase — of Branch operations	5, Table 3, Table 4; Fig 3, Fig 4
Port — see Branch Highway port	
Power-off condition	7.4; A1.10(7)
Priority order — see Graded-L signals	
Read operations — on Branch Highway	5.1.1, 5.1.2, 5.1.3, 5.1.4; 3, Table 3, Fig 3
Read and Write — BH lines $BRW1-BRW24$	4.2.1; A1.5.1
References	See bibliography
Reserved — definition	2
Reserved lines — Branch Highway $BV6$ and $BV7$	4.6; Table 6, Table 7
Response — BH line BQ	4.2.2; Table 9
Return lines — in BH	Table 7; 6, Fig 6
Shaping — of output signals at BH ports	7.2; 4.3, 4.4.1
Should — recommended practice, definition	2
Signal standards — at BH ports	7, Table 8
Skew	5.3; 5, Table 3, Table 4
Station Number — BH lines ($BN1, 2, 4, 8, 16$)	4.1.2
Station Number Codes	Table 2; A1.5.2
Station Number Register (SNR) — in CCA-1	A1.5.2; 4.1.2, Table 9, Fig 7
Strobes — Dataway lines ($S1, S2$)	5; A1.7.1, A1.7, A1.5.3, Table 2
Subaddress — BH lines ($BA1, 2, 4, 8$)	4.1.3; A1.5.2
Termination — of BH lines	7.3; Table 8, Fig 1, Fig 2
— preferred values	Table 8
Time-out	5
Timing, Branch — BH lines ($BTA, BTB1-BTB7$)	4.3; 3, A1.7, A1.10(3), A1.10(4)
Write Operations	5.1.5; 3, Fig 4

ANSI/IEEE
Std 675-1982
(Revision of ANSI/IEEE
Std 675-1979)

An American National Standard

IEEE Standard
Multiple Controllers in a CAMAC*Crate

Sponsor

Instruments and Detectors Committee of
the IEEE Nuclear and Plasma Sciences Society

Approved September 17, 1981

IEEE Standards Board

Approved December 15, 1981

American National Standards Institute

*Computer Automated Measurement and Control

© Copyright 1982 by

The Institute of Electrical and Electronics Engineers, Inc
345 East 47th Street, New York, NY 10017

*No part of this publication may be reproduced in any form,
in an electronic retrieval system or otherwise,
without the prior written permission of the publisher.*

IEEE Standards documents are developed within the Technical Committees of the IEEE Societies and the Standards Coordinating Committees of the IEEE Standards Board. Members of the committees serve voluntarily and without compensation. They are not necessarily members of the Institute. The standards developed within IEEE represent a consensus of the broad expertise on the subject within the Institute as well as those activities outside of IEEE which have expressed an interest in participating in the development of the standard.

Use of an IEEE Standard is wholly voluntary. The existence of an IEEE Standard does not imply that there are no other ways to produce, test, measure, purchase, market, or provide other goods and services related to the scope of the IEEE Standard. Furthermore, the viewpoint expressed at the time a standard is approved and issued is subject to change brought about through developments in the state of the art and comments received from users of the standard. Every IEEE Standard is subjected to review at least once every five years for revision or reaffirmation. When a document is more than five years old, and has not been reaffirmed, it is reasonable to conclude that its contents, although still of some value, do not wholly reflect the present state of the art. Users are cautioned to check to determine that they have the latest edition of any IEEE Standard.

Comments for revision of IEEE Standards are welcome from any interested party, regardless of membership affiliation with IEEE. Suggestions for changes in documents should be in the form of a proposed change of text, together with appropriate supporting comments.

Interpretations: Occasionally questions may arise regarding the meaning of portions of standards as they relate to specific applications. When the need for interpretations is brought to the attention of IEEE, the Institute will initiate action to prepare appropriate responses. Since IEEE Standards represent a consensus of all concerned interests, it is important to ensure that any interpretation has also received the concurrence of a balance of interests. For this reason IEEE and the members of its technical committees are not able to provide an instant response to interpretation requests except in those cases where the matter has previously received formal consideration.

Comments on standards and requests for interpretations should be addressed to:

> Secretary, IEEE Standards Board
> 345 East 47th Street
> New York, NY 10017
> USA

Foreword

(This Foreword is not a part of ANSI/IEEE Std 675-1982, IEEE Standard, Multiple Controllers in a CAMAC Crate.)

This standard makes provision for the use of auxiliary controllers in a CAMAC crate to extend the capabilities and fields of application of the CAMAC modular instrumentation and interface system of ANSI/IEEE Std 583-1982, Modular Instrumentation and Digital Interface System (CAMAC). (Also Report EUR 4100e of the Commissions of the European Communities.) The report on which this document is based (Dept of Energy Report DOE/EV-007), was prepared by the NIM Committee of the United States Department of Energy* and the ESONE Committee of European Laboratories**. Corresponding documents are ESONE Report EUR 6500e and IEC Publication 729 of the International Electrotechnical Commission.

Auxiliary Controllers (ACs), located in several crates on a CAMAC Command Highway (Parallel Branch or Serial), can provide the user with an extremely effective distributed processing capability. This capability is useful in both data acquisition and automatic control systems where parallel processing is desirable. Auxiliary controllers can be of various degrees of sophistication, and some may include processor elements for increased autonomy from central computers.

A further extension of the AC-based CAMAC system may include a special communication channel direct from an AC to another element in the system. High-speed block transfer of data can be effected to and from the AC in such a case.

To permit use of the ACs in CAMAC crates, this report defines the hardware and protocol so as to:

(1) Provide means for granting control of the Dataway to auxiliary controllers that request control and to provide a priority selection scheme to arbitrate in the case of simultaneous requests from multiple controllers.

(2) Provide means for the auxiliary controller to address (a) the dedicated module address lines and (b) the dedicated interrupt request LAM (Look-at-Me) lines, all of which terminate only at the crate controller station and at the module stations to which they are dedicated.

This standard was reviewed and balloted by the Nuclear Instruments and Detectors Committee of the IEEE Nuclear and Plasma Sciences Society. Because of the board applicability of the CAMAC system, this standard was submitted for review to liaison representatives of numerous societies and committees including the Communications Society, Computer Society, Industry Applications Society, Industrial Electronics and Control Instrumentation Group, Instrumentation and Measurement Group, and the Power Generation and Nuclear Power Engineering Committees of the Power Engineering Society.

The revision of this standard was in conjunction with the 1981 review (1982 issue) of the entire family of IEEE CAMAC standards undertaken to incorporate existing addenda and corrections into the standards.

At the time of approval of this standard, the membership of the Nuclear Instruments and Detectors Committee of the IEEE Nuclear and Plasma Sciences Society was as follows:

D. C. Cook, *Chairman*   **Louis Costrell,** *Secretary*

J. A. Coleman	T. R. Kohler	P. L. Phelps
J. F. Detko	H. W. Kraner	J. H. Trainor
F. S. Goulding	W. W. Managan	S. Wagner
F. A. Kirsten	G. L. Miller	F. J. Walter
	D. E. Persyk	

At the time it approved this standard, the American National Standards Committee N42 on Nuclear Instruments had the following personnel:

Louis Costrell, *Chairman*   **David C. Cook,** *Secretary*

Organization Represented   *Name of Representative*

Organization	Representative
American Conference of Governmental Industrial Hygienists	Jesse Lieberman
American Nuclear Society	Frank W. Manning
Health Physics Society	J. B. Horner Kuper
	J. M. Selby (*Alt*)
Institute of Electrical and Electronics Engineers	Louis Costrell
	Julian Forster (*Alt*)
	D. C. Cook (*Alt*)
	A. J. Spurgin (*Alt*)

Organization Represented	Name of Representative
Instrument Society of America	M. T. Slind
	J. E. Kaveckis (*Alt*)
Lawrence Berkeley National Laboratory	E. J. Wagner
Oak Ridge National Laboratory	Frank W. Manning
	D. J. Knowles (*Alt*)
US Department of Energy	Gerald Goldstein
US Department of the Army, Materiel Command	Abraham E. Cohen
US Department of Commerce, National Bureau of Standards	Louis Costrell
US Federal Emergency Management Agency	Carl R. Siebentritt, Jr
US Naval Research Laboratory	D. C. Cook
US Nuclear Regulatory Commission	Robert E. Alexander
Members-at-Large	J. G. Bellian
	O. W. Bilharz, Jr
	John M. Gallagher, Jr
	Voss A. Moore
	R. F. Shea
	E. J. Vallario

*NIM Committee[1]
L. Costrell, *Chairman*

C. Akerlof	N. W. Hill	L. B. Mortara
E. J. Barsotti	D. Horelick	V. C. Negro
B. Bertolucci	M. E. Johnson	L. Paffrath
J. A. Biggerstaff	C. Kerns	D. G. Perry
A. E. Brenner	F. A. Kirsten	I. Pizer
R. M. Brown	P. F. Kunz	E. Platner
E. Davey	R. S. Larsen	S. Rankowitz
W. K. Dawson	A. E. Larsh, Jr	S. J. Rudnick
S. R. Deiss	R. A. LaSalle	G. K. Schulze
S. Dhawan	N. Latner	W. P. Sims
R. Downing	F. R. Lenkszus	D. E. Stilwell
T. F. Droege	R. Leong	J. H. Trainor
C. D. Ethridge	C. Logg	K. J. Turner
C. E. L. Gingell	S. C. Loken	H. Verweij
A. Gjovig	D. R. Machen	B. F. Wadsworth
B. Gobbi	D. A. Mack	L. J. Wagner
D. B. Gustavson	J. L. McAlpine	H. V. Walz
D. R. Heywood		D. H. White

NIM Executive Committee for CAMAC
L. Costrell, *Chairman*

E. J. Barsotti	F. A. Kirsten
J. A. Biggerstaff	R. S. Larsen
S. Dhawan	D. A. Mack
J. H. Trainor	

NIM Dataway Working Group
F. A. Kirsten, *Chairman*

E. J. Barsotti	C. Kerns
J. A. Biggerstaff	P. F. Kunz
L. Costrell	R. S. Larsen
S. Dhawan	D. R. Machen
A. Gjovig	L. Paffrath
D. R. Heywood	S. Rankowitz
D. Horelick	S. J. Rudnick

NIM Software Working Group
D. B. Gustavson, *Chairman*
W. K. Dawson, *Secretary*

A. E. Brenner	R. A. LaSalle
R. M. Brown	F. B. Lenkszus
S. R. Deiss	C. A. Logg
S. Dhawan	J. McAlpine
C. D. Ethridge	L. B. Mortara
M. E. Johnson	D. G. Perry

---

[1] National Instrumentation Methods Committee of the US Department of Energy.

NIM Mechanical and Power Supplies Working Group
L. J. Wagner, *Chairman*

L. Costrell	S. J. Rudnick
C. Kerns	W. P. Sims

NIM Analog Signals Working Group
D. I. Porat, *Chairman*

L. Costrell	N. W. Hill
C. E. L. Gingell	S. Rankowitz

NIM Serial System Subgroup
D. R. Machen, *Chairman*

E. J. Barsotti	R. G. Martin
D. Horelick	L. Paffrath
F. A. Kirsten	S. J. Rudnick

NIM Block Transfer Subgroup
E. J. Barsotti, *Chairman*

W. K. Dawson	F. R. Lenkszus
F. A. Kirsten	R. G. Martin
R. A. LaSalle	R. F. Thomas, Jr

NIM Multiple Controllers Subgroup
P. F. Kunz, *Chairman*

E. J. Barsotti	D. R. Machen
F. A. Kirsten	R. G. Martin

**ESONE Committee[2]

W. Attwenger, Austria, *Chairman* 1980-81   H. Meyer, Belgium, *Secretary*

Representatives of ESONE Member Laboratories

W. Attwenger, Austria	E. Kwakkel, Netherlands	A. C. Peatfield, England
R. Biancastelli, Italy	J. L. Lecomte, France	I. C. Pyle, England
L. Binard, Belgium	J. Lingertat, D.R. Germany	B. Rispoli, Italy
J. Biri, Hungary	M. Lombardi, Italy	M. Sarquiz, France
B. Bjarland, Finland	M. Maccioni, Italy	W. Schoeps, Switzerland
D. A. Boyce, England	P. Maranesi, Italy	R. Schule, F.R. Germany
B. A. Brandt, F.R. Germany	C. H. Mantakas, Greece	P. G. Sjolin, Sweden
F. Cesaroni, Italy	D. Marino, Italy	L. Stanchi, Italy
P. Christensen, Denmark	H. Meyer, Belgium	R. Trechcinski, Poland
W. K. Dawson, Canada	K. D. Muller, F.R. Germany	M. Truong, France
M. DeMarsico, Italy	J. G. Ottes, F.R. Germany	P. Uuspaa, Finland
C. A. DeVries, Netherlands	A. D. Overtoom, Netherlands	H. Verweij, Switzerland
H. Dilcher, F.R. Germany	E. C. G. Owen, England	A. J. Vickers, England
B. V. Fefilov, USSR	L. Panaccione, Italy	S. Vitale, Italy
R. A. Hunt, England	M. Patrutescu, Roumania	M. Vojinovic, Yugoslavia
W. Kessel, F.R. Germany	R. Patzelt, Austria	K. Zander, F.R. Germany
R. Klesse, France		D. Zimmermann, F.R. Germany

ESONE Executive Group (XG)

W. Attwenger, Austria, XG *Chairman* 1980-81

H. Meyer, Belgium, *Secretary*

P. Christensen, Denmark	M. Sarquiz, France
M. Dilcher, F.R. Germany	R. Trechcinski, Poland
B. Rispoli, Italy	A. Vickers, England
H. Verweij, Switzerland	

ESONE Technical Coordination Committee (TCC)

A. C. Peatfield, England, TCC *Chairman*
P. Christensen, Denmark, TCC *Secretary*

W. Attwenger, Austria	W. Kessel, F.R. Germany
R. Biancastelli, Italy	J. Lukacs, Hungary
G. Bianchi, France	R. Patzelt, Austria
H. Dilcher, F.R. Germany	P. J. Ponting, Switzerland
P. Gallice, France	W. Schoeps, Switzerland
S. Vitali, Italy	

---

[2] European Standards on Nuclear Electronics Committee.

ECA/ESONE CAMAC Document Maintenance Study Group (DMSG)

P. Gallice, France, *Chairman*

R. C. M. Barnes, England
L. Besse, Switzerland
J. Davis, England
F. Iselin, Switzerland
H. Meyer, Belgium
H. J. Trebst, F.R. Germany

When this standard was approved on September 17, 1981, the IEEE Standards Board had the following membership:

I. N. Howell, Jr, *Chairman* **Irving Kolodny,** *Vice Chairman*

Sava I. Sherr, *Secretary*

G. Y. R. Allen
J. J. Archambault
J. H. Beall
J. T. Boettger
Edward Chelotti
Edward J. Cohen
Len S. Corey

Jay Forster
Kurt Greene
Loering M. Johnson
Joseph L. Koepfinger
J. E. May
Donald T. Michael*
J. P. Riganati

F. Rosa
R. W. Seelbach
J. S. Stewart
W. E. Vannah
Virginius N. Vaughan, Jr
Art Wall
Robert E. Weiler

*Member emeritus

Contents

SECTION	PAGE
CAMAC and NIM Standards and Reports	9
1. Introduction	11
2. Interpretation	11
3. The Auxiliary Controller Bus and Associated Front Panel Signals	12
4. Use of the Lines on the Auxiliary Controller Bus and Associated Signals	17
4.1 Control Signals	17
4.1.1 Request	18
4.1.2 Grant-In and Grant-Out Signals	18
4.1.3 Request Inhibit	18
4.1.4 Generation of ACL Signal	18
4.1.5 Response to ACL Signal	18
4.2 Encoded-N Signals	19
4.2.1 Generation of Encoded-N Signals	19
4.2.2 Response to Encoded-N Signals	19
4.3 Look-at-Me Signals	19
4.4 Other Signals	19
5. The Auxiliary Controller Bus Connector and Associated Front Panel Connectors	19
6. Signal Standards	21
6.1 Signal Standards on the ACB	21
6.2 Signal Standards for the ACB Associated Front Panel Signals	21
6.3 Signal Standards for AC Dataway Connections	21
6.4 Protection of Critical Signal Inputs	21

FIGURES

Fig 1 Multiple Controllers in a CAMAC Crate	13
Fig 2 Crate Controller, Minimum Configuration	14
Fig 3 Sequence of Signals for an AC to Gain Control of Dataway for an Addressed Command Operation	15
Fig 4 Examples of Sequences of Signals that may Occur with the ACL Signal. A — Example of a Sequence in Which Auxiliary Controller Lockout (ACL) Signal is Generated Too Late to Cause Aborting of Dataway Cycle Started by AC Using Request/Grant; B — Example of Sequence in Which a Dataway Cycle, Started by AC Using Request/Grant, is Caused to Abort by ACL Signal Generated by Another Controller.	16

TABLES

Table 1 Signal Lines at the Auxiliary Controller Bus Connector and Associated Front Panel Connectors	17
Table 2 Contact Assignments on Auxiliary Controller Bus Connector	20
Table 3 Current Signal Standards and Pull-Up Current Sources for the Auxiliary Controller Bus Connector and Associated Front Panel Connectors	20
Table 4 Signal Standards for Q, R, and X at the Auxiliary Controller Dataway Connector	21

APPENDIX

A1. Specification of CAMAC Crate Controller Type A-2	22
A1.1 CAMAC Crate Controller Type A-2	22
A1.2 Other Crate Controllers	22
A1.3 General Features	22
A1.4 Front Panel	23

APPENDIX		PAGE
A1.5	Dataway Signals	23
	A1.5.1 Data Signals	23
	A1.5.2 Command Signals	23
	A1.5.3 Common Control Signals	24
	A1.5.4 Patch Connections	24
	A1.5.5 Arbitration Methods Used by the Crate Controller	25
A1.6	Demand Handling	25
	A1.6.1 Branch Demand	25
	A1.6.2 Graded-L	25
	A1.6.3 Pull-Up for GL and L Lines	25
A1.7	Timing Requirements	25
	A1.7.1 Command Mode Operations with Dataway $S1$, $S2$, and B	26
	A1.7.2 Graded-L Operations	26
	A1.7.3 Command Mode Operations Without Dataway $S1$, $S2$, or B	28
A1.8	Commands Implemented by Crate Controller Type A-2	27
A1.9	LAM-Grader Connector	27
	A1.9.1 Signal Standards	29
	A1.9.2 Timing — Branch Demand	29
	A1.9.3 Timing — Graded-L Operations	29
	A1.9.4 Timing — Command Mode Operations	29
A1.10	Off-Line State	29
A1.11	Dataway Inhibit I in Off-Line State	30
A1.12	Auxiliary Controller Bus Connector	30

APPENDIX FIGURE
Fig A1 CAMAC Crate Controller Type A-2 31

APPENDIX TABLES
Table A1 Branch Commands Implemented by CAMAC Crate Controller Type A-2 28
Table A2 Contact Assignments for LAM-Grader Connector of Crate Controller
          Type A-2 28

INDEX 32

CAMAC and NIM Standards and Reports

Title	IEEE, ANSI Std No	IEC No	DOE No	EURATOM (EUR) or ESONE No
CAMAC Instrumentation and Interface Standards*	SH08482* (Library of Congress No 8185060)	—	—	—
Modular Instrumentation and Digital Interface System (CAMAC)	ANSI/IEEE Std 583-1982	516	TID-25875† and TID-25877†	EUR 4100e
Serial Highway Interface System (CAMAC)	ANSI/IEEE Std 595-1982	640	TID-26488†	EUR 6100e
Parallel Highway Interface System (CAMAC)	ANSI/IEEE Std 596-1982	552	TID-25876† and TID-25877†	EUR 4600e
Multiple Controllers in a CAMAC Crate	ANSI/IEEE Std 675-1982	729	DOE/EV-0007	EUR 6500e
Block Transfers in CAMAC Systems	ANSI/IEEE Std 683-1976 (Reaff 1981)	677	TID-26616†	EUR 4100 suppl
Amplitude Analog Signals within a 50 Ω System	—	—	TID-26614	EUR 5100e
The Definition of IML A Language for Use in CAMAC Systems	—	—	TID-26615	ESONE/IML/01
CAMAC Tutorial Articles	—	—	TID-26618	—
Real-Time BASIC for CAMAC	ANSI/IEEE Std 726-1982	§	TID-26619†	ESONE/RTB/03
Subroutines for CAMAC	ANSI/IEEE Std 758-1979 (Reaff 1981)	713	DOE/EV-0016†	ESONE/SR/01
Recommendations for CAMAC Serial Highway Drivers and LAM Graders for the SCC-L2	—	—	DOE/EV-0006	ESONE/SD/02
Definitions of CAMAC Terms	Included in SH08482	678	DOE/ER-0104	ESONE/GEN/01
Standard Nuclear Instrument Modules NIM	—	547**	TID-20893 (Rev 4)	—

†Superseded by corresponding IEEE Standard listed.

*This is a hard cover book that contains ANSI/IEEE Std 583-1982, ANSI/IEEE Std 595-1982, ANSI/IEEE Std 596-1982, ANSI/IEEE Std 675-1982, ANSI/IEEE Std 683-1976 (Reaff 1981), ANSI/IEEE Std 726-1982 and IEEE Std 758-1979 (Reaff 1981), plus introductory material and a glossary of CAMAC terms.

**Covers only mechanical features and connector pin assignments.

§ In preparation.

NOTE: *Availability of Documents*
ANSI   Sales Department, American National Standards Institute, 1430 Broadway, New York, NY 10018.
IEEE   IEEE Service Center, 445 Hoes Lane, Piscataway, New Jersey 08854, USA.
IEC   International Electrotechnical Commission, 1, rue de Varembé, CH-1211 Geneva 20, Switzerland.
DOE and TID Reports   National Bureau of Standards, Washington, D.C. 20234, USA, Attn: L. Costrell.
EURATOM   Office of Official Publications of the European Communities, P.O. Box 1003, Luxembourg.
ESONE   Commission of the European Communities, CGR-BCMN, B-2440 GEEL, Belgium, Attn: ESONE Secretariat, H. Meyer.

An American National Standard

IEEE Standard
Multiple Controllers in a CAMAC Crate

1. Introduction

This standard defines a method for incorporating more than one source of control into a CAMAC crate. The standard fully conforms to the mechanical and signal standards of the CAMAC system as described in ANSI/IEEE Std 583-1982, Modular Instrumentation and Digital Interface System (CAMAC), or EUR 4100e.

In order to allow more than one Controller to access the Dataway of a CAMAC Crate, an Auxiliary Controller Bus (ACB) and priority arbitration protocol are fully defined. This permits the use of Auxiliary Controllers (ACs) in normal stations in the crate. The ACB carries encoded address information from an AC to the Crate Controller (CC) in the control station of a CAMAC crate, and carries Look-at-Me signals from the CC to the ACs. The ACB, connected between controllers, may also be used to establish priority for control of the CAMAC Dataway.

This standard is fully compatible with ANSI/IEEE Std 595-1982, Serial Highway Interface System (CAMAC), or EUR 6100e, and the ANSI/IEEE Std 596-1982, Parallel Highway Interface System (CAMAC) or EUR 4600e. It may also be used in autonomous systems (systems with no external highways) or in systems with Type U Crate Controllers (such as systems with a computer I/O bus as the highway).

Section A1 of the Appendix to this standard defines a Parallel Highway Crate Controller, Type A-2. This controller is similar to Crate Controller Type A-1 (as defined in Section A1 of the Appendix to ANSI/IEEE Std 596-1982 or EUR 4600e), except for the ACB connector and the priority arbitration protocol. When these two features are not required, Crate Controllers Types A-1 and A-2 are totally interchangeable.

Abbreviations used in this document are as follows:

AC   Auxiliary Controller
ACB   Auxiliary Controller Bus
ACL   Auxiliary Controller Lockout
CC   Crate Controller
R/G   Request Grant

2. Interpretation

This standard is a reference text describing and specifying multisource control within a CAMAC crate. It should be read in conjunction with, and is supplementary to, ANSI/IEEE Std 583-1982, ANSI/IEEE Std 595-1982, and ANSI/IEEE Std 596-1982 or EUR 4100e, EUR 6100e, and EUR 46003, respectively.

No part of this standard is intended to supersede or modify the above-mentioned standards.

In this standard there are mandatory requirements, recommendations, and examples of permitted practice.

> Mandatory clauses of the standard are enclosed in a box, as here, and usually include the word "must."

Definitions of recommended practices (those to be followed unless there are sound reasons to the contrary) include the word "should."

Examples of permitted practice generally include the word "may," and leave freedom of choice to the designer or user.

> In order to "conform" with the specifications of this standard, an equipment or system must satisfy all the mandatory requirements in this standard, excluding the Appendix. If constructed as a CAMAC plug-in unit, the equipment must also satisfy the mandatory requirements of ANSI/IEEE Std 583-1982 or EUR 4100e.

Section A1 of this standard's Appendix defines the Parallel Crate Controller Type A-2 in such a way that Type A-2 Controllers produced by different manufacturers will be operationally interchangeable. The main text to this standard contains a less restrictive definition of Controllers that are not necessarily interchangeable. See Section A1 of the Appendix regarding conformity with the specification of the CAMAC Crate Controller Type A-2.

> In order to be "compatible" with the ACB, equipment need not satisfy all the mandatory requirements, but must not interfere with the full operation of all the features of Controllers which "conform" to this standard.

No part of this standard is intended to exclude the use of equipment that is compatible in the preceding sense, even if it does not conform fully to this standard or is not constructed as CAMAC plug-in units.

No license or permission is needed in order to use this standard.

3. The Auxiliary Controller Bus and Associated Front Panel Signals

The standard CAMAC crate, described in ANSI/IEEE Std 583-1982 or EUR 4100e requires the presence of a Controller to control and coordinate the activities of the crate. During a Dataway addressed command operation, the Controller establishes the necessary signals on the B, N, A, F, $S1$, and $S2$ lines to define the command operation to be performed and to define the timing of the operation. During addressed command operations involving data, the Controller transmits or receives data via the W or R lines, respectively. During unaddressed operations, the Controller establishes the necessary signals on the B, $S1$, $S2$, and C or Z lines. The Controller may establish the state of the I signal and may monitor the state of the L, X, and Q signals.

Each CAMAC crate has one control station, which is the only station providing access to the N and L lines. The control station and a normal station together provide access to all signal lines needed by a Controller to perform the operations described above. The Controller which occupies the control station is designated the CC of the crate. Examples of CCs are the Serial Highway Crate Controller Type L-2 (Section A1 of the Appendix of ANSI/IEEE Std 595-1982 or EUR 6100e) and the Parallel Highway Crate Controller Type A-2 (Section A1 of the Appendix of this standard).

An additional source of control within a CAMAC crate can be provided by an AC, which occupies one or more normal stations. In order to accommodate ACs, two features are required: (1) access to the N and L lines at normal stations, and (2) priority arbitration for control of the Dataway. Access to the N lines is necessary to allow an AC to generate a complete addressed command operation. Access to the L lines is necessary if an AC is to respond to Look-at-Me signals from other modules or controllers, or both. Priority arbitration protocol ensures that at any time only one Controller is permitted to have control of the crate. It also provides the means for assigning control of the Dataway on the basis of a prearranged priority.

Access to the N and L lines is provided by the ACB (see 4.2 and 4.3) via the CC. A Controller which conforms to this standard requires a connector on its rear panel for connection to the ACB, All lines on the ACB are bussed to each Controller as in Fig 1.

When an AC performs an addressed command operation, it generates the 5 bit binary code for the station number associated with the command, and transmits it via the Encoded-N lines of the ACB. The CC receives this Station Number code, decodes it, and places a logic "1" on the appropriate Dataway N line at the control station. The CC receives the 24 L signals at the control station and passes these signals to the ACB connector.

The minimum requirement for a CC which permits the use of ACs is that it links the Dataway L lines to the ACB and contains the N-decoder, the ACB connector, and pull-up current sources as in Fig 2.

The priority arbitration consists of two modes: Request/Grant (R/G) and Auxiliary Controller Lockout (ACL). The preferred arbitration mode is the R/G protocol.

Three signals are involved in this mode: (1) the REQUEST signal which is bussed to each Controller on the ACB and is accessible at a front panel connector on each controller; (2) the Grant signal which is daisy-chained, that is, the Grant-Out of one Controller is connected

Fig 1
Multiple Controllers in a CAMAC Crate

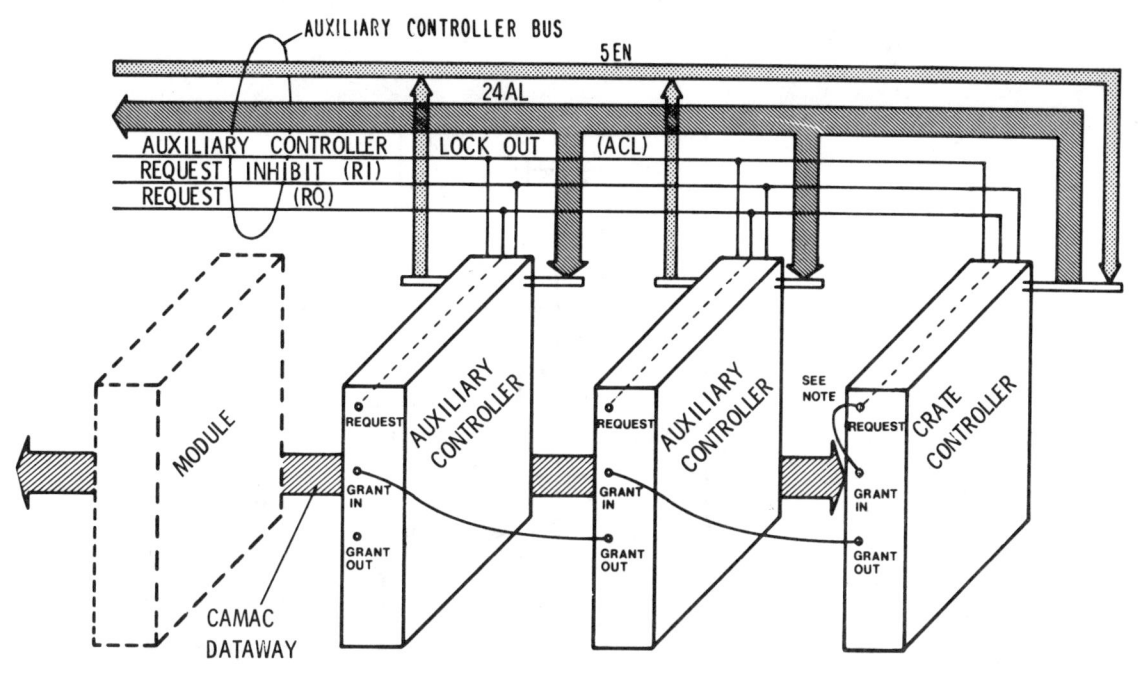

NOTE: FRONT PANEL REQUEST TO GRANT-IN CONNECTION MADE AT HIGHEST PRIORITY CONTROLLER USING R/G PROTOCOL

to the Grant-In of another Controller by front panel connectors; and (3) the Request Inhibit signal which is also bussed on the ACB.

The front panel Request signal output of the highest priority Controller will be connected to its front panel Grant-In signal input. The front panel Grant-Out signal output from this Controller is then connected to the front panel Grant-In signal input of the next highest priority Controller. The connection of Grant-Out to Grant-In is continued from Controller to Controller until it reaches the lowest priority Controller in the CAMAC crate.

The sequence of signals for a Controller to gain control of the crate is shown in Fig 3. A Controller first generates a Request and waits until it receives a Grant-In. Each Controller not generating the Request signal generates a Grant-Out when it receives a Grant-In. The Controller generating the Request signal does not generate a Grant-Out. The chaining of the Grant bus from Controller to Controller ensures that the Grant signal will propagate downstream to the highest priority Controller which is requesting control of the Dataway.

When a Controller requesting control receives Grant-In, it generates and maintains Request Inhibit to indicate that it has control of the crate, and it removes its Request signal. In response to Request Inhibit, any other Controllers also remove their Request signal outputs, thereby causing the Grant signals to be removed. When a Controller has finished its Dataway operations, it removes its Request Inhibit, and control of the crate will be given to the next Controller requesting it. At that time if two or more Controllers request control of the crate at the same time the highest priority Controller will be determined by its position on the Grant chain (Fig 1).

Gain of control of the crate by a Controller is delayed if the Dataway is already in use. If a Controller is connected to an external highway, the interface to the highway is required to accommodate this delay. An example of a highway interface that can accommodate this delay is the CAMAC Parallel Highway (ANSI/IEEE Std 596-1982 or EUR 4600e). The R/G mode is unsuitable for a Controller which can not accommodate this delay. An example is the

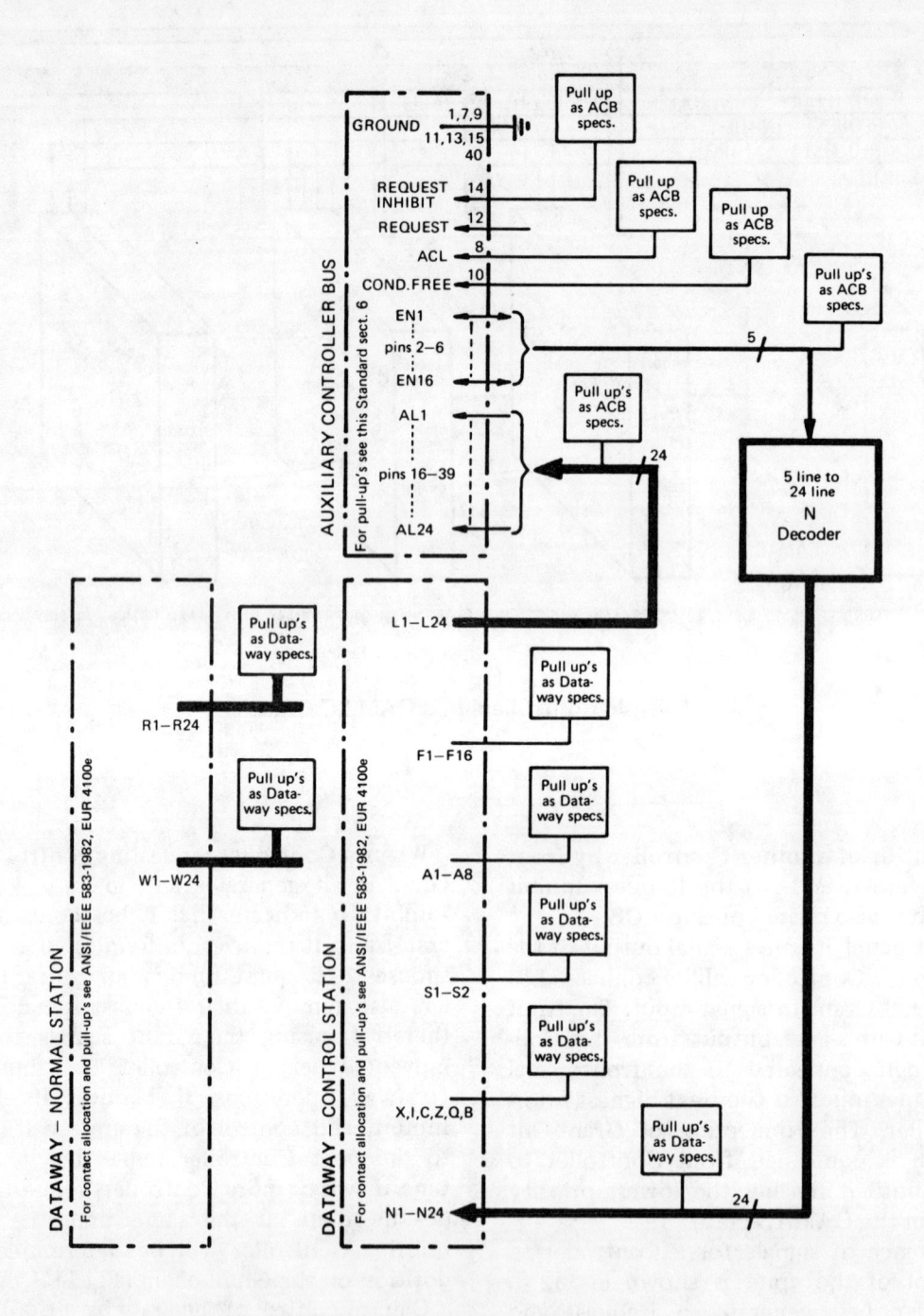

Fig 2
Crate Controller, Minimum Configuration

Serial Crate Controller Type *L*-2 (ANSI/IEEE Std 595-1982 or EUR 6100e). When a Serial Highway Crate Controller Type *L*-2 is addressed by the Serial Highway it will proceed with its Dataway operation independently of the R/G protocol.

The ACL feature is provided to accommodate a Controller which cannot tolerate the delay associated with the R/G protocol. In a given crate, only one Controller (which may be either an AC or the CC) is allowed to use ACL to gain control of the crate. The ACL signal is bussed on the ACB to all other Controllers in the crate. Upon receiving this signal, a Controller which has control of the crate will either abort or complete its operation before the Controller generating ACL starts its Dataway operation (see 4.1.5). Examples of the sequences of signals that may occur with the ACL signal are shown in Fig 4A and B.

The necessary connections of the Request, Grant, Request Inhibit, and ACL lines are illustrated in Fig 1.

The Serial Crate Controller Type *L*-2, as described in the Appendix to ANSI/IEEE Std 595-1982 and EUR 6100e, does not have an ACB connector. However, its SGL-Encoder connector may be used to connect to the ACB since the signals on the ACB are a subset of those on the SGL-Encoder connector. With such an interface, the Serial Crate Controller Type *L*-2 may thus be used as a CC compatible with this standard. However, some *L*-2 controllers may not have a pull-up on the Request Inhibit line. In such instances, it will be necessary to add a pull-up on contact 17 of the SGL encoder connector of the *L*-2 and connect it to the Request Inhibit line.

The unaddressed operations, Dataway Initialize *Z* and Dataway Clear *C*, do not require use of the Encoded-*N* signals of the ACB. However, the Controller still uses one of the priority arbitration modes to gain control of the crate before issuing either of these commands. Care should be taken that a Dataway *Z* or *C* from one Controller does not adversely affect the operations of another Controller.

The Dataway Inhibit *I* is not associated with Dataway operations and may be generated at any time by either Controllers or other plug-in units. In contrast to requirements on other plug-in units (see 5.5.2 of ANSI/IEEE Std 583-1982 or EUR 4100e). Controllers capable of generating and maintaining Dataway *I* do not

Fig 3
Sequence of Signals for an AC to Gain Control of
Dataway for an Addressed Command Operation

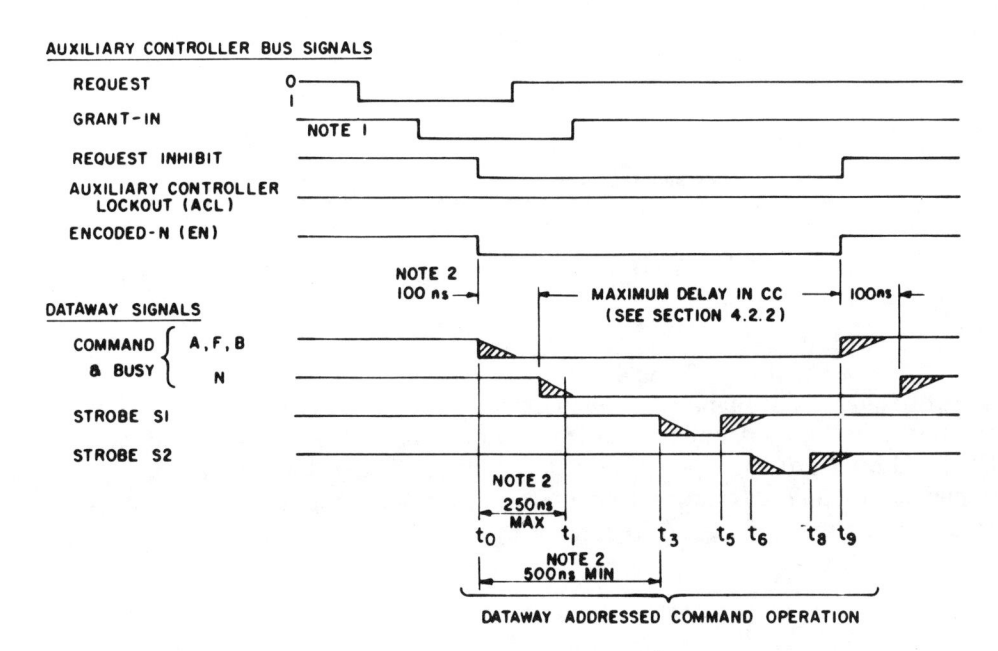

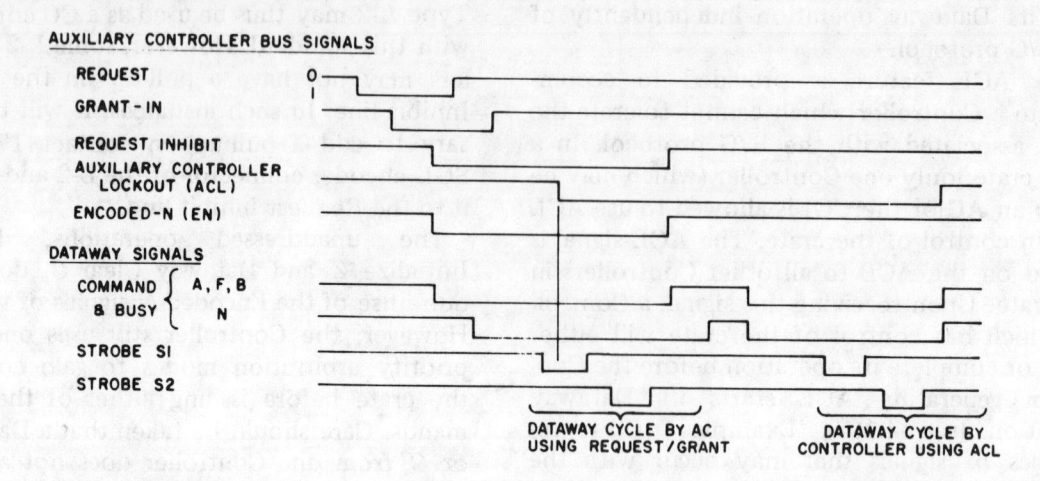

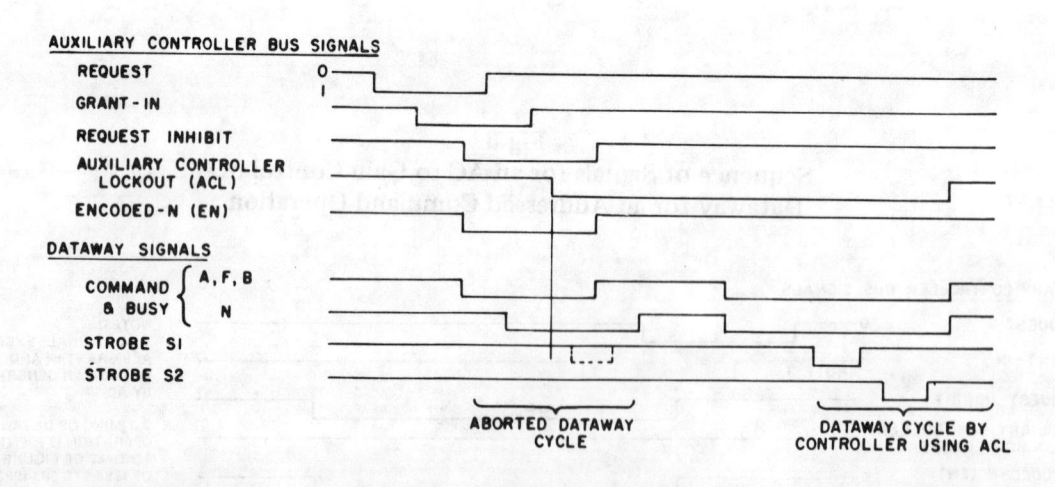

Fig 4
Examples of Sequences of Signals that may Occur with the ACL Signal.
A — Example of a Sequence in Which Auxiliary Controller Lockout (ACL) Signal is Generated Too Late to Cause Aborting of Dataway Cycle Started by AC Using Request Grant;
B — Example of Sequence in Which a Dataway Cycle, Started by AC Using Request/Grant, is Caused to Abort by ACL Signal Generated by Another Controller

IN A CAMAC CRATE

respond to $Z \cdot S2$ by generating and maintaining Dataway I.

4. Use of the Lines on the Auxiliary Controller Bus and Associated Signals

Each line at the ACB connector and the associated front panel signal connectors must be used in accordance with the mandatory requirements detailed in the following sections. Table 1 shows the titles, the standard designations, and the sources of the signals defined in this section.

4.1 Control Signals

A Controller, when used in a CAMAC crate having one or more other Controllers, must not generate any Dataway signals, with the exception of the Dataway I and the L(s) of the station(s) it occupies, unless it has gained control of the crate or is addressed as a module. A Controller must gain control by generating Request in the R/G mode or ACL in the ACL mode. It should preferably gain control by the R/G mode unless there are strong technical reasons to the contrary.

Table 1
Signals Lines at the Auxiliary Controller Bus Connector and
Associated Front Panel Connectors

Title	Location	Designation	Generated by	Signal Lines	Use
Request	ACB and front panel	RQ	CC or AC	1	Indicates request for control
Grant-In	front panel	GI	CC or AC	1	Indicates request is granted
Grant-Out	front panel	GO	CC or AC	1	Issued by Controller when GI is received but Controller is not requesting
Request Inhibit	ACB	RI	CC or AC	1	Indicates Controller has control in Request/Grant mode
Auxiliary Controller Lockout	ACB	ACL	1 CC or AC	1	Indicates lockout control
Encoded-N	ACB	EN1,2,4,8,16	AC	5	Binary coded station number
Auxiliary Look-at-Me	ACB	AL1—AL24	CC	24	24 Look-at-Me lines from modules
Conditional Free line	ACB		CC	1	Line recommended for Byte Clock in CAMAC Serial Highway systems
Ground	ACB	0V	CC or AC	7	System Ground

4.1.1 Request

> In order to gain control of the crate when using the R/G protocol, a Controller must first generate a logic "1" signal on the Request signal line. It must not, however, initiate the 0→1 transition of the Request signal unless Request Inhibit and ACL are both logic "0". If it is generating a Request, it must initiate the 1→0 transition of Request within 50 ns upon receiving either Request Inhibit = 1 or ACL = 1.

4.1.2 Grant-In and Grant-Out Signals

> A Controller participating in the R/G mode must generate a Grant-Out signal as follows:
> (1) It must generate a logic "0" on Grant-Out whenever it receives a logic "0" on Grant-In.
> (2) If it is not generating the Request signal when it receives the 0→1 transition of Grant-In, it must retransmit on Grant-Out the signal it receives on Grant-In.
> (3) If it is generating the Request signal when it receives the 0→1 transition of Grant-In it must maintain a logic "0" on Grant-Out until it receives the next 0→1 transition of the Grant-In signal, and it must generate Request Inhibit to establish control of the crate.

If a Controller retransmits the Grant signal, it should do so with minimum delay.

4.1.3 Request Inhibit

> A Controller gains control of the crate by initiating the 0→1 transition of Request Inhibit, and it maintains control of the crate until it initiates the 1→0 transition of Request Inhibit. It must maintain control of the crate for a minimum of 350 ns unless it receives ACL = 1.

The generation of Request Inhibit by a Controller will establish its control of the crate. If the Controller generates Request Inhibit = 0 between command operations, then it releases its control after each operation, thereby allowing another Controller to gain (and possibly maintain) control. If, on the other hand, the Controller maintains Request Inhibit = 1 between command operations, then the Controller will maintain control of the crate, thus allowing, for example, the execution of a block transfer with minimum delay.

4.1.4 Generation of ACL Signal

> At any one time, the generation of the ACL signal must be reserved to only one Controller in a CAMAC crate. The Controller generating the ACL signal should generate ACL only after it expects to initiate a Dataway operation (for example, on the recognition of the Crate Address in a CAMAC Command addressed to it) in order to allow maximum use of the Dataway by other Controllers. The ACL signal must be maintained until the Dataway operation is complete.
>
> A Controller generating ACL must not initiate its Dataway operation until (1) a minimum of 200 ns has elapsed since generating ACL, and (2) it receives Request Inhibit in a logic "0" state.

The Serial Highway Crate Controller Type L-2, will proceed with its Dataway operation independently of the state of the Request Inhibit signal. After receiving the first byte of a command addressed to it, it will generate ACL in order to gain control of the crate. A Dataway operation may take place after 4 additional bytes have been received. The minimum elapsed time could be as short as 800 ns for a Serial Highway operating at its maximum data rate of 5×10^6 bytes per second in byte-serial mode.

4.1.5 Response to ACL Signal

> A Controller must not initiate a Dataway operation while it is receiving the ACL signal in the logic "1" state. A Controller must complete its Dataway operation if it receives ACL at logic "1" after it has generated strobe $S1$. A Controller must release control of the crate in response to an ACL signal before the Controller generating ACL begins its Dataway operation.

A Controller releases control of the crate by either aborting or completing the Dataway

IN A CAMAC CRATE

operation. When used with the Serial Highway, a Controller may assume that the Serial Highway is operating at maximum speed, in which case 800 ns are allowed for completion. Alternatively, the byte clock information (see 2.3 of ANSI/IEEE Std 595-1982 or EUR 6100e), may be used to extend its control. The latter could permit the Controller to complete several Dataway operations before releasing control.

4.2 Encoded-N Signals
4.2.1 Generation of Encoded-N Signals

In order to execute a Dataway command operation, an AC must generate the binary coded station number of the addressed station on the Encoded-N (EN1 — EN16) lines of the ACB. As with Dataway signals, an AC must not generate the EN signals unless it has gained control of the crate. In the timing of the Dataway command operation of an AC, the time between t_0 and t_1 (See Fig 9 of ANSI/IEEE Std 583-1982 or EUR 4100e) must take into account delays caused by decoding of the EN signals in the CC.

4.2.2 Response to Encoded-N Signals

A CC must repond to the binary coded station numbers on the EN lines whenever it does not have control of the crate. In a CC, each of the Station Number codes $N(1)$ through $N(24)$ must be decoded to produce a signal on the corresponding Dataway line N1 through N24 with a delay of 100 ns maximum.

4.3 Look-at-Me Signals

The CC must retransmit the Look-at-Me ($L1-L24$) signals from the control station of the Dataway to the Auxiliary Look-at-Me (AL1—AL24) contacts of the ACB connector.

4.4 Other Signals

The Conditional Free line is reserved for use as the Byte Clock signal in Serial Highway systems.

5. The Auxiliary Controller Bus Connector and Associated Front Panel Connectors

Each Controller must have a double row post header ACB connector with 40 0.64 mm by 0.64 mm pins on 2.54 mm centers, AMP No 87272-1, AMP No 4-825457-0, 3M No 3495-1002, or an equivalent connector that is fully mateable with AMP No 86896-2 and 3M No 3417-3000. The connector body must be mounted at the rear of the Controller, above the Dataway connector. Contact 1 must be lower right (when facing the tips of the pins), and there must be an indication at the rear of the controller that this is so. Contact 1 must be located 130.4 mm to 133.4 mm above the Dataway plug horizontal datum. Note however that (1) no part of the connector shall extend beyond 188.6 mm above the Dataway plug horizontal datum, and (2) any portion of the connector below 126.6 mm from this datum must be contained within the 290 mm maximum horizontal dimension of the plug-in. (see Figs 4 and 5 of ANSI/IEEE Std 583-1982 or EUR 4100e.) The assignment of the signal lines of the ACB is given in Table 2.

A Controller which uses the Request/Grant protocol to gain control of the crate must have, in addition, three coaxial connectors Type 50CM (see ANSI/IEEE Std 583-1982 or 4.1 of EUR 5100e) on the front panel. The three connectors must be labeled and used as follows:

(1) There must be a connector for the Request signal output. This signal must be the same at all times as the Request signal on the ACB.

(2) There must be a connector for the Grant-In signal input.

(3) There must be a connector for the Grant-Out signal output.

Table 2
Contact Assignments on Auxiliary Controller Bus Connector*

Contact	Usage	Contact	Usage
1	Ground (0V)	2	Encoded-N EN1
3	Encoded-N EN2	4	Encoded-N EN4
5	Encoded-N EN8	6	Encoded-N EN16
7	Ground (0V)	8	ACL
9	Ground (0V)	10	Conditionally Free
11	Ground (0V)	12	Request RQ
13	Ground (0V)	14	Request Inhibit RI
15	Ground (0V)	16	AL1
17	AL2	18	AL3
19	AL4	20	AL5
21	AL6	22	AL7
23	AL8	24	AL9
25	AL10	26	AL11
27	AL12	28	AL13
29	AL14	30	AL15
31	AL16	32	AL17
33	AL18	34	AL19
35	AL20	36	AL21
37	AL22	38	AL23
39	AL24	40	Ground (0V)

*Contact 2 is across from contact 1, contact 4 is across from contact 3, etc.

Table 3
Current Signal Standards and Pull-Up Current Sources for the Auxiliary Controller Bus Connector and Associated Front Panel Connnectors*

Signal standards at connector	Aux Cont Lockout, Request Inhibit	Aux LAM	Request Grant-In/Out	Encoded-N Cond Free
Line at "1" state at 0.5 V Minimum Current Sinking Capability (current drawn from line by unit generating the signal)	For CC 6.4 mA For AC 16.0 mA	3.2 mA	16.0 mA	16.0 mA
Line at "1" state at 0.5 V Maximum Load Current (current fed into line by unit receiving the signal)	0.4 mA per unit (6.4 mA maximum)	0.4 mA per unit (3.2 mA maximum)	12.8 mA	11.2 mA
Line at "0" state at 3.5 V (maximum current drawn from line by CC without sinking pull-up)	100 μA	100 μA	100 μA	100 μA
Line at "0" state at 3.5 V (minimum current fed into line by CC with sinking pull-up)	2.5 mA	2.5 mA	2.5 mA	2.5 mA
Location of pull-up for current, I_p, at 0.5 V: 6 mA $\leq I_p \leq$ 9.6 mA	CC	CC	Grant-In	CC

*A characteristic impedance termination is being considered for the Request line and the Request/Grant connections. This could increase the current sinking requirements noted in this table.

6. Signal Standards

> Signal outputs from Controllers onto all ACB lines must be delivered through intrinsic OR gates. Each line must be provided with an individual pull-up current source to restore the line to the "0" state in the absence of an applied "1" signal. The rise and fall times at signal outputs must not be less than 10 ns, in order that cross-coupling of signals is not excessive.
>
> Signal outputs from Controllers onto all Dataway lines must conform fully with the mandatory requirements of ANSI/IEEE Std 583-1982 or EUR 4100e.

6.1 Signal Standards on the ACB

> All signals on the ACB must conform to the signal voltage standards shown in Table 5 of ANSI/IEEE Std 583-1982 or EUR 4100e and the standards for pull-up current sources in Table 3 of this standard.

The standards for pull-up current sources on the ACB are derived from Table 6 of ANSI/IEEE Std 583-1982 or EUR 4100e and correspond with those for compatible current sinking logic devices (for example, standard TTL and DTL for units generating signals and low power Schottky TTL for units receiving signals). They also correspond with those of the SGL-Encoder connector of Serial Crate Controllers (see Section 14 of ANSI/IEEE Std 595-1982 or EUR 6100e). These standards impose an upper limit on the number of ACs in a single CAMAC crate. This limit is 8 ACs, and it is derived from the current sinking capability and load current on the AL lines. If an SGL-Encoder unit is attached to the SGL-Encoder connector it may impose a more severe restriction on the number of ACs. If some ACs do not receive the AL signals, then the limit on the number of ACs is imposed by the load current on the Dataway (see 6.3).

6.2 Signal Standards for the ACB Associated Front Panel Signals

> The Request, Grant-In, and Grant-Out front panel signals must conform to the signal voltage standards shown in Table 5 of ANSI/IEEE Std 583-1982 or EUR 4100e and the standards for pull-up current sources in Table 3 of this standard.

6.3 Signal Standards for AC Dataway Connections

> All signals at the CAMAC Dataway-interface to an AC must conform to Table 6 of ANSI/IEEE Std 583-1982 or EUR 4100e, with the exception of the signals Q, R, and X. The Q, R, and X signals must conform to standards contained in Table 4 of this standard.

The signal standards for the CAMAC Dataway interface impose a limit of 12 ACs in a single CAMAC crate. However, if other units fully compatible with ANSI/IEEE Std 583-1982 or EUR 4100e receive Q, R, or X signals from the Dataway, then the limit on the number of ACs may be lower. Such a unit may present a current load on the Dataway as much as that of 4 ACs.

6.4 Protection of Critical Signal Inputs.
The Grant-In and Request Inhibit signal inputs should be protected against spikes introduced by cross-coupling from other ACB or Dataway signal lines. This protection should be chosen in accordance with the specific design of the controller. It will usually include an integrating time constant of approximately 50 ns on the Grant-In, and an integrating time constant of approximately 50 ns that is effective on the positive-going 1-to-0 edge of the Request Inhibit input of each controller.

The purpose of this protection is to prevent more than one controller gaining control of the Dataway because of spikes or other unintentional transient signal excursions.

Table 4
Signal Standards for Q, R, and X at the Auxiliary Controller Dataway Connector

Condition at Dataway Connection	Absolute Limit
Line in "1" state at +0.5 V (maximum current fed into line by AC receiving signal)	0.4 mA
Line in "0" state at +3.5 V (maximum current drawn from line by each AC)	100 µA

Appendix

(This Appendix is not a part of ANSI/IEEE Std 675-1982, IEEE Standard, Multiple Controllers in a CAMAC Crate.)

A1. Specification of CAMAC Crate Controller Type *A*-2

A1.1 CAMAC Crate Controller Type *A*-2

> In order to conform with the specifications for CAMAC Controller Type *A*-2, a crate controller must have all the mandatory features defined in this Appendix. It must have no other features that would affect its interchangeability with any other Type *A*-2 Crate Controllers, taking into account the effect of such interchange on both hardware and software. It must be fully interchangeable with one conforming to Fig A1, although it need not have identical structure, internal signals (shown without the prefix B in Fig A1) or logical expressions.

With respect to the communications protocol via the Branch highway, Crate Controller Type *A*-2 (CCA-2) is interchangeable with Crate Controller Type *A*-1 (CCA-1) as defined in Section A1 of the Appendix of ANSI/IEEE Std 596-1982 or EUR 4600e. In practice the Branch Highway cycle times may differ because of the priority arbitration logic used by the CCA-2.

In order to accommodate the use of ACs, the CCA-2 differs from the CCA-1 by having an ACB connector, and its associated features as described in ANSI/IEEE Std 675-1982 or EUR 6500e (Sections 1-6 of this standard). Differences between this Appendix and Section A1 of the Appendix of ANSI/IEEE Std 596-1982 or EUR 4600e are indicated by a vertical bar in the left-hand margin of this Appendix.

A1.2 Other Crate Controllers

It is recommended that other crate controllers should be interchangeable with Crate Controller Type *A*-2 in respect to those features that they have in common, although they need not have all the mandatory features of Crate Controller Type *A*-2 and may have additional features.

A1.3 General Features

> The crate controller must conform fully with the mandatory requirements of ANSI/IEEE Std 583-1982 or EUR 4100e, and ANSI/IEEE Std 596-1982 or EUR 4600e (Sections 1-7 of that standard). It is mandatory that all signal inputs at the Branch Highway ports of Crate Controller Type *A*-2 must satisfy the lower input current standard (± 0.3 mA) shown in Table 8 of ANSI/IEEE Std 596-1982 or EUR 4600e.
>
> Crate Controller Type *A*-2 must not occupy more than three stations. It should preferably be a double-width unit that engages with the Dataway at the control station and the adjacent normal station.
>
> In addition to the two front panel connectors for the Branch Highway ports (see Section A1.4), the CC must have a rear-mounted connector for a link to an optional separate LAM-Grader unit (See Section A1.9).
>
> The CC must conform fully with the mandatory requirements of ANSI/IEEE Std 675-1982 or EUR 6500e (Sections 1-6 of this standard). In addition to the rear-mounted LAM-Grader connector, the CC must have a rear-mounted connector for a link to the ACB.

A1.4 Front Panel

> The CC must have all the following front panel features, and no others that would affect interchangeability (for example, the addition of indicators for test purposes is permitted).
>
> (1) There must be two connectors for Branch Highway ports, as defined in Section 6 of ANSI/IEEE Std 596-1982 or EUR 4600e, mounted with the correct orientation and with all corresponding contacts joined.
>
> (2) There must be a means of indicating the selected crate address (1-7). There may be limited access at or through the front panel to the means of changing the crate address.
>
> (3) There must be a means of selecting off-line status of the CC (see A1.10).
>
> (4) There must be a coaxial connector for the Inhibit signal input. The type of connector is as specified in 4.2.5 of ANSI/IEEE Std 583-1982 or 4.1 of EUR 5100e. The signal standards are as specified in 7.2.1 of ANSI/IEEE Std 583-1982 or EUR 4100e.
>
> (5) There must be two push buttons, or equivalent manual controls, for Initalize and Clear. These are only effective in the off-line state, and the front panel layout or markings should indicate this.
>
> (6) There must be three coaxial connectors for the Request and Grant signals. These connectors must conform fully with the mandatory requirements of ANSI/IEEE Std 675-1982 or EUR 6500e (Section 5 of this standard).
>
> (7) There must be a means of indicating the state of the (R/G)/ACL control option (see A1.7). There may be a limited access at or through the front panel to the means of changing this control option.

A1.5 Dataway Signals

> The Crate Controller Type A-2 must gain control of the crate in accordance with the mandatory requirements of 4.1 of this standard.

A1.5.1 Data Signals

> When the CC is on-line, addressed, and in control of the crate during a Read command operation with a station number other than $N(30)$, it must retransmit the signals from the 24 Dataway Read lines through intrinsic OR outputs to the BRW lines. Crate Controller Type A-2 must have gates between the R and BRW lines so that this transfer of Read data occurs only when the crate controller is on-line, addressed, and in control of the crate, for example when $BCR_i \cdot (BTA + \overline{BTB_i}) = 1$.
>
> During Write operations with station number other than $N(30)$ it must retransmit the signals from the 24 BRW lines to the Dataway Write lines. Crate Controller Type A-2 must have gates between the BRW and W lines so that this transfer of Write data occurs only when the crate controller is addressed, on-line, and in control of the crate.

The gates between the R and BRW lines and between the BRW and W lines may further limit the transfers to Read operations $\overline{BF16} \cdot \overline{BF8} = 1$ and Write operations $BF16 \cdot \overline{BF8} = 1$, respectively. However, the CC is permitted to generate signals on the Dataway Write lines during any operation for which it has control, but other units connected to the Dataway can only rely on the presence of such signals during Dataway Write operations.

A1.5.2 Command Signals

The Branch Highway command signals BN, BA, and BF should be conditioned in the crate controller, for example, by integration or by staticizing at a time related to BTA $0 \rightarrow 1$ (the 0 to 1 transition of BTA), in order to protect the Dataway command lines from the effects of crosstalk into Branch Highway command lines.

> The subaddress and function signals from the BA and BF lines must be retransmitted by the CC on the Dataway A and F lines during all command mode operations when

the controller is on-line, addressed, and in control of the crate.

In a double-width CC each of the Station Number codes $N(1)$ through $N(23)$ must be decoded in the CC to produce a signal on the corresponding Dataway line $N1$ to $N23$. The Station Number code must be decoded from the BN lines of the Branch Highway port connector whenever the CC is in control of the crate in response to a Branch Highway operation. At all other times the Station Number codes must be decoded from the EN lines of the ACB connector.

Station Number codes $N(26)$ through $N(30)$ received from the Branch Highway port connector (but not necessarily the ACB connector) must be decoded to address internal features of the CC.

Command operations with $N(26)$ must generate Dataway signals on all the lines $N1$ through $N23$. Command operations with $N(24)$ generate Dataway signals on $N1$ through $N23$ as determined by the contents of 23 bit Station Number Register (SNR). This register is loaded from BRW1-BRW23 by the command $N(30) \cdot A(8) \cdot F(16)$. The bit of the SNR that is loaded from BRW1 controls the state of $N1$, etc. The register is not reset by the Dataway Initialize signal Z.

A triple-width controller may alternatively have a 22 bit SNR decode $N(1)$ through $N(22)$, and generate signals on Dataway lines $N1$ through $N22$.

A1.5.3 Common Control Signals

The Dataway Initialize signal Z must be generated in response to the command $N(28) \cdot A(8) \cdot F(26)$ and to the Branch Initialize signal (see 4.5.1 of ANSI/IEEE Std 596-1982 or EUR 4600e). It must also be generated in response to the manual Initialize control, but only when the CC is in the off-line state.

The Dataway Clear signal C must be generated in response to the command $N(28) \cdot A(9) \cdot F(26)$. It must also be generated in response to the manual Clear control, but only when the CC is in the off-line state.

The Dataway Initialize Z and Clear C signals must not be generated until the crate controller is in control of the crate and must be generated with the timing specified for unaddressed operations in ANSI/IEEE Std 583-1982 or EUR 4100e, Fig 10. They must be associated with a sequence including B and $S2$ signals, also with the time specified in ANSI/IEEE Std 583-1982 or EUR 4100e, Fig 10. The sequence is permitted to include $S1$ but this is not mandatory, and other units connected to the Dataway must not rely on the generation of $S1$ with Z and C.

The Dataway Initialize Z signal must be generated in response to the Branch Initialize signal only when the CC is in control of the crate and the Branch Initialize signal is in the logic "1" state. (Note that Crate Controller Type A-2 could fail to generate Dataway Initialize Z in response to Branch Initialize while an auxiliary controller is performing continuous Dataway operations.)

The Dataway Inhibit signal I must be initiated when an on-line crate controller generates Dataway Initialize Z, and must reach a maintained "1" state not later than time t_3 (see ANSI/IEEE Std 583-1982 or EUR 4100e, Fig 10). When some other unit generates Initialize (accompanied or not by Inhibit), an on-line crate controller must not generate Inhibit in response to Dataway Z gated by $S2$. The Inhibit signal must be generated in response to the command $N(30) \cdot A(9) \cdot F(26)$. In these cases the Inhibit signal must be maintained by the crate controller until reset by the command $N(30) \cdot A(9) \cdot F(24)$. It must also be generated while the front panel Inhibit signal is present.

The command $N(30) \cdot A(9) \cdot F(27)$ must produce a $Q = 1$ response if there is a "1" state signal on the Dataway Inhibit line.

A1.5.4 Patch Connections

Crate Controller Type A-2 must not use the patch pins of the Dataway stations that it occupies.

A1.5.5 Arbitration Methods Used by the Crate Controller

Crate Controller Type A-2 must provide for both the R/G and ACL methods of gaining control of the crate. Either the ACL method or the R/G method will be used depending on the state of the (R/G)/ACL option control. If the state of this control is "1" the ACL method must be used. If the state of the control is "0" the R/G method must be used. When the crate controller is using the R/G method, it must respond to an ACL signal on the ACB (see 4.1.5 of this standard).

A1.6 Demand Handling
A1.6.1 Branch Demand

The Branch Demand signal BD must be derived, subject to the following conditions, from the OR combination of an External Demand signal from contact 48 of the LAM-Grader connector and an Internal Demand signal which is the OR of the 24 GL signals received via the LAM-Grader connector.

The output of the Branch Demand signal to the BD line must be disabled by the command $N(30) \cdot A(10) \cdot F(24)$ or by the Dataway Initialize signal Z with $S2$ when generated by the CC. It must be enabled by the command $N(30) \cdot A(10) \cdot F(26)$. The command $N(30) \cdot A(10) \cdot F(27)$ must give a $BQ = 1$ response if the output of the BD is enabled. The command $N(30) \cdot A(11) \cdot F(27)$ must give a $BQ = 1$ response if the OR of the Internal and External Demands is in the "1" state, even if the output of BD is disabled.

The Internal Demand signal must be inhibited by the "1" state of the Inhibit Internal D signal from contact 51 of the LAM-Grader connector.

A1.6.2 Graded-L

In response to a Graded-L Request signal BG = 1, accompanied by $BCR_i = 1$, the CC must gain control of the crate. It must generate the Graded-L operation signal on contact 1 of the LAM-Grader connector. It must accept the Graded-L signals GL1-GL24 from the LAM-Grader connector and transmit them to the BRW lines (GL1 to BRW1, etc).

The crate controller must also accept the Graded-L signals from the LAM-Grader connector and transmit them to the BRW lines in response to command mode operations with $N(30) \cdot A(0-7) \cdot F(0)$ (see A1.9.4).

In both cases the GL information must be transferred from the LAM-Grader to the BRW lines with minimum delay, and the signals must not appear on the Dataway Read lines.

A1.6.3 Pull Up for GL and L Lines

Pull up current sources in accordance with ANSI/IEEE Std 583-1982 or EUR 4100e, Table 6, must be provided on all GL lines of the LAM-Grader connector and all AL lines of the ACB connector in the CC. The L lines of the LAM-Grader connector must be isolated from the Dataway L lines and must not be provided with pull-up current sources, thus allowing a simple LAM Grader to form wired-OR combinations of L signals without affecting the AL lines on the ACB.

A1.7 Timing Requirements

The initiation of any operation by the CC is determined by the Branch Highway signals BCR_i, BG, BZ, and BTA and the signals associated with ACB (Request, Grant, ACL, and Request Inhibit).

Before responding to any Branch Highway operation, the crate controller must first gain control of the crate. It must initiate the process of gaining control of the crate when it detects $BCR_i = 1$. It must maintain control of the crate, by generating Request Inhibit or ACL, until it detects $BCR_i = 0$.

If the Branch Driver generates $BCR_i = 0$ between command operations, the CC will release the crate after each operation allowing auxiliary controllers to gain control (and possibly maintain control) of the crate. If on the other hand, the Branch Driver maintains $BCR_i = 1$ between command operations, then the CC will maintain control, thus allowing, for example, the execution of a block transfer with minimum delay.

In command mode operations with Station Number codes other than $N(30)$, the CC generates Dataway Strobe signals $S1$ and $S2$ with timing related to that of the Branch timing signals BTA and BTB as defined in A1.7.1.

Command operations with Station Number code $N(30)$ do not generate, $S1$, $S2$, or B signals on the Dataway lines (see A1.7.3).

In Graded-L operations there are no Dataway Strobe or B signals, and the timing must take into account the signal delays in any non-Dataway connections to a LAM-Grader unit. These timing requirements are defined in A1.7.2 and A1.9.3.

> The internal timing generator of the CC must be protected against spurious signals on the BTA and BCR lines.

One method of protection, shown in Fig A1, is to condition the incoming signals from the BTA line and the selected BCR line by integration with a time constant of 100 ± 50 ns. Another method is to condition the internal signal TA which controls the timing generator. Transitions of the BTA and BCR signals are detected by the CC after a delay (shown in Figs 3 and 4 of IEEE Std 596-1982 or EUR 4600e) due to this protection.

A1.7.1 Command Mode Operations with Dataway $S1$, $S2$, and B.

> The following timing conditions must be satisfied when the crate controller responds to a command mode branch operation which requires a Dataway operation with signals $S1$, $S2$, and B. In this section the times t_0, t_3, t_5, etc, refer to the corresponding key points on Fig 9 of ANSI/IEEE Std 583-1982 or EUR 4100e.
>
> During Phase 1 of the operation, the Branch Driver generates the command signals BF, BA, and BN along with the signals BCR_i of the addressed CCs. The CC must initiate the process of gaining control of the crate with minimum delay when it detects the 0→1 transition of BCR_i (see Section 4 of this standard).
>
> In Phase 2 of the operation, after actions by the Branch Driver and the CC during Phase 1, the CC detects BTA = 1, accompanied by BG = 0, BCR_i = 1, and the appropriate command signals. It must then initiate the required Dataway N and command signals, thus starting the Dataway operation at t_0, either when it detects BTA = 1 or at the appropriate time after it has gained control of the crate, if this occurs later (see Section 4 of this standard.)
>
> At t_3, which is 400(+200/-0) ns after t_0, the CC must initiate the 0→1 transition of the Dataway Strobe $S1$, and the Branch timing signal transition BTB_i 1→0. At t_5, which is 200(+100/-0) ns after t_3, the 1→0 transition of the Strobe $S1$ must be initiated.
>
> In Phase 4, the CC initiates the 0→1 transition of Strobe $S2$ at t_6, which is either when it detects BTA = 0 or when the interval $t_5 - t_6$ is 100(+100/-0) ns, if this occurs later.
>
> At t_8 which is 200(+100/-0) ns after t_6, the 1→0 transition of $S2$ must be initiated.
>
> At t_9, which is 100(+100/-0) ns after t_8, the CC must initiate the 1→0 transitions of the Dataway signals N and B, and must isolate the Dataway Q and R lines from the Branch Highway BQ and BRW lines respectively. It must then initiate the Branch timing signal transition BTB_i 0→1. When the CC detects BCR_i = 0, it must initiate the 1→0 transition of the Request Inhibit signal or the ACL signal if the latter is in use.

A1.7.2 Graded-L Operations

> The CC must satisfy the following timing conditions during Graded-L operations with BG = 1 and BCR_i = 1. In Phase 1 of the operation, the Branch Driver generates the command signal BG along with the signals BCR_i of the addressed CCs. The CC must initiate the process of gaining control of the crate with minimum delay when it detects the 0→1 transition of BCR_i. In Phase 2 it must initiate Branch timing signal transition BTB_i 1→0 within 400(+200/-0) ns after detecting BTA = 1 or 400(+200/-0) ns after gaining control of the crate, if this occurs later. At the same time it must be presenting to its BRW outputs the GL information received via the LAM-Grader connector (see A1.9.3). In Phase 4 it must remove the GL information from its BRW outputs with minimum delay after detecting BTA = 0, and initiate the signal trans-

IN A CAMAC CRATE

ition BTB_i 0→1. The CC must maintain control of the crate until it detects $BCR_i = 0$, and it must then initiate the 1→0 transition of the Request Inhibit signal or the ACL signal if the latter is in use.

A1.7.3 Command Mode Operations without Dataway $S1$, $S2$, or B

Command mode operations addressed to $N(30)$ are concerned with internal features of the crate controller and with reading Graded-L information via the LAM-Grader connector. The crate controller must not generate signals on the Dataway $S1$, $S2$, B, or R lines.

The timing of these operations must follow the requirements for command mode operations (see A1.7.1) with the exception that the $S1$, $S2$, and B signals are not generated on the Dataway lines, although there may be equivalent internal signals.

A1.8 Commands Implemented by Crate Controller Type A-2

Crate Controller Type A-2 must recognize and implement the commands summarized in Table A1 and must not use any other commands. When addressed with any of these commands it must generate BX = 1. The five Function codes $F(0, 16, 24, 26, 27)$ must be fully decoded in the CC.

The CC must generate BQ = 1 in response to all commands that read from or write to its registers, or the LAM-Grader connector. In Crate Controller Type A-2 the two commands to which this applies are $N(30) \cdot A(0\text{-}7) \cdot F(0)$ and $N(30) \cdot A(8) \cdot F(16)$.

A1.9 LAM-Grader Connector

The rear-mounted connector for a link to an optional separate LAM-Grader unit must be a 52 contact Cannon Double-Density fixed member with pins (Type 2DB52P), or equivalent type fully mateable with the free member with sockets, Cannon Type 2DB52S. It must be mounted at the rear of the CC above the Dataway connectors within the area for free access (see ANSI/IEEE Std 583-1982 or EUR 4100e, Fig 3), with Contact 1 lowermost. The 52 contacts are assigned as shown in Table A2.

The LAM Grader accepts L signals from the CC via the LAM-Grader connector. It generates Graded-L GL signals and, optionally, the External Demand signal. It may include gates, mask registers, etc, for processing the L signals, or may merely consist of passive interconnections between the contacts of the LAM-Grader connector. It may interact with the crate controller in the following ways:

(1) Branch Demand. Crate Controller Type A-2 derives the Branch Demand BD signal from the Graded-L signals (and, optionally, the External D signal) which it receives via the LAM-Grader connector.

(2) Graded-L Operations. The CC generates the Graded-L Operations signal on Contact 1 to indicate that it requires Graded-L signals.

If the LAM Grader responds to this signal it must satisfy the timing requirements of A1.9.3.

(3) Command Mode Operations. In response to Branch commands with $N(28)$ or $N(30)$ the CC generates the Controller Addressed signal on Contact 50. This allows the LAM Grader to be treated as a detached part of the CC that can be addressed independently of its location in the crate. The presence of Dataway Busy B distinguishes operations with $N(28)$ from those with $N(30)$. The Controller Addressed signal with Dataway $A(0\text{-}7)$, but without B, indicates that the CC requires Graded-L signals. In conjunction with a Dataway operation and B the Controller Addressed signal may be used, for example, to access registers in the LAM Grader.

If the LAM Grader responds to the Controller Addressed signal it must satisfy the timing requirements of A1.9.4.

The Graded-L Operation signal on Contact 1 must be in the logic "1" state when the CC is on-line, in control of the crate, and $(BTA + \overline{BTB_i}) \cdot BG \cdot BCR_i = 1$.

The Controller Addressed signal on Contact 50 must be in the "1" state during command mode operations to $N(28)$ or $N(30)$ when the CC is on line, in control of the crate, and $[N(28) + N(30)] \cdot (BTA + \overline{BTB_i}) \cdot \overline{BG} \cdot BCR_i = 1$.

Equivalent conditions for the generation of these two signals, Controller Addressed and

ANSI/IEEE
Std 675-1982

IEEE STANDARD: MULTIPLE CONTROLLERS

Table A1
Branch Commands Implemented by CAMAC
Crate Controller Type A-2

Action	Command N	A	F	Response
Generate Dataway Z	28	8	26	BQ=0
Generate Dataway C	28	9	26	BQ=0
Read GL	30	0-7	0	BQ=1
Load SNR	30	8	16	BQ=1
Remove Dataway I	30	9	24	BQ=0
Set Dataway I	30	9	26	BQ=0
Test Dataway I	30	9	27	BQ=1 if I=1
Disable BD output	30	10	24	BQ=0
Enable BD output	30	10	26	BQ=0
Test BD output enabled	30	10	27	BQ=1 if BD enabled
Test demands present	30	11	27	BQ=1 if demands present

Table A2
Contact Assignments for LAM-Grader Connector of
Crate Controller Type A-2

Contact	Usage	Contact	Usage
1	Graded-L operation	2	$L1$
3	GL1	4	$L2$
5	GL2	6	$L3$
7	GL3	8	$L4$
9	GL4	10	$L5$
11	GL5	12	$L6$
13	GL6	14	$L7$
15	GL7	16	$L8$
17	GL8	18	$L9$
19	GL9	20	$L10$
21	GL10	22	$L11$
23	GL11	24	$L12$
25	GL12	26	$L13$
27	GL13	28	$L14$
29	GL14	30	$L15$
31	GL15	32	$L16$
33	GL16	34	$L17$
35	GL17	36	$L18$
37	GL18	38	$L19$
39	GL19	40	$L20$
41	GL20	42	$L21$
43	GL21	44	$L22$
45	GL22	46	$L23$
47	GL23	48	External D
49	GL24	50	Controller Addressed
51	Inhibit Internal D	52	Ground (OV)

Graded-*L* operation, are shown in Fig A1 in terms of the internal (nonmandatory) signals of a particular implementation of Crate Controller Type *A*-2.

All mandatory timing requirements refer to signal conditions at the LAM-Grader connector on the CC. The interval between the initiation of a signal by the CC and the receipt of an established response from the external unit thus includes delays due to both the external unit and its linking cable.

A1.9.1 Signal Standards

All signals via the LAM-Grader connector must satisfy 7.1 of ANSI/IEEE Std 583-1982 or EUR 4100e. This signal standard for *N* signals applies to the Graded-*L* operation signal on Contact 1, and also to the Controller Addressed signal on Contact 50. All other signals including External *D* on Contact 48 and Inhibit Internal *D* on Contact 51, follow the standard for *L* signals.

A1.9.2 Timing — Branch Demand

The maximum overall delay between the time when an *L* signal at the control station of the CC reaches a maintained "1" or "0" state and the time when the BD signal at the Branch Highway port of the same CC reaches a corresponding maintained "1" or "0" state is defined in 4.4.1 of ANSI/IEEE Std 596-1982 or EUR 4600e. When the CC is used in conjunction with an external LAM Grader the component of this delay due to the CC must not exceed 250 ns.

A1.9.3 Timing — Graded-*L* Operations

The interval between the generation of the Graded-*L* operation signal, accompanied by *L* signals, and the establishment of corresponding Graded-*L* signals must not exceed 350 ns.

A1.9.4 Timing — Command Mode Operations

The interval between the generation of the Controller Addressed signal (accompanied by *L* signals, and in conjunction with Dataway signals $F(0)$, $A(0-7)$, $B = 0$) and the establishment of corresponding Graded-*L* signals must not exceed 350 ns.

The external unit must present identical GL information in response to the Graded-*L* operation signal and to the Controller Addressed signal with $A(0)$, $F(0)$, and $B = 0$. Subaddresses $A(1-7)$ may be used to access different selections of GL information.

If the external unit responds to command mode operations with $N(28) \cdot A(0-15)$, $B = 1$, and an *F* code, it must satisfy the normal timing requirements of ANSI/IEEE Std 583-1982 or EUR 4100e for a CAMAC module and is permitted to make data transfers via the Dataway *R* and *W* lines.

A1.10 Off-Line State

The off-line state is selected by means of the off-line manual control on the front panel of the CC. In this state the controller does not respond to command or Graded-*L* Request signals on the Branch Highway, and does not generate Branch timing or demand signals on the Highway.

The following minimum conditions must be observed in the off-line state:

(1) The front panel manual controls for Initialize and Clear must be effective. (They must be ineffective in the on-line state.)

(2) The front panel Inhibit signal input must continue to be effective. Dataway Inhibit must only be generated in response to the front panel Inhibit input.

(3) The CC must not respond to BTA = 1. It must not generate Dataway *B*, *N*, *S*1 or *S*2 signals in response to BTA = 1 with BG = 0, or access the Graded-*L* information in response to BTA = 1 with BG = 1.

(4) The CC must not generate "1" state outputs to the BTB, BD, BRW, BQ, or BX lines. An off-line crate is thus prevented from interfering with Branch operations.

> (5) The CC must not respond to BZ = 1.
>
> (6) The CC must not take any action that would interfere with the control or use of the Dataway by ACs.

The following additional condition is recommended in the off-line state:

(7) In the absence of power to the CC, all inputs and outputs via the Branch Highway ports should be free to assume either the "0" state or the "1" state, as required by other units connected to the Branch, and should not impose abnormal current loadings.

A1.11 Dataway Inhibit I in Off-Line State

It is mandatory in ANSI/IEEE 583-1982 or EUR 4100e (5.5.2) that units generating Initialize Z must also generate I. Units that can generate and maintain I must maintain $I = 1$ until specifically reset. See 5.5.2 of ANSI/IEEE Std 583-1982 or EUR 4100e. Both those requirements are met by Crate Controller Type A-2 in the on-line state. However, A1.10(2) of this standard specifically prohibits the generation of $I = 1$ in the off-line state other than in response to the front panel Inhibit input. The off-line state has been defined in such a way that an AC can control the Inhibit I signal while the CC is off line. Section A1.10(2) is primarily intended to prevent the generation of maintained $I = 1$ by Crate Controller Type A-2 when it is in the off-line state.

To be consistent with this aim:

(1) Crate Controller Type A-2 should generate I as defined in A1.5.3 and A1.10 (and as shown in Fig A1).

(2) Any auxiliary means of generating I (other than ACs) in an off-line crate should conform to 5.5.2 of ANSI/IEEE Std 583-1982 or EUR 4100e by generating $I = 1$ in response to $Z \cdot S2$. It should preferably maintain $I = 1$ and provide a means of resetting it.

A1.12 Auxiliary Controller Bus Connector

> Crate Controller Type A-2 must have a rear mounted ACB connector of the type defined in Section 5 of this standard and with the contact assignment of signal lines given in Table 2 of this standard.

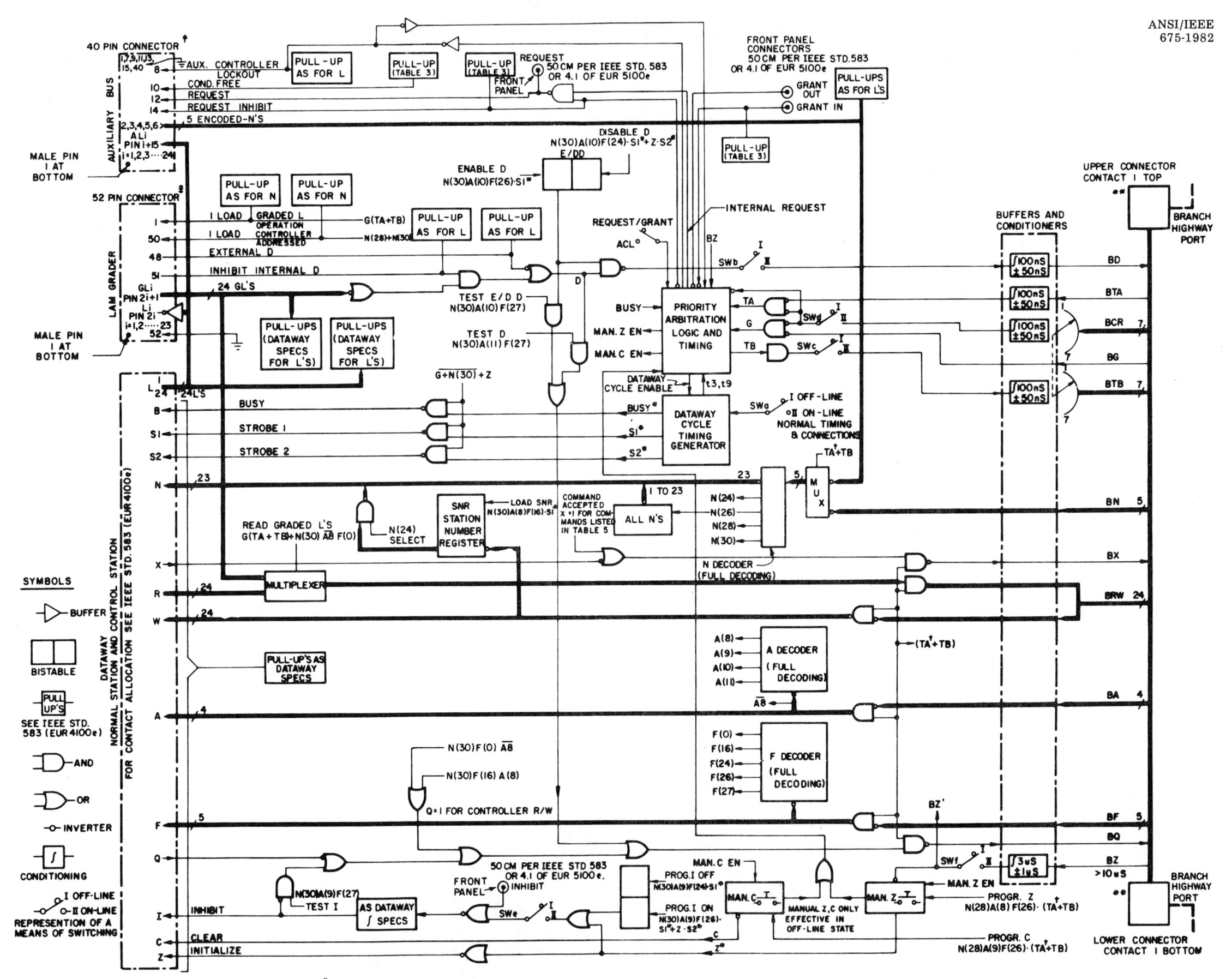

Fig A1
CAMAC Crate Controller Type A-2
(Double Width Unit)

Index

Item	Section
Abbreviations	1
ACL signal	4
generation of	4.14
response to	4.15
Appendix	A1
Auxiliary Controller Bus	3, 4, 5, Tables 1, 2
Auxiliary Controller Bus connector	5, Tables 1, 2
Connector, Auxiliary Controller Bus	5, Tables 1, 2
Contact assignments	Table 2
Current signal standards	Table 3
Connector, LAM-Grader	A1.9
Connectors, front panel	5
Control signals	5
Crate controller	Figs 1, 2
Crate Controller Type A-2	A1, A1.1, Fig A1
Arbitration methods	A1.5.5
Auxiliary Controller Bus connector	A1.12
Branch Demand	A1.6.1, A1.9.1
Command signals	A1.5.2
Commands implemented	A1.8, Table A1
Common control signals	A1.5.3
Connector, Auxiliary Controller Bus	5, Tables 1, 2
Connector, LAM-Grader	A1.9
Data signals	A1.5.1
Dataway Inhibit in Off-Line state	A1.11
Dataway Signal	A1.5
Demand handling	A1.6
Front Panel	A1.4
General features	A1.3
GL Pull Up	A1.6.3
Graded-L	A1.6.2
Inhibit	A1.11
L Pull Up	A1.6.3
LAM-Grader connector	A1.9
LAM-Grader contact assignments	Table A2
LAM-Grader contact, signal standards	A1.9.1
LAM-Grader timing	A1.9.2, A1.9.3, A1.9.4
Off-line state	A1.10, A1.11
Signal standards, LAM-Grader connector	A1.9.1
Crate Controller Type A-2 (Continued)	
Signal standards, timing, Branch Demand	A1.9.2, A1.9.3, A1.9.4
Signals, Command	A1.5.2
Signals, common control	A1.5.3
Timing	A1.7
Timing, Command Mode Operation	A1.7.1, A1.7.3, A1.9.4
Timing, Graded-L operations	A1.7.2, A1.9.3
Crate controllers, other	A1.2
Encoded-N Signals	4.2
Front panel connectors	5
Front panel signals	3, Table 1
Grant-In signals	4.1.2
Grant-Out signals	4.1.2
Interpretation	1
Introduction	2
Look-at-Me signals	4.3
N signals	4.2
Patch connections	A1.5.4
Pull Up current sources	Table 3
Pull Up, GL	A1.6.3
Pull Up, L	A1.6.3
Request	4.1.1
Request Inhibit	4.1.3
Signal Input Protection	6.4
Signal lines at Auxiliary Controller Bus	Table 1
Signal lines at front panel connectors	Table 1
Signal sequence and timing	Figs 3, 4
Signals, control	5
Signals, Encoded-N	4.2
Signals, front panel	3, Table 1
Signals, Grant-In	4.1.2
Signals, Grant-Out	4.1.2
Signals, Look-at-Me	4.3
Signals, other	4.4
Signal standards	6, Table 3, 4
Signal standards, AC Dataway connections	6.3
Signal standards, ACB	6.1, 6.2, Table 1
Signal standards, ACB associated front panel signals	6.2, Table 1

ANSI/IEEE
Std 683-1976
(Reaffirmed 1981)

An American National Standard

IEEE Recommended Practice for Block Transfers in CAMAC* Systems

Sponsor

Instruments and Detectors Committee of
the IEEE Nuclear and Plasma Sciences Society

Approved September 9, 1976
Reaffirmed September 17, 1981

IEEE Standards Board

Approved November 9, 1977
Reaffirmed December 15, 1981

American National Standards Institute

*Computer Automated Measurement and Control

© Copyright 1976 by

The Institute of Electrical and Electronics Engineers, Inc
345 East 47th Street, New York, NY 10017

*No part of this publication may be reproduced in any form,
in an electronic retrieval system or otherwise,
without the prior written permission of the publisher.*

IEEE Standards documents are developed within the Technical Committees of the IEEE Societies and the Standards Coordinating Committees of the IEEE Standards Board. Members of the committees serve voluntarily and without compensation. They are not necessarily members of the Institute. The standards developed within IEEE represent a consensus of the broad expertise on the subject within the Institute as well as those activities outside of IEEE which have expressed an interest in participating in the development of the standard.

Use of an IEEE Standard is wholly voluntary. The existence of an IEEE Standard does not imply that there are no other ways to produce, test, measure, purchase, market, or provide other goods and services related to the scope of the IEEE Standard. Furthermore, the viewpoint expressed at the time a standard is approved and issued is subject to change brought about through developments in the state of the art and comments received from users of the standard. Every IEEE Standard is subjected to review at least once every five years for revision or reaffirmation. When a document is more than five years old, and has not been reaffirmed, it is reasonable to conclude that its contents, although still of some value, do not wholly reflect the present state of the art. Users are cautioned to check to determine that they have the latest edition of any IEEE Standard.

Comments for revision of IEEE Standards are welcome from any interested party, regardless of membership affiliation with IEEE. Suggestions for changes in documents should be in the form of a proposed change of text, together with appropriate supporting comments.

Interpretations: Occasionally questions may arise regarding the meaning of portions of standards as they relate to specific applications. When the need for interpretations is brought to the attention of IEEE, the Institute will initiate action to prepare appropriate responses. Since IEEE Standards represent a consensus of all concerned interests, it is important to ensure that any interpretation has also received the concurrence of a balance of interests. For this reason IEEE and the members of its technical committees are not able to provide an instant response to interpretation requests except in those cases where the matter has previously received formal consideration.

Comments on standards and requests for interpretations should be addressed to:

    Secretary, IEEE Standards Board
    345 East 47th Street
    New York, NY 10017
    USA

Foreword

(This Foreword is not a part of ANSI/IEEE Std 683-1976, Block Transfers in CAMAC Systems.)

A wide variety of block transfer algorithms (or modes) are possible in CAMAC systems. However, a restricted number of such algorithms can satisfy nearly all needs. This document presents recommended algorithms to encourage uniformity in design of CAMAC modules and controllers with resulting increased compatibility. It should be read in conjunction with, and is supplementary to, ANSI/IEEE Std 583-1982, Modular Instrumentation and Digital Interface Systems, ANSI/IEEE 595-1982, Serial Highway Interface Systems, and ANSI/IEEE Std 596-1982, Parallel Highway Interface System (CAMAC). The report on which this document is based, and with which it is technically identical (ERDA† Report TID-26616, February 1976), was prepared by the NIM Committee of the United States Department of Energy* and the ESONE Committee of European Laboratories.** Corresponding documents are ESONE Report EUR 4100 supplement and IEC Publication 677 of the International Electrotechnical Commission.

This standard was reviewed and balloted by the Nuclear Instruments and Detectors Committee of the IEEE Nuclear and Plasma Sciences Society. Because of the broad applicability of this standard, coordination was established with numerous IEEE societies, groups, and committees, including the Communications Society, Computer Society, Industry Applications Society, Industrial Electronics and Control Instrumentation Group, Instrumentation and Measurement Group, and the Power Generation and Nuclear Power Engineering Committees of the Power Engineering Society.

This standard was reviewed and reaffirmed in conjunction with the 1981 revision (1982 issue) of the family of IEEE CAMAC standards, though it differs from the previous issue only in the replacement of ULX by ULS and UDX by UDS in the table of contents (title of 5.4) and in replacement of the dates of the referenced IEEE CAMAC Standards by the current dates.

At the time of approval of this standard, the membership of the Nuclear Instruments and Detectors Committee of the IEEE Nuclear and Plasma Sciences Society was as follows:

D. C. Cook, *Chairman* **Louis Costrell,** *Secretary*

J. A. Coleman	T. R. Kohler	P. L. Phelps
J. F. Detko	H. W. Kraner	J. H. Trainor
F. S. Goulding	W. W. Managan	S. Wagner
F. A. Kirsten	G. L. Miller	F. J. Walter
	D. E. Persyk	

At the time it approved this standard, the American National Standards Committee N42 on Nuclear Instruments had the following personnel:

Louis Costrell, *Chairman* **David C. Cook,** *Secretary*

Organization Represented	Name of Representative
American Conference of Governmental Industrial Hygienists	Jesse Lieberman
American Nuclear Society	Frank W. Manning
Health Physics Society	J. B. Horner Kuper
	J. M. Selby (*Alt*)
Institute of Electrical and Electronics Engineers	Louis Costrell
	Julian Forster (*Alt*)
	D. C. Cook (*Alt*)
	A. J. Spurgin (*Alt*)
Instrument Society of America	M. T. Slind
	J. E. Kaveckis (*Alt*)
Lawrence Berkeley National Laboratory	E. J. Wagner
Oak Ridge National Laboratory	Frank W. Manning
	D. J. Knowles (*Alt*)

---

†Energy Research and Development Administration (now part of the United States Department of Energy).

Organization Represented	Name of Representative
US Department of Energy	Gerald Goldstein
US Department of the Army, Materiel Command	Abraham E. Cohen
US Department of Commerce, National Bureau of Standards	Louis Costrell
US Federal Emergency Management Agency	Carl R. Siebentritt, Jr
US Naval Research Laboratory	D. C. Cook
US Nuclear Regulatory Commission	Robert E. Alexander
Members-at-Large	J. G. Bellian
	O. W. Bilharz, Jr
	John M. Gallagher, Jr
	Voss A. Moore
	R. F. Shea
	E. J. Vallario

*NIM Committee[1]
L. Costrell, *Chairman*

C. Akerlof	N. W. Hill	L. B. Mortara
E. J. Barsotti	D. Horelick	V. C. Negro
B. Bertolucci	M. E. Johnson	L. Paffrath
J. A. Biggerstaff	C. Kerns	D. G. Perry
A. E. Brenner	F. A. Kirsten	I. Pizer
R. M. Brown	P. F. Kunz	E. Platner
E. Davey	R. S. Larsen	S. Rankowitz
W. K. Dawson	A. E. Larsh, Jr	S. J. Rudnick
S. R. Deiss	R. A. LaSalle	G. K. Schulze
S. Dhawan	N. Latner	W. P. Sims
R. Downing	F. R. Lenkszus	D. E. Stilwell
T. F. Droege	R. Leong	J. H. Trainor
C. D. Ethridge	C. Logg	K. J. Turner
C. E. L. Gingell	S. C. Loken	H. Verweij
A. Gjovig	D. R. Machen	B. F. Wadsworth
B. Gobbi	D. A. Mack	L. J. Wagner
D. B. Gustavson	J. L. McAlpine	H. V. Walz
D. R. Heywood		D. H. White

NIM Executive Committee for CAMAC
L. Costrell, *Chairman*

E. J. Barsotti	F. A. Kirsten
J. A. Biggerstaff	R. S. Larsen
S. Dhawan	D. A. Mack

J. H. Trainor

NIM Dataway Working Group
F. A. Kirsten, *Chairman*

E. J. Barsotti	C. Kerns
J. A. Biggerstaff	P. F. Kunz
L. Costrell	R. S. Larsen
S. Dhawan	D. R. Machen
A. Gjovig	L. Paffrath
D. R. Heywood	S. Rankowitz
D. Horelick	S. J. Rudnick

NIM Software Working Group
D. B. Gustavson, *Chairman*
W. K. Dawson, *Secretary*

A. E. Brenner	R. A. LaSalle
R. M. Brown	F. B. Lenkszus
S. R. Deiss	C. A. Logg
S. Dhawan	J. McAlpine
C. D. Ethridge	L. B. Mortara
M. E. Johnson	D. G. Perry

---

[1] National Instrumentation Methods Committee of the US Department of Energy.

NIM Mechanical and Power Supplies Working Group
L. J. Wagner, *Chairman*

L. Costrell	S. J. Rudnick
C. Kerns	W. P. Sims

NIM Serial System Subgroup
D. R. Machen, *Chairman*

E. J. Barsotti	R. G. Martin
D. Horelick	L. Paffrath
F. A. Kirsten	S. J. Rudnick

NIM Analog Signals Working Group
D. I. Porat, *Chairman*

L. Costrell	N. W. Hill
C. E. L. Gingell	S. Rankowitz

NIM Block Transfer Subgroup
E. J. Barsotti, *Chairman*

W. K. Dawson	F. R. Lenkszus
F. A. Kirsten	R. G. Martin
R. A. LaSalle	R. F. Thomas, Jr

NIM Multiple Controllers Subgroup
P. F. Kunz, *Chairman*

E. J. Barsotti	D. R. Machen
F. A. Kirsten	R. G. Martin

**ESONE Committee[2]

W. Attwenger, Austria, *Chairman* 1980-81   H. Meyer, Belgium, *Secretary*

Representatives of ESONE Member Laboratories

W. Attwenger, Austria	E. Kwakkel, Netherlands	A. C. Peatfield, England
R. Biancastelli, Italy	J. L. Lecomte, France	I. C. Pyle, England
L. Binard, Belgium	J. Lingertat, D.R. Germany	B. Rispoli, Italy
J. Biri, Hungary	M. Lombardi, Italy	M. Sarquiz, France
B. Bjarland, Finland	M. Maccioni, Italy	W. Schoeps, Switzerland
D. A. Boyce, England	P. Maranesi, Italy	R. Schule, F.R. Germany
B. A. Brandt, F.R. Germany	C. H. Mantakas, Greece	P. G. Sjolin, Sweden
F. Cesaroni, Italy	D. Marino, Italy	L. Stanchi, Italy
P. Christensen, Denmark	H. Meyer, Belgium	R. Trechcinski, Poland
W. K. Dawson, Canada	K. D. Muller, F.R. Germany	M. Truong, France
M. DeMarsico, Italy	J. G. Ottes, F.R. Germany	P. Uuspaa, Finland
C. A. DeVries, Netherlands	A. D. Overtoom, Netherlands	H. Verweij, Switzerland
H. Dilcher, F.R. Germany	E. C. G. Owen, England	A. J. Vickers, England
B. V. Fefilov, USSR	L. Panaccione, Italy	S. Vitale, Italy
R. A. Hunt, England	M. Patrutescu, Roumania	M. Vojinovic, Yugoslavia
W. Kessel, F.R. Germany	R. Patzelt, Austria	K. Zander, F.R. Germany
R. Klesse, France		D. Zimmermann, F.R. Germany

ESONE Executive Group (XG)

W. Attwenger, Austria, XG *Chairman* 1980-81

H. Meyer, Belgium, *Secretary*

P. Christensen, Denmark	M. Sarquiz, France
M. Dilcher, F.R. Germany	R. Trechcinski, Poland
B. Rispoli, Italy	A. Vickers, England

H. Verweij, Switzerland

ESONE Technical Coordination Committee (TCC)

A. C. Peatfield, England, TCC *Chairman*

P. Christensen, Denmark, TCC *Secretary*

W. Attwenger, Austria	W. Kessel, F.R. Germany
R. Biancastelli, Italy	J. Lukacs, Hungary
G. Bianchi, France	R. Patzelt, Austria
H. Dilcher, F.R. Germany	P. J. Ponting, Switzerland
P. Gallice, France	W. Schoeps, Switzerland

S. Vitali, Italy

---

[2] European Standards on Nuclear Electronics Committee.

ECA/ESONE CAMAC Document Maintenance Study Group (DMSG)

P. Gallice, France, *Chairman*

R. C. M. Barnes, England	F. Iselin, Switzerland
L. Besse, Switzerland	H. Meyer, Belgium
J. Davis, England	H. J. Trebst, F.R. Germany

When this standard was approved September 9, 1976, the IEEE Standards Board had the following membership:

William R. Kruesi, *Chairman* **Irvin N. Howell, Jr,** *Vice Chairman*

Ivan G. Easton, *Secretary*

William E. Andrus	Irving Kolodny	William J. Neiswender
Jean Jacques Archambault	Benjamin J. Leon	Gustave Shapiro
Dale R. Cochran	Anthony C. Lordi	Ralph M. Showers
Warren H. Cook	John P. Markey	Robert A. Soderman
Louis Costrell	Thomas J. Martin	Leonard W. Thomas, Sr
Jay Forster	Donald T. Michael	Charles L. Wagner
Joseph L. Koepfinger	Voss A. Moore	William T. Wintringham
	William S. Morgan	

Contents

SECTION		PAGE
	CAMAC and NIM Standards and Reports	8
1.	Introduction	9
2.	Classification of Block-Transfer Modes	10
	2.1 CAMAC Address Sequencing	10
	2.2 Synchronization Source	10
	2.3 Block-Transfer Termination	11
3.	Block-Transfer Modes Described in IEEE Std 583 (EUR 4100)	11
	3.1 UCS (Stop) Mode	11
	3.2 ACA (Address Scan) Mode	12
	3.3 UQC (Repeat) Mode	12
4.	Additional Block-Transfer Modes	13
	4.1 UCW (Stop-on-Word) Mode	13
	4.2 ULS (LAM Synchronized Stop) Mode	13
	4.3 UDS (Direct Synchronized Stop) Mode	14
	4.4 MCA (Multidevice Action) Mode	14
5.	Compatibility	14
	5.1 Block-Transfer Mode MCA	14
	5.2 Block-Transfer Modes XCX, XLS, XDS	14
	5.3 Block-Transfer Modes ULS, UDS, ULC, UCD, UQC	15
	5.4 Block-Transfer Modes ULS, UDS, ULW, UDW	15
6.	Hardware Design	16
	6.1 Module Design—Q Response	16
	6.2 Module Design—LAM Signal	17
	6.3 Interface Design	17
7.	Software Considerations	18
8.	References	18

APPENDIX

A1.	Other Block-Transfer Modes	19
	A1.1 MCQ (Multiple Test) Mode	19
	A1.2 ECA (Extended Address Scan) Mode	19
	A1.3 UQL (Pause) Mode	19

TABLES

Table 1	Block-Transfer Mode Descriptor	10
Table 2	Block-Transfer Names	10
Table 3	Compatibility Aspects of Stop and Stop-on-Word Block-Transfer Termination Modes	15
Table 4	Single Module, Single Address Block Transfer: Recommended Method for Performing UCS (Stop) Mode and Corresponding LAM Synchronized CAMAC Block Transfers	16
Table 5	Recommended Method for Performing ACA (Address Scan) and UQC (Repeat) Mode CAMAC Block Transfers	16
Table 6	Method for Performing UCW (Stop-on-Word) Mode and Corresponding LAM Synchronized CAMAC Block Transfers	17

FIGURE

| Fig 1 | Recommended Implementation of a Module's LAM Signal Which Is Intended for Block-Transfer Synchronization; Simplified Block Diagram | 18 |

CAMAC and NIM Standards and Reports

Title	IEEE, ANSI Std No	IEC No	DOE No	EURATOM (EUR) or ESONE No
CAMAC Instrumentation and Interface Standards*	SH08482* (Library of Congress No 8185060)	—	—	—
Modular Instrumentation and Digital Interface System (CAMAC)	ANSI/IEEE Std 583-1982	516	TID-25875† and TID-25877†	EUR 4100e
Serial Highway Interface System (CAMAC)	ANSI/IEEE Std 595-1982	640	TID-26488†	EUR 6100e
Parallel Highway Interface System (CAMAC)	ANSI/IEEE Std 596-1982	552	TID-25876† and TID-25877†	EUR 4600e
Multiple Controllers in a CAMAC Crate	ANSI/IEEE Std 675-1982	729	DOE/EV-0007	EUR 6500e
Block Transfers in CAMAC Systems	ANSI/IEEE Std 683-1976 (Reaff 1981)	677	TID-26616†	EUR 4100 suppl
Amplitude Analog Signals within a 50 Ω System	—	—	TID-26614	EUR 5100e
The Definition of IML A Language for Use in CAMAC Systems	—	—	TID-26615	ESONE/IML/01
CAMAC Tutorial Articles	—	—	TID-26618	—
Real-Time BASIC for CAMAC	ANSI/IEEE Std 726-1982	§	TID-26619†	ESONE/RTB/03
Subroutines for CAMAC	ANSI/IEEE Std 758-1979 (Reaff 1981)	713	DOE/EV-0016†	ESONE/SR/01
Recommendations for CAMAC Serial Highway Drivers and LAM Graders for the SCC-L2	—	—	DOE/EV-0006	ESONE/SD/02
Definitions of CAMAC Terms	Included in SH08482	678	DOE/ER-0104	ESONE/GEN/01
Standard Nuclear Instrument Modules NIM	—	547**	TID-20893 (Rev 4)	—

†Superseded by corresponding IEEE Standard listed.

*This is a hard cover book that contains ANSI/IEEE Std 583-1982, ANSI/IEEE Std 595-1982, ANSI/IEEE Std 596-1982, ANSI/IEEE Std 675-1982, ANSI/IEEE Std 683-1976 (Reaff 1981), ANSI/IEEE Std 726-1982 and IEEE Std 758-1979 (Reaff 1981), plus introductory material and a glossary of CAMAC terms.

**Covers only mechanical features and connector pin assignments.

§ In preparation.

NOTE: *Availability of Documents*
ANSI   Sales Department, American National Standards Institute, 1430 Broadway, New York, NY 10018.
IEEE   IEEE Service Center, 445 Hoes Lane, Piscataway, New Jersey 08854, USA.
IEC   International Electrotechnical Commission, 1, rue de Varembé, CH-1211 Geneva 20, Switzerland.
DOE and TID Reports   National Bureau of Standards, Washington, D.C. 20234, USA, Attn: L. Costrell.
EURATOM   Office of Official Publications of the European Communitities, P.O. Box 1003, Luxembourg.
ESONE   Commission of the European Communities, CGR-BCMN, B-2440 GEEL, Belgium, Attn: ESONE Secretariat, H. Meyer.

An American National Standard

IEEE Recommended Practice for Block Transfers in CAMAC Systems

1. Introduction

The basic CAMAC specification (see Ref [1]) defines a single CAMAC operation as the activity which occurs in response to a single CAMAC command. This activity may consist of the transfer of a single data word between a CAMAC module and computer memory or the changing of the status of a module [for example, $F(26)$, $F(24)$] or return of a value for Q as the result of a test made on the module, or any compatible set of the previously named activities. A block transfer is defined as a sequence of single CAMAC operations involving data which the user specifies by a command said to be of a higher level than one which specifies a single CAMAC operation. The higher level command contains all the information required for the specification of the desired sequence of single CAMAC commands and is interpreted by a channel which governs the activity on the CAMAC highway. Control information such as the readiness of the computer to participate in a data transfer, the state of the CAMAC Q line and the state of certain LAMs (Look-at-Me's) or special synchronizing signals must be made available to the channel. The use made of the control information by the channel defines the block-transfer mode. If a module is to influence the sequence of operations within a block transfer, then it must have the features required by the algorithm.

A channel consists of an interface to the CAMAC system as well as a means for selecting and executing the algorithms of the implemented block-transfer modes. An algorithm may be implemented wholly in hardware or wholly in software or by a combination of hardware and software. The possibility of software implementation of any algorithm means that CAMAC block transfers can take place on a system which does not have the hardware (such as direct memory access) required to carry out "computer" block transfers. Note that a module behaves in the same way when it is accessed by a hardware algorithm as it does when accessed by conventional programmed computer input-output. Regardless of the method of channel implementation, the use of the channel results in reductions in both the CPU time required and in the programming effort through the use of pre-defined algorithms.

Many different block-transfer algorithms (or modes) may be defined, all of which are compatible with the CAMAC specifications. It is also possible to have a channel execute a sequence of CAMAC commands which does not involve the transfer of data. An example of such a "Multiple Action" mode is discussed in the Appendix. Many algorithms convey control information by either Q or a LAM signal or both. The requirements placed on these signals by one algorithm may conflict with those placed on them by another algorithm. Hence compatibility problems between modes and channels can occur. This is especially true if no restrictions are placed on the choice of a suitable block-transfer algorithm. However, experience with many different CAMAC systems and extensive analysis of the problem has revealed that a restricted number of block-transfer algorithms can satisfy nearly all needs. To encourage uniformity in future designs of modules and controllers, certain algorithms are recommended to be used whenever possible, and some additional algorithms are suggested for special applications that cannot practically be implemented with the recommended algorithms. The possibility remains for a user to define other block-transfer algorithms to meet special needs.

In the following sections, the recommended block-transfer algorithms are discussed. In Section 3, those given in the basic specification (see Ref [1]) are described. These algorithms are well established and are supported by existing hardware, both in modules and in

[3] Numbers in brackets correspond to those of the references listed in Section 8 of this standard.

CAMAC interfaces available for many small computers. Section 4 discusses the new algorithms.

The CAMAC user is reminded that the block-transfer characteristics of his modules, his controller (Branch Driver or computer-crate interface), and the software in the computer must be matched in order for the block-transfer algorithms to be carried out correctly. The ability to perform block transfers is a feature of the total computer system and its interfaces.

2. Classification of Block-Transfer Modes

The various block-transfer modes can be classified by specifying the nature of each of three fundamental characteristics: how the CAMAC address is determined, the source of the synchronizing signal, and the method used to terminate the block transfer. In the following, these characteristics are described, and a notation permitting a compact specification of the various modes is defined.

The notation is based on the use of a single letter to represent the nature of each characteristic. The letters are written in the order that the characteristics were mentioned above and the resulting three letters completely describe a block-transfer mode. If a certain characteristic need not be specified, then it is denoted by the letter X. Thus the symbol XXX describes a block-transfer mode in which all the characteristics are undefined. Table 1 summarizes the meaning of each letter used and Table 2 lists the modes described below and the IML (see Ref [2]) support for each.

2.1. CAMAC Address Sequencing.
The first letter of the block mode descriptor indicates the method used to determine the CAMAC address of the next CAMAC operation.

A block transfer accessing a CAMAC address which stays constant during the block transfer is typically used to access a buffer within a module or to access an external device (such as a computer peripheral unit) through a CAMAC module. Such Uniaddress block transfers are denoted by the letter U.

A block transfer accessing a sequence of CAMAC addresses is typically used to access a set of registers located at different places

Table 1
Block-Transfer Mode Descriptor

First Letter	— CAMAC Address Sequencing
	U → Uni-Address
	M → Multi-Address (Calculated or Given Array in IML*)
	A → Address Scan
	E → Extended Address Scan
Second Letter	— Synchronizing Source
	C → Controller
	Q → Q Response
	L → LAM Signal
	D → Pseudo-LAM Signal
Third Letter	— Operation Termination
	C → Channel Word Count
	A → Terminal Address Reached
	S → Q = 0 on last + 1
	W → Q = 0 on last
	L → LAM Signal
	D → Pseudo-LAM Signal

*See Ref [2]

Table 2
Block-Transfer Names

Descriptor	IEEE Std 583 EUR 4100	IML*
UCS	Stop Mode	UBC
UCW	none	UBC
UQC	Repeat Mode	UBR
ACA	Address Scan Mode	MAD
MCA	none	MA
ULS	none	UBL
UDS	none	UBL
ULW	none	UBL
UDW	none	UBL

within a CAMAC system but containing related data. The algorithm for determining the next CAMAC address may depend on the state of Q resulting from the last operation.

The ACA (Address Scan) Mode described in Ref [1] (Section 5.4.3.1) is the main example of this technique, although there does exist a variant, the ECA (Extended Address Scan) Mode, which is described in the Appendix.

The sequence of CAMAC addresses may also be determined either by a list of all the addresses or by a starting address, increments to be applied to each part of the address and a final address. These correspond to the Given and Calculated Arrays of CAMAC addresses in IML (see Ref [2]), and both are indicated by the letter M for Multiaddress.

2.2 Synchronization Source.
The second letter of a block-transfer mode descriptor indicates

the source of synchronization of individual transfers. In some cases the module is continuously ready to effect a transfer, and the channel controller may execute the CAMAC commands at any convenient rate. The block transfer is then said to be "controller synchronized," and the descriptor for all such modes is of the form XCX.

In other cases the module is not continuously ready. Following a particular transfer a certain time must elapse before it can effect another. The rate at which the block transfer proceeds is controlled by the module (or modules). Hence the module must provide synchronization information to the channel so that it can correctly execute its part of the block-transfer process. For such *module synchronized* block transfers three sources of synchronizing signal have been identified. They are (1) the Q response as in the Repeat Mode (Ref [1], Section 5.4.3.2) denoted by Q; (2) a specific LAM signal denoted by L; (3) a special signal (Pseudo-LAM) sent directly from the module to the controller denoted by D.

2.3 Block-Transfer Termination. The third and final letter of a block transfer mode descriptor indicates the means by which the block transfer is terminated. A given block transfer may be halted by either the channel or the module depending on the conditions existing at the time of the action. It is assumed that all block transfers will be executed with a limit on the number of transfers permitted; in addition to this channel word-count limit, there may be other conditions which occur either in the channel or the module which will terminate the transfer. The first such condition to occur terminates the process.

If a block transfer can be terminated only by exhausting a word count within the channel, it is designated by the letter C. If a block transfer can be terminated by the channel either because a word count is exhausted or because a sequence of addresses in a multiaddress transfer is exhausted, it is designated by the letter A.

A block transfer can also be terminated by a status signal from the module. While either a LAM signal or a special signal sent directly to the channel could be used, all the module terminated modes described in this document use the Q response associated with each CAMAC command of the block transfer.

In the CAMAC specification given in Ref [1] (Section 5.4.3.3) such a mode is described. A $Q = 1$ response is interpreted as meaning that the operation took place within the block. The first $Q = 0$ response indicates that the end-of-block occurred on the previous operation. In Ref [1] this is called the Stop mode and is designated here by the letter S.

Another possible interpretation of a $Q = 0$ response is that it accompanies the last operation in the block. For write commands the data word has been accepted by the module and for read commands the data word has been transferred to the computer. This is the Stop-on-Word mode and is designated by the letter W.

3. Block-Transfer Modes Described in ANSI/IEEE Std 583-1982, Modular Instrumentation and Digital Interface System (CAMAC) (EUR 4100)

The CAMAC specification places no restrictions on the modes which may be used for block transfers. However, three modes which depend on defined uses of the Q response signal are given as examples. These modes have been widely implemented, particularly for high-energy physics applications, and many modules are available with the appropriate features incorporated.

3.1 UCS (Stop) Mode (Uni-address Block Transfer, Controller Synchronized, Q Response Terminated). The algorithm for this block transfer mode is defined for the module in Ref [1], Section 5.4.3.3. This mode enables the channel to transfer a sequence of data words between a single fixed CAMAC address and computer memory at a rate determined by the computer system but with the block size determined by the CAMAC module.

The response $Q = 1$ indicates that a data word has been transmitted by the module in a Read operation or accepted by the module in a Write operation, that is, the transfer occurred within the block of data words.

The response $Q = 0$ indicates a transfer attempt beyond the end of the block, and hence is the method used for terminating a UCS mode block transfer. In a Read operation no valid data has been transmitted by the module, and the sequence is to be stopped by closing the channel. Since the transfer of a nonexistent (or dummy) data word to computer memory would require some form of recovery, the channel should not complete the transfer.

A $Q = 0$ response in a Write operation indicates that the module has not accepted the data word. This mode is supported in IML by the UBC mode (Section 5.9 of Ref [2]).

3.2 ACA (Address Scan) Mode (Multiaddress Block Transfer, Controller Synchronized and Terminated).

The algorithm for this block-transfer mode is defined for the module in Ref [1], Section 5.4.3.1. Its object is to access registers at successive subaddresses in a sequence of modules. The channel controller, either hardware or software, initiates the transfer by establishing an initial CAMAC address, final CAMAC address, and function code with which registers included in the address scan are to be accessed. The channel controller accesses each register in turn by sequentially incrementing through the CAMAC subaddresses. A $Q = 1$ response from a module indicates that a register included in the scan is present. The response $Q = 0$ indicates register absent. A $Q = 0$ response from any subaddress or a $Q = 1$ response from Subaddress $A(15)$ causes the channel controller to set the subaddress to zero and increment the station number before the next transfer attempt.

For both Read and Write operations the channel must attempt to execute the CAMAC operation. The Q response from this attempt is used by the channel to determine the next CAMAC address to be used as well as to indicate to the computer what has to be done concerning data transfers. The response $Q = 1$ indicates that the module either transmitted or accepted a data word. A $Q = 0$ response to a Read operation means that no data word was available at the CAMAC address accessed, and the channel does not initiate a transfer to memory. A $Q = 0$ response to a Write operation means that the data word was not accepted at the CAMAC address accessed and the channel does not allow the computer to step to the next data word. The block transfer is terminated when the channel controller accesses the final CAMAC address or when a specific number of data transfers have taken place.

Since the response combination $Q = 0$, $X = 0$ can mean that no module was present, it must be expected. Because of this the ACA mode is potentially dangerous and is usually implemented with special precautions, for example, unique and known data words at various points of the sequence to allow verification that all modules are present. The combination $Q = 1$, $X = 0$ is an error condition.

This mode is supported in IML by the MAD mode (Section 5.13 of Ref [2]).

3.3 UQC (Repeat) Mode (Uniaddress Block Transfer, Q Response Synchronized, Controller Terminated).

The algorithm for this block transfer mode is defined for the module in Ref [1], Section 5.4.3.2. The channel addresses one fixed CAMAC address and is able to transfer data words only when the module is ready, that is, with the maximum rate determined by the module.

The response $Q = 1$ indicates that a data word has been transmitted in a Read operation or accepted in a Write operation, that is, the module was ready. For Read operations $Q = 0$ indicates that the module was unable to supply a data word. For Write operations $Q = 0$ indicates that the data word was not accepted by the module. The command should, therefore, be repeated with the same data word until $Q = 1$ is obtained.

In this mode the Q response cannot indicate the end of the sequence and hence some other mechanism is required, such as the sequence being terminated when a specified number of data transfers have been achieved. Similarly the module has no method of indicating that it has no data; there is, therefore, a risk that the CAMAC operation will be repeated indefinitely while waiting for $Q = 1$. Since this situation may also be encountered in a fault condition, a channel should include an error-detecting feature such as specifying a maximum number of consecutive $Q = 0$ operations that will be attempted or implementation of a time-out feature.

This mode is particularly useful in both Read and Write operations for communication with a buffer or auxiliary equipment whose data-transfer rate is marginally slower than the transfer rate of the computer system. It is not recommended when there is a wide discrepancy between module and channel speeds because transfer efficiency may become low. This mode is supported in IML by the UBR mode (Section 5.10 of Ref [2]).

4. Additional Block-Transfer Modes

Since the basic CAMAC specification in Ref [1] was published additional block-transfer modes have been identified. These are not intended to replace the modes discussed in Section 3 but rather to complement them. In the following sections some of these additional modes are described, and typical applications are given in order to illustrate the special conditions which require the modes.

4.1 UCW (Stop-on-Word) Mode (Uniaddress Block Transfer, Controller Synchronized, Q Response Terminated). The UCW mode is used for the transfer of a block of data between a single CAMAC address and computer memory at a speed determined by the channel. It is identical to the UCS (Stop) mode except for the interpretation of the data that accompanies the first $Q = 0$ response. This different interpretation of Q means that the UCW mode cannot be used with many early interfaces and modules unless special precautions are taken. Hence its use should be restricted to those cases which require all its features.

In a UCS block transfer of N data words $Q = 0$ is returned on the $(N + 1)$th transfer; that is, no significant data is transferred in the operation returning $Q = 0$. Under certain circumstances more efficient operation is obtained if $Q = 0$ is returned on the Nth transfer; that is, accompanying the last word actually transferred. If $Q = 0$ accompanies the last word transferred, then the mode is described as UCW or Stop-on-Word.

Its implementation for Read operations requires the channel controller to pass the word received with $Q = 0$ to computer memory. For Write operations the module responds with $Q = 0$ in the operation that fills the available space.

With modules acting as computer peripherals (for example, for bulk storage) fixed length blocks are often transferred in both directions. By sending $Q = 0$ with the last successful transfer (UCW mode) a mechanism (compatible with standard computer peripherals) is obtained whereby it can be determined that the required number of transfers has been achieved.

UCS (Stop) mode has the advantage that it can distinguish between a module containing a single data word and no data word on the first transfer attempt. In UCW mode a preliminary test operation is required if this is a problem for a particular module.

Both UCS mode and UCW mode are supported in IML by the UBC mode. (Section 5.9 of Ref [2].) Whether the transfer accompanied by $Q = 0$ is completed or not is determined by the channel specified for the transfer.

4.2 ULS (LAM Synchronized Stop) Mode (Uniaddress Block Transfer, LAM Synchronized, Q Response Terminated). Modules designed to work in UCS (Stop) mode can be upgraded to LAM synchronized mode by adding a LAM feature. This mode allows the transfer of a block of data between a single CAMAC address and computer memory with each individual data-word transfer requested by a LAM signal from the module. This LAM signal may be used as a "sync" to the hardware or software channel, that is, not necessarily generating a computer interrupt. The mode is particularly appropriate for low-speed devices (for example, paper tape I/O or detector/sensor scanning) which would otherwise make repeated demands for an interrupt-service routine.

LAM synchronized modes requiring a LAM to initiate each transfer can make use of the normal LAM handling mechanism of the local implementation to identify the specific LAM signal and hence the channel required. In an implementation using the "Graded-L" mechanism (see Ref [3], Section 5.2) for LAM handling it simplifies the controller activity if a specific Graded-L number is dedicated to the block transfer at the time of its execution. The operation performing the requested transfer should also reset the LAM signal. This gives more efficient operation where the module is accessed by a block-transfer channel or an interrupt-service routine. Because each channel is individually requested, many different block transfer channels can operate concurrently provided that each has a unique LAM signal. In addition these block-transfer channels can operate in parallel with program controlled CAMAC operations using normal input-output facilities.

The ULS mode block transfer is terminated either by a predetermined number of data words having been transferred or by a $Q = 0$

response from the module. For Read or Write operations in the ULS mode the module issues another LAM after the last word of the block has been transferred in order that the $Q = 0$ end-of-block indication be obtained.

One example of the use of this mode is the input of a message from a teletypewriter. Each character is transferred to a message buffer in computer memory in response to a LAM request from the interface module. The CAMAC operation which reads the character also resets the LAM signal. The block transfer continues until the buffer contains a predetermined number of characters or the interface module identifies a specific character, for example, a carriage return, and returns $Q = 0$ in response to the Read operation initiated by the LAM.

For special applications the ULW (LAM Synchronized Stop-on-Word) mode may be used. Note that, like the UCW mode (Section 4.1), this mode cannot be used with many earlier interfaces and modules.

The ULS mode is supported in IML by the UBL mode (Section 5.8 of Ref [2]).

4.3 UDS (Direct Synchronized Stop) Mode (Uniaddress Block Transfer, Special Signal Synchronized, Q Response Terminated). This mode is identical to the ULS mode described in Section 4.2 with the exception of the way the synchronizing signal reaches the channel. In the ULS mode the synchronizing signal is processed by the LAM handling mechanism before it is used by the channel to initiate the required CAMAC operation. In the UDS mode the synchronizing signal has the same properties as the LAM signal in the ULS mode except that the signal is sent directly to the channel, bypassing the LAM handling mechanism of the channel. Hence the UDS mode can for some systems have a more rapid response to a transfer request than the ULS mode. The synchronizing signal can be sent to the channel via a front panel connector or a patch pin (or a free use Branch Highway line if appropriate, or both) and is called a "Pseudo-LAM."

For special applications the UDW (Direct Synchronized Stop-on-Word) mode may be used. Note that, like the UCW mode (Section 4.1), this mode cannot be used with many earlier interfaces and modules.

The UDS mode is supported in IML by the UBL mode (Section 5.8 of Ref [2]).

4.4 MCA (Multidevice Action) Mode (Multi-address Block Transfer, Controller Synchronized and Terminated). This mode permits a random distribution [a Given Array in IML (see Ref [2])] or a regular pattern [a Calculated Array in IML (see Ref [2])] of CAMAC addresses to be treated as a single hardware array by the programmer. The Q response from the modules is ignored, and no special module features are required. The mode is both synchronized and terminated by the channel controller. Termination occurs when a specified word count is reached or the sequence of CAMAC addresses is exhausted.

This mode is supported in IML by the MA mode (Section 5.11 of Ref [2]).

5. Compatibility

The various rules which exist for the use of the Q response and LAM signals in block transfers raise certain questions of compatibility of modules and channel controllers used together in certain block-transfer modes (see Table 3). Ideally all modules and controllers would be compatible with any block-transfer mode, but existing equipment and conflicting requirements prevent this. Among the modes described in Sections 3 and 4, two significant sources of incompatibility of equipment exist. The Q response to a given CAMAC function clearly can have only one meaning; it cannot be used both for synchronization and for indicating end-of-block, and the modes XXS and XXW require different behavior on the part of the controller and module when the end of the block occurs.

5.1 Block-Transfer Mode MCA. Since neither the Q response nor the LAM is used in the MCA mode, no compatibility problems can arise, given that each module is always ready to respond to a CAMAC command.

5.2 Block-Transfer Modes XCX, XLX, XDX. If a module is not continuously ready to effect a transfer, then the XCX mode cannot be used in place of XLX or XDX. If a module is continuously ready to effect a transfer, then any of the XCX, XLX, or XDX modes can be used provided the LAM or Pseudo-LAM be continuously asserted in order to get a transfer rate comparable to the controller synchronized one.

The XLX and XDX modes are compatible

Table 3
Compatibility Aspects of Stop and Stop-on-Word Block-Transfer Termination Modes

Direction of Transfer	Module Mode	Controller Mode XXS	Controller Mode XXW
READ	XXS	Correct Transfer	Additional dummy word stored at end of block
READ	XXW	Irrecoverable loss of last word from module	Correct Transfer
WRITE	XXS	Correct Transfer	Last word transmitted from memory not accepted by module
WRITE	XXW	Last word accepted by module, but controller assumes that last word has not been accepted	Correct Transfer

provided the LAM of the XLX mode and the Pseudo-LAM of the XDX mode have exactly the same meaning.

5.3 Block-Transfer Modes ULS, UDS, ULC, UDC, UQC. The modes ULC and UDC have not been specifically described because of the assumption stated in Section 2 that every channel contains a word count limit which terminates a block transfer if it is exhausted. Thus, from the point of view of the channel, these two modes are subsets of their respective module-terminated modes - ULS and UDS. In practice these modes are common ones, and a useful level of compatibility among modules can be achieved by proper specification of the Q response to transfer commands. Thus the response of a module designed to operate in the ULS or UDS modes, or both, to a transfer command should be $Q = 1$ if ready (that is, transfer took place) and $Q = 0$ if not ready (that is, transfer did not take place). Then, if the block length is known, such a module can also be operated in the UQC mode since the synchronizing signal in ULS and UDS modes indicates the presence of information in the module. This information may be that the block has terminated ($Q = 0$) or that a successful data transfer has taken place ($Q = 1$).

5.4 Block-Transfer Modes ULS, UDS, ULW, UDW. System designers should be aware that very serious compatibility problems can arise if channels and modules are not matched in modes XXS and XXW. Table 3 shows the results for each direction of transfer. Taking into account the different situations which arise depending on the direction of transfer, there are four possible mismatches.

5.4.1 *Write Via XXS Controller to XXW Module.* The controller in this case assumes that the last word transmitted was not accepted and sets the address pointer and word count accordingly. The software must be aware that one more word has been transferred than is indicated and perform any necessary recovery.

5.4.2 *Write Via XXW Controller to XXS Module.* The controller assumes that the last word transmitted was accepted, whereas the module has actually rejected it. In this case also the software must be aware of the situation and make an appropriate recovery.

5.4.3 *Read Via XXS Controller from XXW Module.* In this case the channel controller discards the last word read (the data word which accompanies the $Q = 0$ response) and does not transfer it to the computer memory. The word is irretrievably lost.

5.4.4 *Read Via XXW Controller from XXS Module.* The last word is transferred to computer memory and the channel word count and address pointer set accordingly. Since the module expects the word to be discarded by the controller, the word contains no information and should be discarded by the software.

6. Hardware Design

Sections 6.1 and 6.2 detail recommendations for the Q response and LAM signal design in modules which may be addressed during a block-transfer operation. Section 6.3 lists recommendations for the hardware design of a computer interface with block-transfer capabilities.

6.1 Module Design—Q Response. The general character of a module will largely determine the block-transfer mode which is appropriate, for example, with many addressable registers to be read in sequence at computer speeds a module Q response accommodating ACA mode is clearly indicated. Tables 4 and 5 show the recommended implementation of the Q response as a function of the modes indicated. Table 6 describes the Q response for special application modes only. When there is an option, the UCS (Stop) mode should be adopted unless there are clear and specific technical reasons for adopting an alternative mode. This recommendation is dictated by the incompatibilities discussed in Section 5 and the widespread use of UCS mode controllers in existing systems. The module may operate in more than one mode, for example, depending on the direction of transfer or selectable by the user. In the latter case the selection should be made or at least tested via program control.

Table 4
Single Module, Single Address Block Transfer:
Recommended Method for Performing UCS (Stop) Mode and
Corresponding LAM Synchronized CAMAC Block Transfers

Q	Sequence in Block	Transfer Direction	Module	Interface	Computer
$Q = 1$	First to last data word	Read	Data word transmitted	Pass data word; Keep channel open	Store data word in computer memory
		Write	Data word accepted	Data word already passed; Keep channel open	Data word already delivered
$Q = 0$	Last + 1 data word	Read	No data word transmitted	Do not pass data word; Close channel	No action required
		Write	Data word not accepted	Close channel; Data word already passed	Do not update data word pointer

Table 5
Recommended Method for Performing ACA (Address Scan) and
UQC (Repeat) Mode CAMAC Block Transfers

Q	Sequence in Block	Transfer Direction	Module	Interface	Computer
$Q = 1$	First to last data word at a station*	Read	Data word transmitted	Pass data word; Keep channel open	Store data word in computer memory
		Write	Data word accepted	Data word already passed; Keep channel open	Data word already delivered
$Q = 0$	Last + 1 data word at a station*	Read	No data word transmitted	Do not pass data word; Keep channel open	No action required
		Write	Data word not accepted	Retain data word; Keep channel open	Data word already delivered

*For the Repeat mode the phase "at a station" does not apply.

Table 6
Method for Performing UCW (Stop-on-Word) Mode and Corresponding LAM Synchronized CAMAC Block Transfers*

Q	Sequence in Block	Transfer Direction	Module	Interface	Computer
Q = 1	First to (last −1) data word	Read	Data word transmitted	Pass data word; Keep channel open	Store data word in computer memory
		Write	Data word accepted	Data word already passed; Keep channel open	Data word already delivered
Q = 0	Last data word	Read	Data word transmitted	Pass data word Close channel	Store data word in computer memory
		Write	Data word accepted	Close channel Data word already passed	Data word already delivered

*This is a special application mode only and cannot be used with many earlier interfaces and modules.

All modules that may be used in the ACA mode must have their registers sequentially subaddressable starting with Subaddress zero. If a module has, for example, four registers to be accessed during an address scan, the Q and X responses to commands accessing these registers must be as follows: For Subaddresses 0 through 3 both the Q and X response must be one. At Subaddress 4 the Q response must be zero and the X response may be a one or zero depending upon whether another register not intended to be accessed by Address Scan is present.

6.2 Module Design—LAM Signal. All CAMAC modules incorporating one or more LAM signals should adhere to all mandatory requirements of Ref [1], Section 5.4.1. Implementations are illustrated in Figs. K5.4.1A, K5.4.1B, K5.4.1C, and 11 of ANSI/IEEE Std 583-1982.

In a block transfer LAM signals may be used in two ways. First, a LAM signal may be used to request the transfer of each data word in the block transfer, that is, one LAM per data transfer. The other use of the LAM signal is to initiate a block transfer. In this case the LAM signal occurs only once, that is, at the beginning of the block transfer. Note that a module may issue a number of different LAM signals.

If a module is being designed to be used in a LAM-synchronized mode, the following recommendations concerning routing of the LAM signal should be considered.

As shown in Fig 1 there are three possible routings of a module's LAM signal. In the conventional approach the LAM signal used to request the transfer of a word is ORed with all other LAMs in the module before arriving at the L line of the dataway. This signal is then transmitted to the crate controller and eventually to the channel controller.

The second and third routings are similar in that they allow for quick accessing and interpretation by the channel controller. These two are classed as "Pseudo-LAMs." For example, in a Branch Highway System the LAM signal can be transmitted via a crate patch pin and a BV line to a Branch Driver, and may directly request a computer DMA cycle. As another example, the LAM signal can be sent directly to the channel controller via a front or rear panel connector. Modules designed with these two LAM routing implementations must have all the properties of LAM signals referred to in the first paragraph of this section.

6.3 Interface Design. Modules which implement the different modes of block transfer require corresponding interfaces or channels. The features required in the interface, with the exception of CAMAC address manipulations, which are related to the Q response, are relatively straightforward. In particular, the distinction between UCS and UCW modes is that the word accompanied by $Q = 0$ in Read operations is transferred to memory for the latter mode. The control and data registers in the channel controller should not be updated

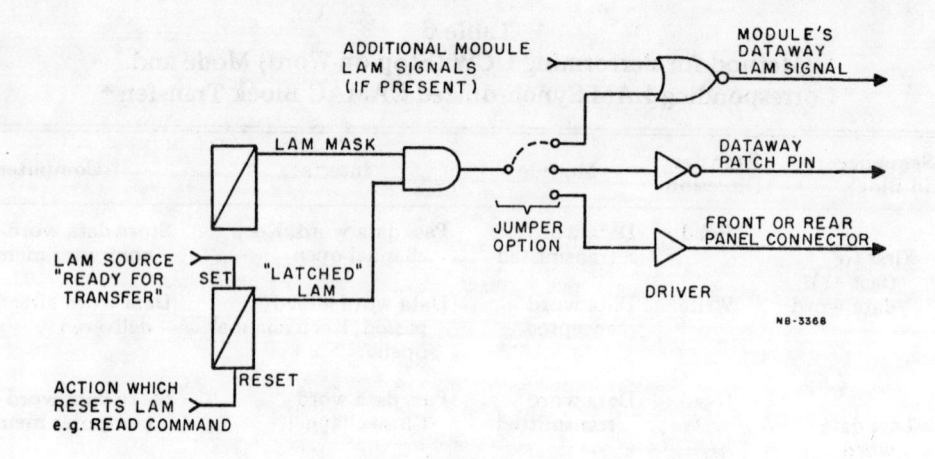

**Fig 1
Recommended Implementation of a Module's LAM Signal
Which Is Intended for Block-Transfer Synchronization;
Simplified Block Diagram**

in the UCS mode until a $Q = 1$ response is received from the module. Interface designers should incorporate a software controllable and testable switch to control whether or not this last word is transferred. This eases the problem of the user with a module designed for a specific application as described in Section 4.1.

All channel controllers should incorporate a word count register as a means to terminate transfers.

As suggested in Section 6.2, an interface can be designed to accept a LAM signal either indirectly or directly and either interrupt the computer or cause the controller to sequence to the next operation in the block transfer.

It is important to note that a channel controller's design need not be restricted to one type of block transfer. For most practical systems several types of block transfers would need to be implemented in the controller.

7. Software Considerations

To the maximum extent practicable, the software for implementing block transfers should be independent of the various modes, as detailed mechanisms for accomplishing these transfers should not have to be considered at the level of application software. From the software point of view, the modes ULS, ULW, UDS, and UDW are identical, if modules and channel controllers are correctly matched. Ref [2], Section 5.8, describes the UBL mode in IML, which is intended to correspond to all four of these modes. The modes UCS and UQC differ at the software level only in the necessity, in the case of UCS, of providing for the possibility that the module will terminate the transfer. Ref [2] describes them in Sections 5.9 and 5.10 as UBC and UBR modes, respectively. Mode MCA corresponds to MA mode, described in Ref [2], Section 5.11. Table 2 gives the correspondences among the names used for block-transfer modes in Ref [1], Ref [2], and this document.

8. References

[1] ANSI/IEEE Std 583-1982, Modular Instrumentation and Digital Interface System (CAMAC). EURATOM Report EUR 4100e, CAMAC: A Modular Instrumentation System for Data Handling; Description and Specification Revised version 1972.

[2] US Energy Research and Development Administration Report TID-26615, Jan 1975 and CAMAC: The Definition of IML, A Language for Use in CAMAC Systems ESONE Secretariat Report ESONE/IML/01, Oct 1974.

[3] ANSI/IEEE Std 596-1982, Parallel Highway Interface System (CAMAC) and EURATOM Report EUR 4600e, CAMAC: Organization of Multi-Crate Systems; Specification of the Branch Highway and CAMAC Crate Controller Type A.

Appendix

(This Appendix is not a part of ANSI/IEEE Std 683-1976, IEEE Recommended Practice for Block Transfers in CAMAC Systems.)

A.1 Other Block-Transfer Modes

This appendix contains descriptions of three other block-transfer modes which, while not having received the same amount of use and recognition as the modes described in the document, have provided CAMAC solutions to particular system problems. They are included to illustrate the wide latitude available to the system designer while still conforming to the CAMAC standard.

A1.1 MCQ (Multiple Test) Mode (Multiaddress Block Transfer, Controller Synchronized, Q Response Terminated). This multiple action mode does not involve the transfer of data but rather provides an efficient method of determining which of a number of different CAMAC addresses has a LAM request or LAM status set. Each of a predefined sequence of modules is addressed with either a Test Look-at-Me $F(8)$ or a Test Status $F(27)$ function. The sequence is stopped on the first occurrence of $Q = 1$. The number of operations performed then identifies the first address in the sequence with the tested flag set. This mode is appropriate to those installations in which many different LAM signals are joined to a common signal (for example, a common Graded-L number or a computer interrupt). No specific hardware is required additionally in the modules used, and the channel may well be implemented in software.

This mode is supported in IML (see Ref [2]) by the MNQ mode (Section 5.12 of Ref [2]).

A1.2 ECA (Extended Address Scan) Mode (Extended Multiaddress Block Transfer, Controller Synchronized and Terminated). This block-transfer mode is an extension of the ACA mode. Its objective is to access registers at successive subaddresses in a sequence of modules or crates, or both.

The length of the block to be transferred is determined by a word count register in the channel controller or by accessing the final address. The channel controller action is identical to that used for an ACA mode block transfer with the exception that when the controller receives a $Q = 1$ response from Subaddress $A(15)$ it resets the subaddress to zero but does not increment the station number. The module typically ignores the subaddress. Data continues to be transferred between a module and the channel controller as long as the module's Q response equals one. Only when a $Q = 0$ response is received does the channel controller increment the station number and set the value of the subaddress A to zero. The channel controller keeps the channel open until the selected number of transfers has occurred or address limit is reached.

The response of $Q = 1$ implies that a data word was transmitted from or accepted by the module. With a $Q = 0$ response for Read operations no data has been passed by the channel controller to computer memory. With a $Q = 0$ response for Write operations, the data word was not accepted by the module and hence is retained by the channel controller for use in the next transfer attempt.

A1.3 UQL (Pause) Mode (Uniaddress Block Transfer, Q Response Synchronized, LAM Terminated). The UQL (Pause) Mode is an extension of the UQC (Repeat) Mode. Its objective is to retain the features of the UQC mode while at the same time providing a means for determining when the end-of-block has been reached.

A $Q = 1$ response to a Read or Write operation means the requested operation took place and that another such operation should be tried immediately. A $Q = 0$ response means the module was not ready to participate in the requested data transfer and that the next CAMAC operation be an examination of the module LAM status register to determine the status of the end-of-block indicator. If this LAM status is set, then the block transfer is considered complete. If it is not set, then another attempt is immediately made to carry out the data transfer and the resulting state of Q used as before to determine both what happened and what should be done next.

Like the Repeat mode, the Pause mode makes continuous use of the CAMAC highway. Hence any channel implementing the Pause mode should incorporate a time-out feature to avoid the possibility of waiting indefinitely for a block to be transferred.

ANSI/IEEE
Std 726-1982
(Revision of
IEEE Std 726-1979)

An American National Standard

IEEE Standard Real-Time BASIC for CAMAC*

Sponsor

**Instruments and Detectors Committee of the
IEEE Nuclear and Plasma Sciences Society**

Approved September 17, 1981

IEEE Standards Board

Approved December 15, 1981

American National Standards Institute

*Computer Automated Measurement and Control

© Copyright 1982 by

The Institute of Electrical and Electronics Engineers, Inc
345 East 47th Street, New York, NY 10017

*No part of this publication may be reproduced in any form,
in an electronic retrieval system or otherwise,
without the prior written permission of the publisher.*

IEEE Standards documents are developed within the Technical Committees of the IEEE Societies and the Standards Coordinating Committees of the IEEE Standards Board. Members of the committees serve voluntarily and without compensation. They are not necessarily members of the Institute. The standards developed within IEEE represent a consensus of the broad expertise on the subject within the Institute as well as those activities outside of IEEE which have expressed an interest in participating in the development of the standard.

Use of an IEEE Standard is wholly voluntary. The existence of an IEEE Standard does not imply that there are no other ways to produce, test, measure, purchase, market, or provide other goods and services related to the scope of the IEEE Standard. Furthermore, the viewpoint expressed at the time a standard is approved and issued is subject to change brought about through developments in the state of the art and comments received from users of the standard. Every IEEE Standard is subjected to review at least once every five years for revision or reaffirmation. When a document is more than five years old, and has not been reaffirmed, it is reasonable to conclude that its contents, although still of some value, do not wholly reflect the present state of the art. Users are cautioned to check to determine that they have the latest edition of any IEEE Standard.

Comments for revision of IEEE Standards are welcome from any interested party, regardless of membership affiliation with IEEE. Suggestions for changes in documents should be in the form of a proposed change of text, together with appropriate supporting comments.

Interpretations: Occasionally questions may arise regarding the meaning of portions of standards as they relate to specific applications. When the need for interpretations is brought to the attention of IEEE, the Institute will initiate action to prepare appropriate responses. Since IEEE Standards represent a consensus of all concerned interests, it is important to ensure that any interpretation has also received the concurrence of a balance of interests. For this reason IEEE and the members of its technical committees are not able to provide an instant response to interpretation requests except in those cases where the matter has previously received formal consideration.

Comments on standards and requests for interpretations should be addressed to:

    Secretary, IEEE Standards Board
    345 East 47th Street
    New York, NY 10017
    USA

Foreword

(This Foreword is not a part of ANSI/IEEE Std 726-1982, IEEE Standard Real-Time BASIC for CAMAC.)

This standard deals with software for the CAMAC modular instrumentation and interface system of ANSI/IEEE Std 583-1982, Modular Instrumentation and Digital Interface System (CAMAC) (also Report EUR 4100e of the Commission of the European Communities). The report on which this standard was originally based, ERDA Report TID-26619, was prepared by the NIM Committee* of the United States Energy Research and Development Administration (now the Department of Energy) and the ESONE Committee** of European Laboratories. The corresponding ESONE document was ESONE/RTB/02.

This standard differs from these original documents in several respects, particularly to be consistent with the new standard for BASIC[1] that has been developed jointly by ANSI X3J2, ECMA TC21 and EWICS TC2. Enhancements to the BASIC[1] standard include a real-time module that provides input-output to process peripherals, response to asynchronous events, synchronization and communication between concurrent activities, and the handling of exceptions.

The BASIC[1] standard is general, and specific parts of the syntax and semantics are left open to be defined appropriately for particular process peripheral systems. The purpose of this standard is to define specific syntax and semantics for use with CAMAC in the areas left open in the standard.

The revision of this standard was in conjunction with the 1981 review (1982 issue) of the entire family of IEEE CAMAC standards undertaken to incorporate existing addenda and corrections into the standards.

This standard was reviewed and balloted by the Nuclear Instruments and Detectors Committee of the IEEE Nuclear and Plasma Sciences Society. At the time of approval of this standard, the membership was as follows:

D. C. Cook, Chairman   **Louis Costrell,** Secretary

J. A. Coleman	T. R. Kohler	P. L. Phelps, Jr
J. F. Detko	H. W. Kraner	J. H. Trainor
F. S. Goulding	W. W. Managan	S. Wagner
F. A. Kirsten	G. L. Miller	F. J. Walter
	D. E. Persyk	

At the time it approved this standard, the American National Standards Committee N42 on Nuclear Instruments had the following personnel:

Louis Costrell, Chairman   **David C. Cook,** Secretary

Organization Represented — *Name of Representative*

American Conference of Governmental Industrial Hygienists . . . . Jesse Lieberman
American Nuclear Society. . . . . Frank W. Manning
Health Physics Society . . . . J. B. Horner Kuper
    J. M. Selby (*Alt*)
Institute of Electrical and Electronics Engineers . . . . Louis Costrell
    Julian Forster (*Alt*)
    D. C. Cook (*Alt*)
    A. J. Spurgin (*Alt*)
Instrument Society of America . . . . M. T. Slind
    J. E. Kaveckis (*Alt*)
Lawrence Berkeley National Laboratory . . . . L. J. Wagner
Oak Ridge National Laboratory . . . . Frank W. Manning
    D. J. Knowles (*Alt*)
US Department of Energy. . . . . Gerald Goldstein
US Department of the Army, Materiel Command . . . . Abraham E. Cohen
US Department of Commerce, National Bureau of Standards. . . . . Louis Costrell
US Federal Emergency Management Agency. . . . . Carl R. Siebentritt, Jr

---

[1] Standard ECMA-55 Minimal BASIC (1978), also published as: ANSI X3.60, American National Standard for Minimal BASIC (1978), ISO Minimal BASIC (1980) ref DIS 6376, ANSI X3J2/80-64, Draft American National Standard for BASIC (1980), and EWICS-TC2 81/7, IRTB Industrial Real-Time BASIC (1981), published by the CEC. This is the standard referred to in this document when reference is made to "the ANSI standard".

Organization Represented	Name of Representative
US Naval Research Laboratory	D. C. Cook
US Nuclear Regulatory Commission	Robert E. Alexander
Members-at-Large	J. G. Bellian
	O. W. Bilharz, Jr
	John M. Gallagher, Jr
	Voss A. Moore
	R. F. Shea
	E. J. Vallario

*NIM Committee[2]
L. Costrell, Chairman

C. Akerlof	N. W. Hill	L. B. Mortara
E. J. Barsotti	D. Horelick	V. C. Negro
B. Bertolucci	M. E. Johnson	L. Paffrath
J. A. Biggerstaff	C. Kerns	D. G. Perry
A. E. Brenner	F. A. Kirsten	I. Pizer
R. M. Brown	P. F. Kunz	E. Platner
E. Davey	R. S. Larsen	S. Rankowitz
W. K. Dawson	A. E. Larsh, Jr	S. J. Rudnick
S. R. Deiss	R. A. LaSalle	G. K. Schulze
S. Dhawan	N. Latner	W. P. Sims
R. Downing	F. R. Lenkszus	D. E. Stilwell
T. F. Droege	R. Leong	J. H. Trainor
C. D. Ethridge	C. Logg	K. J. Turner
C. E. L. Gingell	S. C. Loken	H. Verweij
A. Gjovig	D. R. Machen	B. F. Wadsworth
B. Gobbi	D. A. Mack	L. J. Wagner
D. B. Gustavson	J. L. McAlpine	H. V. Walz
D. R. Heywood		D. H. White

NIM Executive Committee for CAMAC
L. Costrell, Chairman

E. J. Barsotti	F. A. Kirsten
J. A. Biggerstaff	R. S. Larsen
S. Dhawan	D. A. Mack
J. H. Trainor	

NIM Block Transfer Subgroup
E. J. Barsotti, Chairman

W. K. Dawson	F. R. Lenkszus
F. A. Kirsten	R. G. Martin
R. A. LaSalle	R. F. Thomas, Jr

NIM Dataway Working Group
F. A. Kirsten, Chairman

E. J. Barsotti	C. Kerns
J. A. Biggerstaff	P. F. Kunz
L. Costrell	R. S. Larsen
S. Dhawan	D. R. Machen
A. Gjovig	L. Paffrath
D. R. Heywood	S. Rankowitz
D. Horelick	S. J. Rudnick

NIM Software Working Group
D. B. Gustavson, Chairman
W. K. Dawson, Secretary

A. E. Brenner	R. A. LaSalle
R. M. Brown	F. B. Lenkszus
S. R. Deiss	C. A. Logg
S. Dhawan	J. McAlpine
C. D. Ethridge	L. B. Mortara
M. E. Johnson	D. G. Perry

NIM Mechanical and Power Supplies Working Group
L. J. Wagner, Chairman

L. Costrell	S. J. Rudnick
C. Kerns	W. P. Sims

NIM Analog Signals Working Group
D. I. Porat, Chairman

L. Costrell	N. W. Hill
C. E. L. Gingell	S. Rankowitz

NIM Serial System Subgroup
D. R. Machen, Chairman

E. J. Barsotti	R. G. Martin
D. Horelick	L. Paffrath
F. A. Kirsten	S. J. Rudnick

NIM Multiple Controllers Subgroup
P. F. Kunz, Chairman

E. J. Barsotti	D. R. Machen
F. A. Kirsten	R. G. Martin

---

[2] National Instrumentation Methods Committee of the US Department of Energy.

**ESONE Committee[3]

W. Attwenger, Austria, *Chairman* 1980-81   H. Meyer, Belgium, *Secretary*

Representatives of ESONE Member Laboratories

W. Attwenger, Austria	E. Kwakkel, Netherlands	A. C. Peatfield, England
R. Biancastelli, Italy	J. L. Lecomte, France	I. C. Pyle, England
L. Binard, Belgium	J. Lingertat, D.R. Germany	B. Rispoli, Italy
J. Biri, Hungary	M. Lombardi, Italy	M. Sarquiz, France
B. Bjarland, Finland	M. Maccioni, Italy	W. Schoeps, Switzerland
D. A. Boyce, England	P. Maranesi, Italy	R. Schule, F.R. Germany
B. A. Brandt, F.R. Germany	C. H. Mantakas, Greece	P. G. Sjolin, Sweden
F. Cesaroni, Italy	D. Marino, Italy	L. Stanchi, Italy
P. Christensen, Denmark	H. Meyer, Belgium	R. Trechcinski, Poland
W. K. Dawson, Canada	K. D. Muller, F.R. Germany	M. Truong, France
M. DeMarsico, Italy	J. G. Ottes, F.R. Germany	P. Uuspaa, Finland
C. A. DeVries, Netherlands	A. D. Overtoom, Netherlands	H. Verweij, Switzerland
H. Dilcher, F.R. Germany	E. C. G. Owen, England	A. J. Vickers, England
B. V. Fefilov, USSR	L. Panaccione, Italy	S. Vitale, Italy
R. A. Hunt, England	M. Patrutescu, Roumania	M. Vojinovic, Yugoslavia
W. Kessel, F.R. Germany	R. Patzelt, Austria	K. Zander, F.R. Germany
R. Klesse, France		D. Zimmermann, F.R. Germany

ESONE Executive Group (XG)

W. Attwenger, Austria, XG *Chairman* 1980-81

H. Meyer, Belgium, *Secretary*

P. Christensen, Denmark   M. Sarquiz, France
M. Dilcher, F.R. Germany   R. Trechcinski, Poland
B. Rispoli, Italy   A. Vickers, England
          H. Verweij, Switzerland

ESONE Technical Coordination Committee (TCC)

A. C. Peatfield, England, TCC *Chairman*

P. Christensen, Denmark, TCC *Secretary*

W. Attwenger, Austria   W. Kessel, F.R. Germany
R. Biancastelli, Italy   J. Lukacs, Hungary
G. Bianchi, France   R. Patzelt, Austria
H. Dilcher, F.R. Germany   P. J. Ponting, Switzerland
P. Gallice, France   W. Schoeps, Switzerland
          S. Vitali, Italy

ECA/ESONE CAMAC Document Maintenance Study Group (DMSG)

P. Gallice, France, *Chairman*

R. C. M. Barnes, England   F. Iselin, Switzerland
L. Besse, Switzerland   H. Meyer, Belgium
J. Davis, England   H. J. Trebst, F.R. Germany

When this standard was approved on September 17, 1981, the IEEE Standards Board had the following membership:

I. N. Howell, Jr, *Chairman*   **Irving Kolodny,** *Vice Chairman*

Sava I. Sherr, *Secretary*

G. Y. R. Allen	Jay Forster	F. Rosa
J. J. Archambault	Kurt Greene	R. W. Seelbach
J. H. Beall	Loering M. Johnson	J. S. Stewart
J. T. Boettger	Joseph L. Koepfinger	W. E. Vannah
Edward Chelotti	J. E. May	Virginius N. Vaughan, Jr
Edward J. Cohen	Donald T. Michael*	Art Wall
Len S. Corey	J. P. Riganati	Robert E. Weiler

*Member emeritus

---

[3] European Standards on Nuclear Electronics Committee.

Contents

SECTION	PAGE
CAMAC and NIM Standards and Reports	8
1. Introduction	9
2. Real-Time Capabilities	9
3. Declarations	9
3.1 Data Structure Declarations	9
3.2 CAMAC Declarations	10
3.3 Dynamic Address Modification	11
4. Parallel Activities	12
5. CAMAC Input and Output	12
5.1 Simple Data Transfers	12
5.2 Block Transfers	12
5.3 CAMAC Control Actions	13
6. The CAMAC 'Q' and 'X' Signals	13
7. CAMAC LAM Handling	13
7.1 LAM Control Actions	13
7.2 LAM Servicing	14
7.3 Dynamic GL Modification	14
8. Message Passing	14
9. Shared Data	15
10. Bit Manipulation	15
APPENDIXES	
Appendix 1 Method of Syntax Specification	16
Appendix 2 The Formal Definitions	16
Appendix 3 CAMAC Keywords, Functions and Statements	18

CAMAC and NIM Standards and Reports

Title	IEEE, ANSI Std No	IEC No	DOE No	EURATOM (EUR) or ESONE No
CAMAC Instrumentation and Interface Standards*	SH08482* (Library of Congress No 8185060)	—	—	—
Modular Instrumentation and Digital Interface System (CAMAC)	ANSI/IEEE Std 583-1982	516	TID-25875† and TID-25877†	EUR 4100e
Serial Highway Interface System (CAMAC)	ANSI/IEEE Std 595-1982	640	TID-26488†	EUR 6100e
Parallel Highway Interface System (CAMAC)	ANSI/IEEE Std 596-1982	552	TID-25876† and TID-25877†	EUR 4600e
Multiple Controllers in a CAMAC Crate	ANSI/IEEE Std 675-1982	729	DOE/EV-0007	EUR 6500e
Block Transfers in CAMAC Systems	ANSI/IEEE Std 683-1976 (Reaff 1981)	677	TID-26616†	EUR 4100 suppl
Amplitude Analog Signals within a 50 Ω System	—	—	TID-26614	EUR 5100e
The Definition of IML A Language for Use in CAMAC Systems	—	—	TID-26615	ESONE/IML/01
CAMAC Tutorial Articles	—	—	TID-26618	—
Real-Time BASIC for CAMAC	ANSI/IEEE Std 726-1982	§	TID-26619†	ESONE/RTB/03
Subroutines for CAMAC	ANSI/IEEE Std 758-1979 (Reaff 1981)	713	DOE/EV-0016†	ESONE/SR/01
Recommendations for CAMAC Serial Highway Drivers and LAM Graders for the SCC-L2	—	—	DOE/EV-0006	ESONE/SD/02
Definitions of CAMAC Terms	Included in SH08482	678	DOE/ER-0104	ESONE/GEN/01
Standard Nuclear Instrument Modules NIM	—	547**	TID-20893 (Rev 4)	—

†Superseded by corresponding IEEE Standard listed.

*This is a hard cover book that contains ANSI/IEEE Std 583-1982, ANSI/IEEE Std 595-1982, ANSI/IEEE Std 596-1982, ANSI/IEEE Std 675-1982, ANSI/IEEE Std 683-1976 (Reaff 1981), ANSI/IEEE Std 726-1982 and IEEE Std 758-1979 (Reaff 1981), plus introductory material and a glossary of CAMAC terms.

**Covers only mechanical features and connector pin assignments.

§ In preparation.

NOTE: *Availability of Documents*
ANSI Sales Department, American National Standards Institute, 1430 Broadway, New York, NY 10018.
IEEE IEEE Service Center, 445 Hoes Lane, Piscataway, New Jersey 08854, USA.
IEC International Electrotechnical Commission, 1, rue de Varembé, CH-1211 Geneva 20, Switzerland.
DOE and TID Reports National Bureau of Standards, Washington, D.C. 20234, USA, Attn: L. Costrell.
EURATOM Office of Official Publications of the European Communities, P.O. Box 1003, Luxembourg.
ESONE Commission of the European Communities, CGR-BCMN, B-2440 GEEL, Belgium, Attn: ESONE Secretariat, H. Meyer.

An American National Standard

IEEE Standard
Real-Time BASIC for CAMAC

1. Introduction

Real-Time BASIC for CAMAC is ANSI standard Real-Time BASIC in which the declarations and real-time statements are defined for use with CAMAC hardware. The ANSI standard[4] defines 'process declarations' and 'process input-output' in a general way, independently of any particular interface hardware. Specific parts of the syntax and semantics of the language are left 'implementation defined' so that an implementation for a particular hardware system can define these areas in the most appropriate way. The purpose of this standard is to provide a standard for reference, to achieve maximum compatibility between different implementations of ANSI BASIC for use with CAMAC.

The language is defined by formal syntax definitions and symbols. An explanation of these definitions and symbols is given in Appendix A.

It is assumed that the reader is familiar with BASIC and CAMAC.

2. Real-Time Capabilities

A real-time program is written as a number of concurrent activities which can communicate and synchronize as necessary to achieve the overall objective of the application.

Each activity has local variables, and communicates with its environment through 'ports' which are of three types:

(1) Message ports which transfer data between two activities

(2) Shared-data ports which allow access to shared data that exists outside the activities

(3) Process ports which provide CAMAC input-output

Also, statements are defined to schedule concurrent activities in real-time and in response to externally and internally generated events.

---

[4] See footnote 1 on page 3.

3. Declarations

Declaration statements are used to define the attributes of ports and the structure of data transferred through them. The declarations must be at the head of the program before any executable statements. They apply to the whole real-time program, unlike DIM statements which are local to a concurrent activity.

3.1 Data Structure Declarations. A data structure is a list of the data types real, integer and string. It is used to define valid lists of variables and arrays in statements transferring data through ports, and to define units of access to shared data. For example, the declaration for a data structure called STAT comprising 3 real numbers, an integer array and 2 strings could be:

200 STRUCTURE STAT: 3 OF REAL, INTEGER(100), 2 OF STRING

The formal syntax of data structure declarations is:

data-structure-dec	= STRUCTURE structure-name colon repeat-count? type (comma repeat-count? type)*
structure-name	= letter (letter/digit)*
repeat-count	= integer OF
integer	= digit digit*
†type	= (REAL/INTEGER/STRING) dimensioning?
dimensioning	= left-parenthesis bounds right-parenthesis
bounds	= integer (comma integer)?

†NOTE: The ANSI standard defines only two data types, numeric (decimal floating-point) and string. For many CAMAC applications the numeric format would use excessive memory and would cause calculations to be too slow. For use with CAMAC 'numeric' is subdivided into two types: 'real' using decimal or binary internal format, and 'integer' corresponding to CAMAC 24-bit words. When used in numeric operations, integers are treated as 24-bit two's complement numbers.

3.2 CAMAC Declarations. CAMAC declarations are the 'process port declarations' in the ANSI standard, in the specific case when the process peripheral system is CAMAC.

Examples of CAMAC declarations are:

```
400  PROCESS INPUT WEIGHT "CAMAC
     (1,3,17,0) (F2) (B 10)"
410  PROCESS OUTPUT PANEL "CAMAC
     (,,2,4)(C4)"
420  PRODIM MPX(4)
421  PROCESS INPUT MPX(1) "CAMAC
     (,,5,0)"
422  PROCESS INPUT MPX(2) "CAMAC
     (,,5,1)"
423  PROCESS INPUT MPX(3) "CAMAC
     (,,5,2)"
424  PROCESS OUTIN MPX(4) "CAMAC
     (,,7,0) (F1)"
430  PROCESS EVENT FULL "CAMAC
     WEIGHT GL3"
```

Line 400 declares an 'input only' device called WEIGHT which is at Branch 1, Crate 3, Station 17 and Sub-Address 0. It is read by Function code 2, and its data is in the form binary sign and magnitude with 10 magnitude bits.

The second example in line 410 declares an 'output only' device called PANEL which is in Station 2 and is accessed at Sub-Address 4 (the Branch and Crate numbers are defaulted — this declaration would be appropriate for single-crate systems). The default function code 16 is used for writing, and the date is interpreted as a BCD number with four digits.

Lines 420—424 show the declaration of an array of CAMAC ports. Line 420 is the dimension statement, and lines 421—424 declare the attributes of each element of the array. In the example the elements MPX(1), MPX(2) and MPX(3) are the first three sub-addresses in Station 5, while MPX(4) is Sub-Address 0 in Station 7.

The final example on line 430 declares a CAMAC LAM emanating from the device WEIGHT, and connected to GL3. The LAM is given the name FULL and it is used as an 'event' in WAIT statements (see Section 4).

The part of a declaration inside quotation marks is 'implementation defined' in the standard, and is defined here for use with CAMAC.

The formal definitions are as follows:

process-dim-statement	= PRODIM process-array-dec (comma process-array-dec)*
process-array-dec	= process-array-name dimensioning
process-port-dec	= PROCESS (process-clause / event-clause) access-information
process-clause	= io-qualifier camac-port-name bounds? (OF structure-name)? (TIMEOUT numeric-rep)
io-qualifier	= INPUT / OUTPUT / OUTIN
camac-port-name	= letter (letter/digit)*
numeric-rep	= significand xrad?
significand	= integer full-stop? / integer? fraction
fraction	= full-stop integer
xrad	= E (plus-sign / minus-sign)? integer
event-clause	= EVENT lam-name
lam-name	= letter (letter / digit)*
‡ acccess-information	= quotation-mark CAMAC (register-dec / lam-dec) quotation-mark
‡ register-dec	= bcna access? format?
‡ bcna	= left-parenthesis integer? comma integer? comma integer comma integer right-parenthesis
‡ access	= left-parenthesis acess-function (comma access-function)* right-parenthesis
‡ access-function	= (F integer) / NX / ch-name
‡ ch-name	= implementation-defined
‡ format	= left-parenthesis (B / C / I) integer right-parenthesis
‡ lam-dec	= camac-port-name GL integer ((P / A) integer)?

‡ = 'implementation defined' in the standard, defined here for use with CAMAC.

The keyword CAMAC is used to distinguish CAMAC declarations in systems with more than one type of process peripheral.

In the 'bcna' field, the four values represent

B, C, N and A, that is, the Branch, Crate, Station and Sub-Address parts of the CAMAC address respectively. If the last two parameters are zero, the address represents a crate controller for use in crate actions (see 5.3), for example, (1,3,0,0) represents Crate 3 on Branch 1.

Up to two function codes can be specified in the 'access' field, one from each group 'read' and 'write'. Note that the function code is given in the declaration — it is not a parameter of the input or output statement. The default codes for read and write are F0 and F16. The keywords INPUT, OUTPUT or OUTIN indicate that the device provides input only, accepts output only, or supports both input and output.

If a system contains a mixture of modules with and without the X-error facility, either the X-error trap must be permanently disabled, or it must be disabled when a module without 'X' is accessed. The keyword NX allows this attribute to be associated with the device rather than stated explicitly in each input or output statement. The ch-name is a system parameter identifying a block-transfer data-channel. It specifies the mode of operation and, in the case of a LAM synchronized channel, the synchronizing LAM. The default data channel is programmed input-output. A possible method of implementation is to use an implementation supplied 'environment description file' containing the definitions of channels and other relatively fixed facilities.

For process input-output where the internal reference is numeric, the format field specifies the numeric interpretation of the data. B represents binary sign and magnitude, with the integer indicating the number of magnitude bits; C represents binary coded decimal (8-4-2-1 code), with the integer indicating the number of decimal digits in the CAMAC word; and I represents two's complement integer (the 'integer' gives the total number of bits, including sign). If no format is specified, the default interpretation is a 24-bit integer.

Some CAMAC devices, such as scanning multiplexers, have complex data structures. The inclusion of a structure-name in the process-clause allows the language processor to check the validity of variable-lists in input-output statements referring to such devices.

In the case of a lam-declaration, the camac-port-name is the CAMAC device that is the source of the LAM, the integer following the keyword GL is the Graded-LAM line to which it is connected. If the LAM is accessed at a Sub-Address other than that given in the CAMAC declaration, the LAM Sub-Address is given by an integer following the keyword A. If the LAM is accessed at a bit position in the group 2 registers A12, A13 and A14, the bit position is given by the integer following the keyword P. Bit positions are numbered from 1 to 24 starting with the least significant.

3.3 Dynamic Address Modification. Some applications require the ability to dynamically re-assign the address part of a CAMAC device. An example is the requirement in a high-integrity system to activate a standby input-output device by a keyboard command. For this purpose, 8 functions are defined specifically to examine and modify the components of a CAMAC address. Note that these functions are specifically for use with CAMAC; they have no equivalent in the ANSI standard.

The address examine functions are BEX, CEX, NEX and AEX, which return the address parts B, C, N and A respectively of a CAMAC device, for example:

520 LET G = AEX(MPX(2))

would load the variable G with the Sub-Address corresponding to device 2 in the vector of CAMAC devices MPX (this would be the value 1 following the declaration in line 422).

The address modify functions are BMY, CMY, NMY and AMY, which modify the address parts B, C, N and A of a CAMAC device. They are used in assignment statements, for example:

530 LET MPX(2) = NMY(4)

would change the station number used to access the process object MPX(2) from the value 5 (declared in statement 422) to the value 4. Note that these are special functions that can only be used to assign values to the address part of a CAMAC device and that this is the only statement in which a process object can appear on the left-hand side of an assignment.

4. Parallel Activities

A parallel activity is defined by the BASIC statements between the keywords PARACT and END PARACT. The PARACT line gives the name of the parallel activity for use in scheduling statements, and an 'urgency' value. An example of a parallel activity could be:

```
550 PARACT WORK 3
555     START RIG1
560     SIGNAL E6
565     WAIT TIME 17*60*60
570     PRINT "TIME TO GO HOME"
575 END PARACT
```

The parallel activity is called WORK and it has urgency 3. The interpretation of the urgency value is implementation defined. It could, for example, represent a relative priority or a deadline but whatever the interpretation a lower value effectively indicates a higher priority.

A parallel activity is started by a START statement. It can be suspended by a WAIT statement, and stopped by a PARSTOP statement. A PARSTOP statement terminates the parallel activity in which it is executed, while a STOP statement terminates the whole real-time program. If control reaches the END PARACT line the effect is the same as a PARSTOP statement.

The formal definitions are as follows:

scheduling-statement	= start-statement / wait-statement / signal-statement / parstop-statement
start-statement	= START activity-name
activity-name	= letter (letter / digit)*
wait-statement	= WAIT (wait-interval / wait-time / wait-event)
wait-interval	= DELAY time-expression
time-expression	= numeric-expression / string-expression
wait-time	= TIME time-expression
wait-event	= EVENT (event-name / lam-name)
event-name	= letter (letter / digit)*
parstop-statement	= PARSTOP
signal-statement	= SIGNAL event-name

The value of the numeric-expression in a time-expression is the number of seconds past the previous midnight. The string expression in a time-expression is an alternative way of expressing time in the form hours:minutes:seconds on a 24-hour clock.

An event-name is the name of a 'software event'. A software event is declared implicitly by its occurrence in a SIGNAL statement. A software event is cleared automatically when it is 'consumed' by a WAIT EVENT statement.

5. CAMAC Input and Output

5.1 Simple Data Transfers. Simple data transfers from and to the process peripheral system use IN FROM and OUT TO statements as follows:

IN FROM camac-port-name TO variable
OUT TO camac-port-name FROM expression

For example, assuming the declaration in line 400, the statement:

600 IN FROM WEIGHT TO G

would read the data from Branch 1, Crate 3, Station 17 Sub-Address 0, convert 10 bits with the sign in bit 11 to floating-point form, and store the result in the variable G.

The formal definitions are as follows:

process-io-statement	= in-statement / out-statement
in-statement	= IN FROM camac-port-name TO in-structure
in-structure	= in-structure-element (comma in-instructure-element)*
in-structure-element	= variable / formal-array
out-statement	= OUT TO camac-port-name FROM out-structure
out-structure	= out-structure-element (comma out-structure-element)*
out-structure-element	= expression / formal-array
formal-array	= array left-parenthesis comma? right-parenthesis

5.2 Block Transfers. Block transfers can be specified by using a formal array and a numeric variable as an in-structure or an out-structure. The formal array contains the block of data and the numeric variable contains, after the block transfer statement has terminated, the number of transfers actually achieved. This may be fewer than the number of elements in the array if the module signals 'end of block' by setting Q=0, or if an exception occurs. The maximum number of transfers will be equal to the number of elements in the array. The following lines of program illustrate an input block transfer:

```
700  PROCESS INPUT M "CAMAC (, , 5,0)
     (FO, CHAN3)"
710  DIM D(100)
720  IN FROM M TO D(), C
730  IF C <> 100 THEN 790
     —
     —
     —
790  PRINT C; "TRANSFERS ACCOM-
     PLISHED"
     —
```

NOTE: Block transfers are not formally defined in the ANSI standard. An implicit block transfer occurs when a formal-array appears in an in-structure or an out-structure, but no mechanism is defined for the process-object to signal end-of-block or for the number of transfers achieved to be available to the program.

5.3 CAMAC Control Actions. These actions are specific to CAMAC and no equivalent statements are defined in the ANSI standard. They concern the CAMAC 'operate' functions, which refer to the status of devices but do not transfer data over the data lines. They operate at the hardware level in the CAMAC system. The action to be executed is specified in the statement. A control action statement has the following form:

CONTROL camac-port-name control-action

where:

control action	= (F op-code) / module-op / crate-op
module-op	= ENB / DIS / CL1 /CL2
crate-op	= CZ / ENCD / DISCD / CC / CLRCI / SETCI

The op-code is a decimal constant representing one of the 'operate' group of function codes. A crate-op is a control action addressed to a crate controller, that is, a process-object for which the A and N parameters are zero (see 3.2).

An example of a control action is:

800 CONTROL MPX(4) ENB

The meaning of the various keywords is as follows:

ENB	= function code 26
DIS	= function code 24
CL1	= function code 9 (clear group 1 register)
CL2	= function code 11 (clear group 2 register)
CZ	= activate the crate 'Z' bus
ENCD	= enable crate demands
DISCD	= disable crate demands
CC	= activate the crate 'clear' bus
CLRCI	= remove 'crate inhibit'
SETCI	= set 'crate inhibit'

6. The CAMAC 'Q' and 'X' Signals

The values of Q and X returned for the last CAMAC operation in each activity are stored in the variable QCAM and XCAM. These are program status variables of which a local pair are created for each activity. They can be tested with IF statements, for example:

800   IF QCAM = 0 THEN 900

7. CAMAC LAM Handling

A CAMAC LAM is handled at the BASIC level as a lam-name. Statement 430 in 3.2 is an example of a lam-declaration. Two aspects of LAM handling must be considered: how to enable it, disable it and test its status by control actions, and how the program makes use of its occurrence.

7.1 LAM Control Actions. LAM CONTROL actions operate at the hardware level in the CAMAC system. The LAM address is supplied by the CAMAC device associated with the event in the declaration.

The syntax of a LAM CONTROL action is as follows:

CONTROL lam-name lam-action

where:

lam-action=ENL/DISL/TEST/CLRL/ MENL/MDISL/MTEST/MCLRL

The meaning of the various keywords is as follows:

ENL = enable LAM, either F26 at a sub-address or set the appropriate bit in the group 2 mask register A13, according to the declaration

DISL = disable LAM, either F24 at a sub-address or clear the appropriate bit in the group 2 mask register A13, according to the declaration

TEST = test LAM request, either F8 at a Sub-Address or test the bit position in the group 2 register A14. This test leaves QCAM=1 if the LAM is set, QCAM-0 if clear

CLRL = clear LAM, either F10 at a Sub-Address or clear the bit in the group 2 register A12

MENL = module enable LAM using the Sub-Address for overall control, see 2.3

MDISL = module disable LAM, overall control

MTEST = test LAM in the module, overall test

MCLRL = clear all LAMs in the module

NOTES: (1) Only modules built with the facility of overall LAM control at a sub-address will respond to the control actions MENL, MDISL, MTEST and MCLRL.
(2) ENL and DISL are CAMAC statements equivalent to the general statements CONNECT and DISCONNECT defined in the ANSI standard.

An example of a LAM CONTROL action is:

830 CONTROL FULL ENL

which enables the LAM in Station 17, Sub-Address 0 given the declaration in line 430, of 3.2.

7.2 LAM Servicing. A program interrupt mechanism is not provided at the BASIC level. A LAM service routine is programmed as a parallel activity with a wait-statement naming the LAM. When the LAM occurs the activity proceeds from the wait statement, which then clears the LAM automatically. The activity is normally programmed to return to the wait statement after servicing the LAM.

7.3 Dynamic GL Modification. Systems using a programmable LAM grader must be able to change the GL number associated with an event. The function GMY is used for this purpose:

LET lam-name = GMY left-parenthesis
                numeric-expression
                right-parenthesis

for example:

940 LET FULL = GMY(7)

This is a special assignment statement similar to the CAMAC address assignments defined in Section 3. It is the only statement in which an event-name can appear on the left hand side of an assignment. The function GMY can be used only in this statement.

8. Message Passing

Send-statements and receive-statements are used to transmit data over message-paths between concurrent-activities. Message-paths are established at run-time by the execution of a send-statement in one activity and a receive-statement in another, referring to the same message-port-name.

These statements also synchronize the two activities. Both activities wait at their respective send-statement and receive-statement until the data are transferred successfully to the variable-list in the receiving activity.

Message-ports must be declared in the declaration section at the beginning of the program. The declaration gives the name of the port and the structure of the data transferred through it. Using the structure declared in 3.1, line 200, an example of a message port declaration is:

210 MESSAGE LINK OF STAT

Examples of valid message statements conforming to this declaration are:

1100 SEND TO LINK FROM A, X/2, 17.35, RESULTS(), COND$, "FIRST"
1200 RECEIVE FROM LINK TO P(1), P(2), Q, I(), COND$, A$

Note that these statements must be in different concurrent activities, and that concurrent activities have their own local variables so that COND$ in line 1100 and COND$ in line 1200 are not the same variable.

The formal definitions are:

message-port-dec = MESSAGE message-port-name OF structure-name
message-port-name = letter (letter / digit)*

send-statement	= SEND TO message-port-name FROM out-structure (TIMEOUT time-expression)?
receive-statement	= RECEIVE FROM message-port-name TO in-structure (TIMEOUT time-expression)?

9. Shared Data

Get-statements and put-statements are used to transmit data through shared-data ports to collections of shared data. The ports must be declared in the declaration section at the head of the program. The declaration gives the name of the port, the structure of the data transferred through it, and also defines a real section of shared data that exists outside the concurrent activities accessing it.

Get-statements and put-statements are indivisible operations. Indivisible means that when a section of shared data is accessed by a get-statement or a put-statement, no other get-statement or put-statement in another concurrent activity can access that section of shared data until the original access is complete, regardless of any rescheduling of activities carried out by the operating system.

Example of these statements are:

```
140  SHARED FLIGHT(10) OF STAT
1110 GET FROM FLIGHT(3) to X,Y,ALL,
     MOD(), ID$, IQ$
1210 PUT TO FLIGHT(I) FROM 3,Y,Z*1.5,
     CON(), "NEXT", A$
```

The formal definitions are:

data-port-dec	= SHARED data-port-name dimensioning? OF structure-name
data-port-name	= letter (letter / digit)*
get-statement	= GET FROM data-port-name TO in-structure
put-statement	= PUT TO data-port-name FROM out-structure

10. Bit Manipulation

The ANSI standard defines bit manipulation in terms of 'bit strings'. A bit string is a sequence of the characters '0' and '1'. An application program uses the normal string operations of segmentation and concatenation to set and test bits, for example if it is required to test bit 5 of a bit pattern stored in the variable B$, statements such as:

IF B$(5:5) = '1' THEN 2000

or

IF SEG$(B$,5,5) = '1' THEN 2000

are used depending on the implementation. For many CAMAC applications this approach will be too slow and occupy too much memory, in which case the following bit-functions are recommended:

AND	= left-parenthesis numeric-expression comma numeric-expression right-parenthesis
OR	= left-parenthesis numeric-expression comma numeric-expression right-parenthesis
XOR	= left-parenthesis numeric-expression comma numeric-expression right-parenthesis
NOT	= left-parenthesis numeric-expression right parenthesis

The numeric-expressions must contain only numeric constants and integers.

Appendixes

(These Appendixes are not a part of ANSI/IEEE 726-1982, IEEE Standard Real-Time BASIC for CAMAC.)

Appendix 1
Method of Syntax Definition

The conventions used in the formal definitions are those employed in the relevant ECMA, ANSI and ISO standards. The conventions are explained fully in those documents, but a brief description of the method of syntax definition is given below.

The syntactic metalanguage used is derived from BNF. The syntax is defined by a series of 'production rules' that define syntactic elements in terms of other syntactic elements in a hierarchical manner, until a 'terminal symbol' is reached. A terminal symbol is an element of the language being defined, that is, Real-Time BASIC for CAMAC. Certain special symbols are used whose meaning is defined below:

=   separates the left part from the right part in a production rule
?   the preceding syntactic element is optionally present
*   the preceding syntactic element is optionally present an arbitrary number of times (including zero times)
(and)   are used to group syntactic elements into a single unit
/   separates alternatives

Spaces and new lines are used to improve legibility of the definitions; they have no syntactic significance.

The following example illustrates the use of some of these symbols:

out-structure	= out-structure-element (comma out-structure-element)*
out-structure-element	= expression / formal-array

This means that an out-structure is a list of out-structure-elements. If there is more than one item in the list, the items are separated by commas. Each item can be either an expression or a formal array. An example of an out-structure satisfying this definition is:

A + 2, B(), C$

Appendix 2
The Formal Definitions

This appendix lists all the formal definitions in alphabetical order, with a reference in the left hand margin to the section in which they are introduced.

3.2	access	= left-parenthesis access-function (comma access-function)* right parenthesis
3.2	access-function	= (F integer) / NX / ch-name
3.2	access-information	= quotation-mark CAMAC (register-dec / lam-dec) quotation-mark
4.	activity-name	= letter (letter / digit)*
3.2	bcna	= left-parenthesis integer? comma integer? comma integer comma integer right-parenthesis
3.1	bounds	= integer (comma integer)?
3.2	camac-port-name	= letter, (letter / digit)*
3.2	ch-name	= implementation-defined

5.3	control-action	= (F op-code) / module-op / crate-op	
5.3	crate-op	= CZ/ENCD/ DISCD/CC/ CLRCI/SETCI	
9.	data-port-dec	= SHARED data-port-name dimensioning? OF structure-name	
9.	data-port-name	= letter (letter / digit)*	
3.1	data-structure-dec	= STRUCTURE structure-name colon repeat-count? type (comma repeat-count? type)*	
3.1	dimensioning	= left-parenthesis bounds right-parenthesis	
3.2	event-clause	= EVENT lam-name	
4.	event-name	= letter (letter / digit)*	
5.	formal-array	= array left-parenthesis comma? right-parenthesis	
3.2	format	= left-parenthesis (B / C / I) integer right-parenthesis	
3.2	fraction	= full-stop integer	
9.	get-statement	= GET FROM data-port-name TO in-structure	
5.	in-statement	= IN FROM camac-port-name TO in-structure	
5.	in-structure	= in-structure-element (comma in-structure-element)*	
5.	in-structure-element	= variable / formal-array	

3.1	integer	= digit digit*	
3.2	io-qualifier	= INPUT / OUTPUT / OUTIN	
7.1	lam-action	= ENL / DISL / TEST / CLRL / MENL / MDISL / MTEST / MCLRL	
3.2	lam-dec	= camac-port-name GL integer ((P / A) integer)?	
3.2	lam-name	= letter (letter / digit)*	
8.	message-port-dec	= MESSAGE message-port-name OF structure-name	
8.	message-port-name	= letter (letter / digit)*	
5.3	module-op	= ENB / DIS / CL1 / CL2	
3.2	numeric-rep	= significand xrad	
5.	out-statement	= OUT TO camac-port-name FROM out-structure	
5.	out-structure	= out-structure-element (comma out-structure-element)*	
5.	out-structure-element	= expression / formal-array	
4.	parstop-statement	= PARSTOP	
3.2	process-array-dec	= process-array-name dimensioning	
3.2	process-clause	= io-qualifier camac-port-name bounds? (OF structure-name)? (TIMEOUT numeric-rep)?	
3.2	process-dim-statement	= PRODIM process-array-dec (comma process-array-dec)*	

5.	process-io-statement	= in-statement / out-statement	3.2	significand	= integer full-stop?/integer? fraction
3.2	process-port-dec	= PROCESS (process-clause/ event-clause)	4.	start-statement	= START activity-name
8.	put-statement	= PUT TO data-port-name FROM out-structure	3.1	structure-name	= letter (letter / digit)*
			4.	timeout-expression	= numeric-expression/ string-expression
8.	receive-statement	= RECEIVE FROM message-port-name TO in-structure (TIMEOUT time-expression)?	3.1	type	= (REAL / INTEGER / STRING) dimensioning?
			4.	wait-event	= EVENT(event-name / lam-name)
3.2	register-dec	= bcna access? format?			
3.1	repeat-count	= integer OF	4.	wait-interval	= DELAY time-expression
4.	scheduling-statement	= start-statement / wait-statement / signal-statement/ parstop-statement	4.	wait-statement	= WAIT (wait-interval / wait-time / wait-event)
8.	send-statement	= SEND TO message-port-name FROM out-structure (TIMEOUT time-expression)?	4.	wait-time	= TIME time-expression
			3.2	xrad	= E (plus-sign / minus-sign)? integer
4.	signal-statement	= SIGNAL event-name			

Appendix 3

CAMAC Keywords, Functions and Statements

This Appendix lists all the new keywords and statements in alphabetical order, with a reference in the left-hand margin to the section in which they are introduced.

3.2	A	the sub-address for overall LAM control
3.3	AEX	a function to examine a CAMAC address
3.3	AMY	a function to modify a CAMAC address
10.	AND	a logical function
3.2	B	binary sign and magnitude
3.3	BEX	a function to examine a CAMAC address
3.3	BMY	a function to modify a CAMAC address
3.2	C	BCD format

BASIC FOR CAMAC   ANSI/IEEE Std 726-1982

3.2	CAMAC	signifies a CAMAC entity	8.	GET	keyword for a shared-data input statement
5.3	CC	a CAMAC CONTROL action	3.2	GL	introduces a CAMAC 'GL' number
3.3	CEX	a function to examine a CAMAC address	7.3	GMY	a function to change a LAM GL number
5.3	CLRCI	a CAMAC CONTROL action	3.2	I	two's complement.. integer format
7.1	CLRL	a LAM CONTROL action	5.	IN FROM	keyword for a CAMAC input statement
5.3	CL1	a CAMAC CONTROL action			
5.3	CL2	a CAMAC CONTROL action	3.2	INPUT	a CAMAC register provides input only
3.3	CMY	a function to modify a CAMAC address	3.1	INTEGER	a data-type in a structure declaration
5.3, 7.1	CONTROL	keyword for a CONTROL statement	7.1	MCLRL	a LAM CONTROL action
5.3	CZ	a CAMAC CONTROL action	7.1	MDISL	a LAM CONTROL action
4.	DELAY	introduces a time delay in a wait-statement	7.1	MENL	a LAM CONTROL action
			8.	MESSAGE	introduces a message-port declaration
5.3	DIS	a CAMAC CONTROL action			
5.3	DISCD	a CAMAC CONTROL action	7.1	MTEST	a LAM CONTROL action
7.1	DISL	a LAM CONTROL action	3.3	NEX	a function to examine a CAMAC address
5.3	ENB	a CAMAC CONTROL action	3.3	NMY	a function to modify a CAMAC address
5.3	ENCD	a CAMAC CONTROL action	10.	NOT	a logical function
4.	END PARACT	terminates the code for a parallel activity	3.2	NX	a keyword indicating that no X error check should be performed when the module is accessed
7.1	ENL	a LAM CONTROL action			
3.2, 4	EVENT	introduces an event-name	10.	OR	a logical function
			3.2	OUTIN	a CAMAC register supports both input and output
3.2	F	introduces a CAMAC function code			
			3.2	OUTPUT	a CAMAC register accepts input only

5.	OUT TO	a CAMAC output statement	5.3	SETCI	a CAMAC CONTROL action
3.2	P	introduces the bit position of a LAM	8.	SHARED	defines a section of data shared by parallel activities
4.	PARACT	introduces a section of code for a parallel activity	4.	SIGNAL	keyword for a signal-statement
4.	PARSTOP	a scheduling statement	4.	START	statement to start a parallel activity
3.2	PROCESS	introduces a CAMAC I/O port declaration	3.1	STRUCTURE	introduces a list of data-types defining a structure
3.2	PRODIM	introduces a CAMAC array declaration	3.1	STRING	a data-type in a structure declaration
8.	PUT	keyword for a shared-date output statement	7.1	TEST	a LAM CONTROL action
6.	QCAM	a system variable for staticizing the CAMAC 'Q' signal	4.	TIME	introduces a time-expression in a wait-statement
			3.2, 8	TIMEOUT	introduces a time-out value for I/O through ports
3.1	REAL	a data-type in a structure declaration	4.	WAIT	keyword for a wait-statement
8.	RECEIVE	a message input statement	6.	XCAM	a system variable for staticizing the CAMAC 'X' signal
8.	SEND	keyword for a message output statement	10.	XOR	a logical function

ANSI/IEEE
Std 758-1979
(Reaffirmed 1981)

An American National Standard

IEEE Standard Subroutines for CAMAC*

Sponsor

**Instruments and Detectors Committee of the
IEEE Nuclear and Plasma Sciences Society**

Approved December 14, 1978
Reaffirmed September 17, 1981

IEEE Standards Board

Approved December 15, 1981

American National Standards Institute

*Computer Automated Measurement and Control

© Copyright 1979 by

The Institute of Electrical and Electronics Engineers, Inc
345 East 47th Street, New York, NY 10017

*No part of this publication may be reproduced in any form,
in an electronic retrieval system or otherwise,
without the prior written permission of the publisher.*

IEEE Standards documents are developed within the Technical Committees of the IEEE Societies and the Standards Coordinating Committees of the IEEE Standards Board. Members of the committees serve voluntarily and without compensation. They are not necessarily members of the Institute. The standards developed within IEEE represent a consensus of the broad expertise on the subject within the Institute as well as those activities outside of IEEE which have expressed an interest in participating in the development of the standard.

Use of an IEEE Standard is wholly voluntary. The existence of an IEEE Standard does not imply that there are no other ways to produce, test, measure, purchase, market, or provide other goods and services related to the scope of the IEEE Standard. Furthermore, the viewpoint expressed at the time a standard is approved and issued is subject to change brought about through developments in the state of the art and comments received from users of the standard. Every IEEE Standard is subjected to review at least once every five years for revision or reaffirmation. When a document is more than five years old, and has not been reaffirmed, it is reasonable to conclude that its contents, although still of some value, do not wholly reflect the present state of the art. Users are cautioned to check to determine that they have the latest edition of any IEEE Standard.

Comments for revision of IEEE Standards are welcome from any interested party, regardless of membership affiliation with IEEE. Suggestions for changes in documents should be in the form of a proposed change of text, together with appropriate supporting comments.

Interpretations: Occasionally questions may arise regarding the meaning of portions of standards as they relate to specific applications. When the need for interpretations is brought to the attention of IEEE, the Institute will initiate action to prepare appropriate responses. Since IEEE Standards represent a consensus of all concerned interests, it is important to ensure that any interpretation has also received the concurrence of a balance of interests. For this reason IEEE and the members of its technical committees are not able to provide an instant response to interpretation requests except in those cases where the matter has previously received formal consideration.

Comments on standards and requests for interpretations should be addressed to:

> Secretary, IEEE Standards Board
> 345 East 47th Street
> New York, NY 10017
> USA

Foreword

(This Foreword is not a part of ANSI/IEEE Std 758-1979, Standard Subroutines for CAMAC.)

This standard presents a recommended set of software subroutines for use with CAMAC modular instrumentation and interface system of ANSI/IEEE Std 583-1982. These subroutines provide a general capability for communicating with CAMAC systems. They will be of primary interest to those who wish to write their own data-processing programs in a high-level programming language, such as Fortran. The achievable data transfer rate is, of course, dependent on a number of factors, including the language used, the operating system, the compiler, and the method and level of subroutines implementation and the computer.

The report on which this standard is based, and with which it is technically identical (DOE/EV-0016), was prepared by the NIM (National Instrumentation Methods) Committee* of the US Department of Energy and the ESONE (European Standards on Nuclear Electronics) Committee** of European Laboratories. Corresponding documents are ESONE Report SR/01 and IEC Publication 713 of the International Electrotechnical Commission.

This standard was reviewed and balloted by the Nuclear Instruments and Detectors Committee of the IEEE Nuclear and Plasma Sciences Society.

This standard was reviewed and reaffirmed in conjunction with the 1981 revision (1982 issue) of the family of IEEE CAMAC standards. It differs from the previous issue only as follows:

Page 3: Foreword — Dates of referenced IEEE standards replaced
Page 8: Replaced by current list of CAMAC specifications and reports
Page 13: 2.3.5, in list of parameters inta corrected to read intc
Page 16: Dates of referenced IEEE standards replaced

At the time of approval of this standard, the membership of the Nuclear Instruments and Detectors Committee of the IEEE Nuclear and Plasma Sciences Society was as follows:

D. C. Cook, *Chairman*   **Louis Costrell,** *Secretary*

J. A. Coleman	T. R. Kohler	P. L. Phelps
J. F. Detko	H. W. Kraner	J. H. Trainor
F. S. Goulding	W. W. Managan	S. Wagner
F. A. Kirsten	G. L. Miller	F. J. Walter
	D. E. Persyk	

At the time it approved this standard, the American National Standards Committee on Nuclear Instruments, N42, had the following personnel:

Louis Costrell, *Chairman*   **David C. Cook,** *Secretary*

Organization Represented	*Name of Representative*
American Conference of Governmental Industrial Hygienists	Jesse Lieberman
American Nuclear Society	Frank W. Manning
Health Physics Society	J. B. Horner Kuper
	J. M. Selby (*Alt*)
Institute of Electrical and Electronics Engineers	Louis Costrell
	Julian Forster (*Alt*)
	D. C. Cook (*Alt*)
	A. J. Spurgin (*Alt*)
Instrument Society of America	M. T. Slind
	J. E. Kaveckis (*Alt*)
Lawrence Berkeley National Laboratory	L. J. Wagner
Oak Ridge National Laboratory	Frank W. Manning
	D. J. Knowles (*Alt*)
US Department of Energy	Gerald Goldstein
US Department of the Army, Materiel Command	Abraham E. Cohen
US Department of Commerce, National Bureau of Standards	Louis Costrell
US Federal Emergency Management Agency	Carl R. Siebentritt, Jr
US Naval Research Laboratory	D. C. Cook
US Nuclear Regulatory Commission	Robert E. Alexander
Members-at-Large	J. G. Bellian
	O. W. Bilharz, Jr
	John M. Gallagher, Jr
	Voss A. Moore
	R. F. Shea
	E. J. Vallario

*NIM Committee[1]
L. Costrell, Chairman

C. Akerlof	N. W. Hill	L. B. Mortara
E. J. Barsotti	D. Horelick	V. C. Negro
B. Bertolucci	M. E. Johnson	L. Paffrath
J. A. Biggerstaff	C. Kerns	D. G. Perry
A. E. Brenner	F. A. Kirsten	I. Pizer
R. M. Brown	P. F. Kunz	E. Platner
E. Davey	R. S. Larsen	S. Rankowitz
W. K. Dawson	A. E. Larsh, Jr	S. J. Rudnick
S. R. Deiss	R. A. LaSalle	G. K. Schulze
S. Dhawan	N. Latner	W. P. Sims
R. Downing	F. R. Lenkszus	D. E. Stilwell
T. F. Droege	R. Leong	J. H. Trainor
C. D. Ethridge	C. Logg	K. J. Turner
C. E. L. Gingell	S. C. Loken	H. Verweij
A. Gjovig	D. R. Machen	B. F. Wadsworth
B. Gobbi	D. A. Mack	L. J. Wagner
D. B. Gustavson	J. L. McAlpine	H. V. Walz
D. R. Heywood		D. H. White

NIM Executive Committee for CAMAC
L. Costrell, Chairman

E. J. Barsotti	F. A. Kirsten
J. A. Biggerstaff	R. S. Larsen
S. Dhawan	D. A. Mack
J. H. Trainor	

NIM Dataway Working Group
F. A. Kirsten, Chairman

E. J. Barsotti	C. Kerns
J. A. Biggerstaff	P. F. Kunz
L. Costrell	R. S. Larsen
S. Dhawan	D. R. Machen
A. Gjovig	L. Paffrath
D. R. Heywood	S. Rankowitz
D. Horelick	S. J. Rudnick

NIM Software Working Group
D. B. Gustavson, Chairman
W. K. Dawson, Secretary

A. E. Brenner	R. A. LaSalle
R. M. Brown	F. B. Lenkszus
S. R. Deiss	C. A. Logg
S. Dhawan	J. McAlpine
C. D. Ethridge	L. B. Mortara
M. E. Johnson	D. G. Perry

NIM Mechanical and Power Supplies Working Group
L. J. Wagner, Chairman

L. Costrell	S. J. Rudnick
C. Kerns	W. P. Sims

NIM Analog Signals Working Group
D. I. Porat, Chairman

L. Costrell	N. W. Hill
C. E. L. Gingell	S. Rankowitz

NIM Serial System Subgroup
D. R. Machen, Chairman

E. J. Barsotti	R. G. Martin
D. Horelick	L. Paffrath
F. A. Kirsten	S. J. Rudnick

NIM Block Transfer Subgroup
E. J. Barsotti, Chairman

W. K. Dawson	F. R. Lenkszus
F. A. Kirsten	R. G. Martin
R. A. LaSalle	R. F. Thomas, Jr

NIM Multiple Controllers Subgroup
P. F. Kunz, Chairman

E. J. Barsotti	D. R. Machen
F. A. Kirsten	R. G. Martin

---
[1] National Instrumentation Methods Committee of the US Department of Energy.

**ESONE Committee[2]

W. Attwenger, Austria, *Chairman* 1980-81   **H. Meyer**, Belgium, *Secretary*

Representatives of ESONE Member Laboratories

W. Attwenger, Austria	E. Kwakkel, Netherlands	A. C. Peatfield, England
R. Biancastelli, Italy	J. L. Lecomte, France	I. C. Pyle, England
L. Binard, Belgium	J. Lingertat, D.R. Germany	B. Rispoli, Italy
J. Biri, Hungary	M. Lombardi, Italy	M. Sarquiz, France
B. Bjarland, Finland	M. Maccioni, Italy	W. Schoeps, Switzerland
D. A. Boyce, England	P. Maranesi, Italy	R. Schule, F.R. Germany
B. A. Brandt, F.R. Germany	C. H. Mantakas, Greece	P. G. Sjolin, Sweden
F. Cesaroni, Italy	D. Marino, Italy	L. Stanchi, Italy
P. Christensen, Denmark	H. Meyer, Belgium	R. Trechcinski, Poland
W. K. Dawson, Canada	K. D. Muller, F.R. Germany	M. Truong, France
M. DeMarsico, Italy	J. G. Ottes, F.R. Germany	P. Uuspaa, Finland
C. A. DeVries, Netherlands	A. D. Overtoom, Netherlands	H. Verweij, Switzerland
H. Dilcher, F.R. Germany	E. C. G. Owen, England	A. J. Vickers, England
B. V. Fefilov, USSR	L. Panaccione, Italy	S. Vitale, Italy
R. A. Hunt, England	M. Patrutescu, Roumania	M. Vojinovic, Yugoslavia
W. Kessel, F.R. Germany	R. Patzelt, Austria	K. Zander, F.R. Germany
R. Klesse, France		D. Zimmermann, F.R. Germany

ESONE Executive Group (XG)

W. Attwenger, Austria, XG *Chairman* 1980-81

H. Meyer, Belgium, *Secretary*

P. Christensen, Denmark	M. Sarquiz, France
M. Dilcher, F.R. Germany	R. Trechcinski, Poland
B. Rispoli, Italy	A. Vickers, England

H. Verweij, Switzerland

ESONE Technical Coordination Committee (TCC)

A. C. Peatfield, England, TCC *Chairman*

P. Christensen, Denmark, TCC *Secretary*

W. Attwenger, Austria	W. Kessel, F.R. Germany
R. Biancastelli, Italy	J. Lukacs, Hungary
G. Bianchi, France	R. Patzelt, Austria
H. Dilcher, F.R. Germany	P. J. Ponting, Switzerland
P. Gallice, France	W. Schoeps, Switzerland

S. Vitali, Italy

ECA/ESONE CAMAC Document Maintenance Study Group (DMSG)

P. Gallice, France, *Chairman*

R. C. M. Barnes, England	F. Iselin, Switzerland
L. Besse, Switzerland	H. Meyer, Belgium
J. Davis, England	H. J. Trebst, F.R. Germany

When this standard was approved December 14, 1978, the IEEE Standards Board had the following membership:

Joseph L. Koepfinger, *Chairman*   **Irvin N. Howell, Jr**, *Vice Chairman*

Ivan G. Easton, *Secretary*

William E. Andrus	Jay Forster	Donald T. Michael
C. N. Berglund	Ralph I. Hauser	Voss A Moore
Edward J. Cohen	Loering M. Johnson	William S. Morgan
Warren H. Cook	Irving Kolodny	Robert L. Pritchard
David B. Dobson	William R. Kruesi	Blair A. Rowley
R. O. Duncan	Thomas J. Martin	Ralph M. Showers
Charles W. Flint	John E. May	B. W. Whittington

---

[2] European Standards on Nuclear Electronics Committee.

Contents

SECTION	PAGE
CAMAC and NIM Standards and Reports	8
1. Introduction	9
2. Functional Specifications	9
2.1 Primary Subroutines	10
2.1.1 Declare CAMAC Register	10
2.1.2 Perform Single CAMAC Action	10
2.2 Single-Action Subroutines	10
2.2.1 Generate Dataway Initialize	11
2.2.2 Generate Crate Clear	11
2.2.3 Set or Clear Dataway Inhibit	11
2.2.4 Test Dataway Inhibit	11
2.2.5 Enable or Disable Crate Demand	11
2.2.6 Test Crate Demand Enabled	11
2.2.7 Test Crate Demand Present	11
2.2.8 Declare LAM	11
2.2.9 Enable or Disable LAM	11
2.2.10 Clear LAM	11
2.2.11 Test LAM	11
2.2.12 Link LAM to Service Procedure	12
2.3 Block Transfers, Multiple Actions, and Inverse Declarations	12
2.3.1 General Multiple Action	12
2.3.2 Address Scan	12
2.3.3 Controller-Synchronized Block Transfer	13
2.3.4 LAM-Synchronized Block Transfer	13
2.3.5 Repeat Mode Block Transfer	13
2.3.6 Analyze LAM Identifier	14
2.3.7 Analyze Register Identifier	14
3. Definitions of Parameters	14
3.1 *ext* (external address)	14
3.2 *b* (branch number)	14
3.3 *c* (crate number)	14
3.4 *n* (station number)	14
3.5 *a* (subaddress)	14
3.6 *f* (function code)	14
3.7 *int* (CAMAC data word)	14
3.8 *q* (Q response)	15
3.9 *l* (logical truth value)	15
3.10 *lam* (LAM identifier)	15
3.11 *m* (LAM access specifier)	15
3.12 *inta* (integer array)	15
3.13 *label* (entry point identifier)	15
3.14 *fa* (function codes)	15
3.15 *exta* (CAMAC external address)	15
3.16 *intc* (CAMAC data array)	15
3.17 *qa* (Q responses)	15
3.18 *cb* (control block)	15
3.19 *extb* (external addresses)	16
4. Bibliography	16

SECTION			PAGE
Appendix A		System-Dependent Subroutines.	17
A1		Introduction	17
A2		Access to Special Signals	17
	A2.1	Branch Initialize	17
	A2.2	Test Status of Preceding Action	17
A3		Channel Identifier	17
	A3.1	Declare Channel	18
	A3.2	Analyzer Channel Declaration	18
A4		Short Data-Word Transfers	18
	A4.1	Perform Single CAMAC Action	18
	A4.2	General Multiple Action	18
	A4.3	Address Scan	18
	A4.4	Controller-Synchronized Block Transfer	18
	A4.5	LAM-Synchronized Block Transfer	18
	A4.6	Repeat Mode Block Transfer	18
A5		Define Crate Identifier	19
A6		Definitions of Parameters	19
	A6.1	*k* (status code)	19
	A6.2	*chan* (channel identifier)	19
	A6.3	*ints* (truncated CAMAC data word)	19
	A6.4	*intt* (truncated CAMAC data array)	19
	A6.5	*intb* (integer array)	19
Appendix B		Fortran Implementation	20
B1		General	20
B2		Description of Subroutines	20
	B2.1	Primary Subroutines	20
		B2.1.1 Declare CAMAC Register	20
		B2.1.2 Perform Single CAMAC Action	20
	B2.2	Single Action Subroutines	21
		B2.2.1 Generate Dataway Initialize	21
		B2.2.2 Generate Crate Clear	21
		B2.2.3 Set or Clear Dataway Inhibit	21
		B2.2.4 Test Dataway Inhibit	21
		B2.2.5 Enable or Disable Crate Demands	21
		B2.2.6 Test Crate Demand Enabled	21
		B2.2.7 Test Demand Present	21
		B2.2.8 Declare LAM	21
		B2.2.9 Enable or Disable LAM	21
		B2.2.10 Clear LAM	21
		B2.2.11 Test LAM	21
		B2.2.12 Link LAM to Service Procedure	21
	B2.3	Block Transfers, Multiple Actions, and Inverse Declarations	21
		B2.3.1 General Multiple Action	21
		B2.3.2 Address-Scan	22
		B2.3.3 Controller-Synchronized Block Transfer	23
		B2.3.4 LAM-Synchronized Block Transfer	23
		B2.3.5 Repeat-Mode Block Transfer	23
		B2.3.6 Analyze LAM Identifier	23
		B2.3.7 Analyze Register Identifier	23
B3		Parameter Types	23
	B3.1	Single Integers	23
	B3.2	Single Logical Values	24
	B3.3	Integer Arrays	24
	B3.4	Logical Array	24
	B3.5	CAMAC Data Word	24
	B3.6	CAMAC Data Array	24
	B3.7	Label	24
Appendix C		Function-Code Mnemonic Symbols	25

CAMAC and NIM Standards and Reports

Title	IEEE, ANSI Std No	IEC No	DOE No	EURATOM (EUR) or ESONE No
CAMAC Instrumentation and Interface Standards*	SH08482* (Library of Congress No 8185060)	—	—	—
Modular Instrumentation and Digital Interface System (CAMAC)	ANSI/IEEE Std 583-1982	516	TID-25875† and TID-25877†	EUR 4100e
Serial Highway Interface System (CAMAC)	ANSI/IEEE Std 595-1982	640	TID-26488†	EUR 6100e
Parallel Highway Interface System (CAMAC)	ANSI/IEEE Std 596-1982	552	TID-25876† and TID-25877†	EUR 4600e
Multiple Controllers in a CAMAC Crate	ANSI/IEEE Std 675-1982	729	DOE/EV-0007	EUR 6500e
Block Transfers in CAMAC Systems	ANSI/IEEE Std 683-1976 (Reaff 1981)	677	TID-26616†	EUR 4100 suppl
Amplitude Analog Signals within a 50 Ω System	—	—	TID-26614	EUR 5100e
The Definition of IML A Language for Use in CAMAC Systems	—	—	TID-26615	ESONE/IML/01
CAMAC Tutorial Articles	—	—	TID-26618	—
Real-Time BASIC for CAMAC	ANSI/IEEE Std 726-1982	§	TID-26619†	ESONE/RTB/03
Subroutines for CAMAC	ANSI/IEEE Std 758-1979 (Reaff 1981)	713	DOE/EV-0016†	ESONE/SR/01
Recommendations for CAMAC Serial Highway Drivers and LAM Graders for the SCC-L2	—	—	DOE/EV-0006	ESONE/SD/02
Definitions of CAMAC Terms	Included in SH08482	678	DOE/ER-0104	ESONE/GEN/01
Standard Nuclear Instrument Modules NIM	—	547**	TID-20893 (Rev 4)	—

†Superseded by corresponding IEEE Standard listed.

*This is a hard cover book that contains ANSI/IEEE Std 583-1982, ANSI/IEEE Std 595-1982, ANSI/IEEE Std 596-1982, ANSI/IEEE Std 675-1982, ANSI/IEEE Std 683-1976 (Reaff 1981), ANSI/IEEE Std 726-1982 and IEEE Std 758-1979 (Reaff 1981), plus introductory material and a glossary of CAMAC terms.

**Covers only mechanical features and connector pin assignments.

§ In preparation.

NOTE: *Availability of Documents*
ANSI   Sales Department, American National Standards Institute, 1430 Broadway, New York, NY 10018.
IEEE   IEEE Service Center, 445 Hoes Lane, Piscataway, New Jersey 08854, USA.
IEC   International Electrotechnical Commission, 1, rue de Varembé, CH-1211 Geneva 20, Switzerland.
DOE and TID Reports   National Bureau of Standards, Washington, D.C. 20234, USA, Attn: L. Costrell.
EURATOM   Office of Official Publications of the European Communitities, P.O. Box 1003, Luxembourg.
ESONE   Commission of the European Communities, CGR-BCMN, B-2440 GEEL, Belgium, Attn: ESONE Secretariat, H. Meyer.

An American National Standard

IEEE Standard Subroutines for CAMAC

1. Introduction

This standard describes a set of standard subroutines to provide access to CAMAC facilities in a variety of computer-programming languages. It is specifically intended that the subroutines be suitable for use with FORTRAN, although they are not restricted to that language. Appendix B describes a recommended implementation of the subroutines explicitly for FORTRAN.

The present approach is based largely on IML and IML-M1 [6].[3] A distinction is made between *declarations*, which are used to name and specify computer and CAMAC entities, and *actions*, which are used to implement the various data movements and condition tests which make up the CAMAC-related portion of a program. As far as possible, the nomenclature of IML-M1 has been followed in order to take advantage of existing familiarity with that system and to provide as uniform a terminology and style as possible among the various CAMAC software documents.

Because of the widespread use of CAMAC on computers with a word length of less than 24 bits and because of the great differences in computer and operating-system features, special system-dependent features are often required to provide greater efficiency or to make appropriate use of the features of particular systems. The main body of this standard describes subroutines which, at the user interface, depend only on the features of CAMAC and therefore, when implemented in any standard procedural language, should be computer-independent. Appendix A describes subroutines which depend not only on CAMAC, but to some extent on individual computers. They cannot be made independent of the system on which they are implemented, and the user should take special precautions when it is necessary to incorporate them into a program. Such a program may not be transportable from one computer to another without modification.

The subroutines have been grouped into three subsets in order to provide different standard levels of implementation. The lowest level requires only two subroutines, but, nevertheless, gives access to most of the facilities which can be found in CAMAC. In higher levels of implementation subroutines are added which permit procedures to be written in more mnemonic terminology, provide better handling of LAM's, permit procedures to be independent of the type of CAMAC highway used, and provide efficient block-transfer capability.

2. Functional Specifications

This section introduces and describes in detail all the recommended subroutines. Since many implementations will not require the complete set, the subroutines are grouped into three subsets corresponding to the recommended implementation levels. Level A, the simplest, requires only the subroutines from 2.1. Level B, an intermediate level, requires the subroutines from 2.1 and 2.2. Level C is

---

[3] Numbers in brackets correspond to those in the Bibliography, Section 4, of this standard.

the highest level and requires implementation of the subroutines from 2.1, 2.2, and 2.3.

Two CAMAC facilities are not available through the use of these subroutines: the X response from an action and the BZ command. These facilities are available through the use of system-dependent subroutines described in Appendix A.

Naming conventions, compatible with the requirements of many computer languages, have been adopted. In order to make it simple for a user to avoid name conflicts, the name of each recommended subroutine begins with the letter C. The second letter of the subroutine name is coded to indicate the general function of the subroutine. Six letters have been used for this purpose:

C indicates that the subroutine performs a control function

D indicates that the subroutine is a declaration of a CAMAC entity

F indicates that the subroutine transfers full-length (24-bit) data words

G indicates that the subroutine analyzes a named CAMAC entity into its address components

S indicates that the subroutine transfers short (less than 24-bit) data words

T indicates that the subroutine tests the state of a signal or status indication

The remaining letters of each subroutine name (to a maximum of six) are chosen for their mnemonic value in identifying which function the subroutine performs.

Since no particular language is assumed in the body of this standard, no syntax can be defined for a subroutine call. The subroutines are described in terms of a subroutine name and an ordered sequence of parameters. As far as possible every implementation should retain the designated name and the order of the parameters. For the same reason no specific form can be defined for the subroutine parameters. CAMAC data words are bit strings with a length of twenty-four. An implementation must have the capacity to represent such strings. Other subroutine parameters are represented either as integer values or as the logical values *true* or *false*. In this standard the parameters of subroutines are described as variables or arrays of types CAMAC word, integer, or logical. For each implementation appropriate storage units and data formats must be chosen for each of these generalized entities.

2.1 Primary Subroutines. These two subroutines, which are required in all implementations, make up level A. The first provides the capability to define the address of a CAMAC entity and to access it. The second is used to perform CAMAC operations on the defined entities. In principle any CAMAC entity for which there is a defined standard mode of access can be accessed through the use of these two subroutines. In practice some systems may contain restrictions on the use of crate controllers or other system modules.

2.1.1 Declare CAMAC Register.
Name: CDREG
Parameters:
  ext (external address, see 3.1)
  b (branch number, see 3.2)
  c (crate number, see 3.3)
  n (station number, see 3.4)
  a (subaddress, see 3.5)
Function: CDREG combines the branch number b, the crate number c, the station number n, and the subaddress a into a convenient system-dependent form and stores the result in ext. Since the method of encoding depends on the implementation, the contents of ext should not be modified by the program. Some subroutines (see 2.2) require only a crate address; if the parameters n and a are both zero, CDREG encodes a crate address and stores it in ext.

2.1.2 Perform Single CAMAC Action.
Name: CFSA
Parameters:
  f (function code, see 3.6)
  ext (external address, see 3.1)
  int (CAMAC data word, see 3.7)
  q (Q response, see 3.8)
Function: CFSA causes the CAMAC action specified by the function code f to be performed at the CAMAC address specified by ext. If f contains a read or write code, a twenty-four-bit data transfer occurs between the CAMAC register addressed by ext and the computer storage location int. Otherwise int is ignored. The state of Q resulting from the operation is stored in q, *true* if Q=1, *false* if Q=0.

2.2 Single-Action Subroutines. These subroutines, together with those described in 2.1, form the level B implementation, which provides a complete facility for specifying single CAMAC actions in a way which is mnemonic, compact, and independent of the type of highway or crate controller. Facilities are provided for declaring LAMs and performing LAM actions using constructions which are independent of the LAM access mode (that is, subaddress or register access).

2.2.1 Generate Dataway Initialize.
Name: CCCZ
Parameter:
   ext   (external address, see 3.1)
Function: CCCZ causes Dataway Initialize (Z) to be generated in the crate specified by *ext*.

2.2.2 Generate Crate Clear.
Name: CCCC
Parameter:
   ext   (external address, see 3.1)
Function: CCCC causes Dataway Clear (C) to be generated in the crate specified by *ext*.

2.2.3 Set or Clear Dataway Inhibit.
Name: CCCI
Parameters:
   ext   (external address, see 3.1)
   l   (logical truth value, see 3.9)
Function: CCCI causes Dataway Inhibit (I) to be set in the crate specified by *ext* if the value of *l* is *true* and to be reset if the value of *l* is *false*.

2.2.4 Test Dataway Inhibit.
Name: CTCI
Parameters:
   ext   (external address, see 3.1)
   l   (logical truth value, see 3.9)
Function: CTCI sets the value of *l* to *true* if Dataway Inhibit is set in the crate specified by *ext* and sets the value of *l* to *false* if Dataway Inhibit is not set.

2.2.5 Enable or Disable Crate Demand.
Name: CCCD
Parameters:
   ext   (external address, see 3.1)
   l   (logical truth value, see 3.9)
Function: CCCD causes Crate Demand to be enabled in the crate specified by *ext* if the value of *l* is *true* and causes Crate Demand to be disabled if the value of *l* is *false*.

2.2.6 Test Crate Demand Enabled.
Name: CTCD
Parameters:
   ext   (external address, see 3.1)
   l   (logical truth value, see 3.9)
Function: CTCD sets the value of *l* to *true* if Crate Demand is enabled in the crate specified by *ext* and sets the value of *l* to *false* if Crate Demand is disabled.

2.2.7 Test Crate Demand Present.
Name: CTGL
Parameters:
   ext   (external address, see 3.1)
   l   (logical truth value, see 3.9)
Function: CTGL sets the value of *l* to *true* if any demand is present in the crate specified by *ext* and sets the value of *l* to *false* if no demand is present.

2.2.8 Declare LAM.
Name: CDLAM
Parameters:
   lam   (LAM identifier, see 3.10)
   b   (branch number, see 3.2)
   c   (crate number, see 3.3)
   n   (station number, see 3.4)
   m   (LAM access specifier, see 3.11)
   inta   (integer array, see 3.12)
Function: CDLAM encodes the branch number *b*, the crate number *c*, the station number *n* and all other necessary information concerning a LAM into a system-dependent, integer form and stores the result in *lam*. The parameter *m* is interpreted as a subaddress if its value is greater than or equal to zero. It is interpreted as the negative of a bit position if its value is less than zero. This information is encoded in the value assigned to *lam* and used by other subroutines to determine whether the LAM is accessed via special functions for LAM access or via reading and writing group 2 registers. The information contained in the array *inta* is completely implementation-dependent. It contains any information which may be required by the computer system or the CAMAC interface to enable the program to access the LAM. The use of *inta* by CDLAM is not required, but *inta* must appear in the parameter string regardless.

2.2.9 Enable or Disable LAM.
Name: CCLM
Parameters:
   lam   (LAM identifier, see 3.10)
   l   (logical truth value, see 3.9)
Function: CCLM causes the LAM specified by *lam* to be enabled if the value of *l* is *true* and causes it to be disabled if the value of *l* is *false*.

2.2.10 Clear LAM.
Name: CCLC
Parameter:
   lam   (LAM identifier, see 3.10)
Function: CCLC causes the LAM specified by *lam* to be cleared.

2.2.11 Test LAM.
Name: CTLM
Parameters:
- lam (LAM identifier, see 3.10)
- l (logical truth value, see 3.9)

Function: CTLM sets *l* to the value *true* if the LAM specified by *lam* is asserted, and sets *l* to the value *false* if the LAM is not asserted.

2.2.12 Link LAM to Service Procedure.
Name: CCLNK
Parameters:
- lam (LAM identifier, see 3.10)
- label (entry point identifier, see 3.13)

Function: CCLNK performs an association between the LAM specified by *lam* and a procedure identified by the parameter *label*. As a result of this association the procedure will be executed whenever the LAM is recognized by the system.

2.3 Block Transfers, Multiple Actions, and Inverse Declarations.
These subroutines together with those described in 2.1 and 2.2, form a comprehensive CAMAC support system, level C, with very general features and a potential for efficient execution of block transfers and multiple actions [5].

All the action subroutines in this section employ a *control block*, *cb*, which is an integer array containing four elements as follows:

element 1	Repeat count
element 2	Tally
element 3	LAM identification
element 4	Channel identification

The repeat count specifies the number of CAMAC actions or the maximum number of data words to be transferred. The tally is returned by the subroutine and indicates the number of actions executed or the number of data words actually transferred. Thus the calling program can detect and analyze a premature termination of an activity. The LAM identification is a coded identifier of the same form and interpretation as that returned by the subroutine CDLAM. The channel identification is a system-dependent or implementation-dependent parameter which may not be required in all implementations. It is used to identify any computer-related facilities necessary for the execution of the specified CAMAC action. Examples of facilities which might be required to be identified include computer input/output channels and operating system facilities such as unit numbers.

2.3.1 General Multiple Action.
Name: CFGA
Parameters:
- fa (function codes, see 3.14)
- exta (external address, see 3.15)
- intc (CAMAC data array, see 3.16)
- qa (Q responses, see 3.17)
- cb (control block, see 3.18)

Function: CFGA causes a sequence of CAMAC functions specified in successive elements of *fa* to be performed at a corresponding sequence of CAMAC addresses specified in successive elements of *exta*. Any read or write function in *fa* causes a CAMAC data word to be transferred between the corresponding element of *intc* and the specified CAMAC register. The Q response for each individual CAMAC action is stored in the corresponding element of *qa*. The number of actions performed and the number of elements required in the arrays *fa*, *exta*, *intc*, and *qa* is given by the value contained in the first element of *cb*. If the third element of *cb* contains the value zero, the specified sequence of actions is executed immediately; if it contains a LAM identification, then the sequence of actions is not initiated until the LAM is recognized.

2.3.2 Address Scan.
Name: CFMAD
Parameters:
- f (function code, see 3.6)
- extb (external addresses, see 3.19)
- intc (CAMAC data array, see 3.16)
- cb (control block, see 3.18)

Function: CFMAD causes a single CAMAC function specified by the value of *f* to be executed at a succession of addresses computed using the Address Scan algorithm described in [1], 5.4.3.1 and [5], 3.2. In the Address Scan mode the specified function is executed first at the address given by the first element of *extb*. Then if the Q response is 1, the subaddress is incremented by 1 and the index into the data array *intc* incremented by 1, and the function is executed at this new subaddress. If the subaddress is incremented beyond 15, it is set to zero, and the station number is incremented. If the Q response is 0, the subaddress is set to zero and the station number is incremented, but the index into *intc* is not changed. Execution of the specified function is attempted at the resulting new address, and the process is repeated until either the requested number of actions (given in the first element of *cb*) have been performed or the address computed by the procedure described above exceeds the address contained in the second element of *extb*. If the third element of *cb* contains the value 0, execution of the CAMAC actions is begun immediately. If it contains a LAM identifier, execution does not begin until the specified LAM is recognized. Address Scan is defined within a single crate only in [1]. For the purpose of this standard Address Scan is extended to other address components as follows:

(1) If the station number is incremented beyond the number of stations in a crate, it is set to 1, and the crate number is incremented.

(2) If the crate number is incremented beyond the number of crates in the branch, the crate number is set to 1, and the branch number is incremented.

2.3.3 Controller-Synchronized Block Transfer.
Name: CFUBC
Parameters:

   f   (function code, see 3.6)
   ext   (external address, see 3.1)
   intc   (CAMAC data array, see 3.16)
   cb   (control block, see 3.18)

Function: CFUBC causes the single CAMAC function given by the value of *f* to be executed at the CAMAC address specified by the value of *ext*. In this mode the CAMAC address is never changed, but the single register is expected to supply or accept many words of data. It is assumed able to supply or accept a data word whenever the controller addresses it until the block is exhausted or the controller terminates the process because the number of data transfers exceeds the limit given by the first element of *cb*. If the third element of *cb* contains the value 0, execution of the CAMAC actions is begun immediately. If it contains a LAM identifier, execution does not begin until the specified LAM is recognized.

The module indicates that the block is exhausted by its Q response. Two methods for doing this are described in [5], that is, Stop Mode (3.1) and Stop-on-Word Mode (4.1). In general both methods are not implemented in a given hardware system and no confusion should occur. However, if both methods are permissible, then the fourth element of *cb* which contains system dependent facilities must be used to specify for each case which method the module is using.

2.3.4 LAM-Synchronized Block Transfer.
Name: CFUBL
Parameters:

   f   (function code, see 3.6)
   ext   (external address, see 3.1)
   intc   (CAMAC data array, see 3.16)
   cb   (control block, see 3.18)

Function: CFUBL causes the single CAMAC function specified by the contents of *f* to be executed at the CAMAC address specified by the contents of *ext*, with the usage of the Q response and the LAM signal defined as in [5], 4.2 for the ULS mode. In the ULS mode the CAMAC address is not changed, but the single address is expected to supply or accept many words of data. In the ULS mode the module asserts a LAM whenever it is ready to participate in a data transfer, and the controller responds with the appropriate command to effect the transfer. Any data words transferred are placed into or taken from the array *intc*. The controller terminates the process if the number of executions exceeds the limit given by the contents of the first element of *cb*, or the module may terminate the process by responding with Q=0. The response Q=1 indicates that the specified function was properly executed. The response Q=0 indicates that the attempted data transfer was not completed and that the process should be terminated. The LAM which synchronizes the process is specified by the contents of the third element of *cb*. Upon termination the number of Q=1 responses is stored in the second element of *cb*.

2.3.5 Repeat Mode Block Transfer.
Name: CFUBR
Parameters:

   f   (function code, see 3.6)
   ext   (external address, see 3.1)
   intc   (CAMAC data array, see 3.16)
   cb   (control block, see 3.18)

Function: CFUBR causes the single CAMAC function specified by the contents of *f* to be executed at the CAMAC address specified by the contents of *ext* with the usage of the Q response defined as in [1], 5.4.3.2 and [5], 3.3 for the Repeat mode. In the Repeat mode the CAMAC address is never changed, but the single address is expected to supply or accept many words of data. Q is used as a timing signal. Q=1 indicates that the previously executed function succeeded; Q=0 indicates that the module was not ready to execute the function and that the controller should try again.

Any data words transferred are placed into or taken from the array *intc*. If the response is Q=0, no transfer took place, and the index into the array *intc* is not changed. The number of Q=1 responses expected is given by the contents of the first element of *cb*. If the third element of *cb* contains zero, the process is initiated immediately; if it contains a LAM identifier, the process is initiated only when the specified LAM is recognized.

2.3.6 Analyze LAM Identifier.
Name: CGLAM
Parameters:
- *lam*   (LAM identifier, see 3.10)
- *b*   (branch number, see 3.2)
- *c*   (crate number, see 3.3)
- *n*   (station number, see 3.4)
- *m*   (LAM access identifier, see 3.11)
- *inta* (integer array, see 3.12)

Function: CGLAM decodes the LAM identifier *lam* into its component parts, consisting of the branch number *b*, the crate number *c*, the station number *n*, the subaddress or bit-position *m*, and the system-dependent information *inta*. This subroutine exactly reverses the process performed by CDLAM, and all parameters have the same interpretation and form.

2.3.7 Analyze Register Identifier.
Name: CGREG
Parameters:
- *ext* (external address, see 3.1)
- *b*   (branch number, see 3.2)
- *c*   (crate number, see 3.3)
- *n*   (station number, see 3.4)
- *a*   (subaddress, see 3.5)

Function: CGREG decodes the CAMAC address identifier *ext* into its component parts, consisting of the branch number *b*, the crate number *c*, the station number *n*, and the subaddress *a*. This subroutine exactly reverses the process performed by CDREG, and all parameters have the same interpretation and form.

3. Definitions of Parameters

The meanings of the parameters used in the subroutine definitions are given briefly in the subroutine function descriptions. Since similar parameters are used in many subroutines, the various symbols are defined and discussed more completely in the subsections which follow. This information is intended particularly as a guide to implementors in the hope that it will result in the greatest possible degree of uniformity among different implementations.

3.1 *ext* **(external address).** The symbol *ext* represents an integer which is used as an identifier of an external CAMAC address. The address may represent a register which can be read or written, a complete CAMAC address which can be accessed by control or test functions, or a crate address. The value of *ext* is explicitly defined to be an integer. Normally it can be expected to be an encoded version of the address components, in which the coding has been selected for the most efficient execution of CAMAC actions on the interface to which the implementation applies. Other possibilities are allowed, however. For example, *ext* may be an index or a pointer into a data structure in which the actual CAMAC address components are stored.

3.2 *b* **(branch number).** The symbol *b* represents an integer which is the branch number component of a CAMAC address. It may represent a physical highway number in multiple highway systems, or it may represent sets of crates grouped together for functional or other reasons. In some systems it may be ignored, although it must be included in the parameter list for the sake of compatibility.

3.3 *c* **(crate number).** The symbol *c* represents an integer which is the crate number component of a CAMAC address. *Crate number* in this context can be either the physical crate number or it can be an integer symbol which is interpreted by the computer system software to produce appropriate hardware access information.

3.4 *n* **(station number).** The symbol *n* represents an integer which is the station number component of a CAMAC address.

3.5 *a* **(subaddress).** The symbol *a* represents an integer which is the subaddress component of a CAMAC address.

3.6 *f* **(function code).** The symbol *f* represents an integer which is the function code for a CAMAC action.

3.7 *int* **(CAMAC data word).** The symbol *int* represents a CAMAC data word stored in computer memory. The form is not specified, but the word must be stored in an addressable storage entity capable of containing twenty-four bits. In a computer or programming system which does not have an addressable unit of storage which can contain twenty-four bits, multiple units must be used.

3.8 *q* **(Q response).** The symbol *q* represents a logical truth value which corresponds to the CAMAC Q response. It is set to *true* if the Q response is 1, to *false* if the Q response is 0.

3.9 *l* **(logical truth value).** The symbol *l* represents a logical truth value which can be either *true* or *false*.

3.10 *lam* **(LAM identifier).** The symbol *lam* represents an integer which is used as the identifier of a CAMAC LAM signal. The information associated with the identifier must include not only the CAMAC address but also information about the means of accessing and controlling the LAM, that is, whether it is accessed via dataless functions at a subaddress or via read/write functions in group 2 registers. The value of *lam* is explicitly defined to be a non-zero integer; whether it is an encoded representation of the information required to describe the LAM or simply provides a key for accessing the information in a system-data structure is an implementation decision. The value 0 is used to indicate, where appropriate, that no LAM is being specified.

3.11 *m* **(LAM access specifier).** The symbol *m* represents an integer which is used to indicate the mode of access of a LAM and the lowest-order address component for LAM addressing. If *m* is zero or positive, it is interpreted as a subaddress and the LAM is assumed to be accessed via dataless functions at this subaddress. If *m* is negative, it is interpreted as the negative of a bit position for a LAM which is accessed via reading, setting, or clearing bits in the group 2 registers at subaddresses 12, 13, or 14.

3.12 *inta* **(integer array).** The symbol *inta* represents an integer array, the length and contents of which are not defined in this standard. It is intended to contain system-dependent or implementation-dependent information associated with the definition of a LAM. If no such information is required, the array need not be used. This information can include parameters necessary for interrupt linkage, event specification, etc. The documentation for an implementation must describe the requirements for any parameters contained in this array.

3.13 *label* **(entry point identifier).** The symbol *label* represents an entry point into a programmed procedure. Such a procedure will typically be executed in response to the recognition of a LAM, and it may interrupt the process being executed at the time of recognition of the LAM. Under these circumstances the procedure must be capable of saving and restoring the state of the computer so that the interrupted process can be resumed. At least one value of labels should identify a system error procedure which deals with LAMs not linked to user processes.

3.14 *fa* **(function codes).** The symbol *fa* represents an array of integers, each of which is the function code for a CAMAC action. The length of *fa* is given by the value of the first element of *cb* (see 3.18) at the time the subroutine is executed.

3.15 *exta* **(CAMAC external addresses).** The symbol *exta* represents an array of integers, each of which is a CAMAC register address. The form and information content of each element of *exta* must be identical to the form and information content of the quantity *ext* (see 3.1). The length of *exta* is given by the value of the first element of *cb* (see 3.18) at the time the subroutine is executed.

3.16 *intc* **(CAMAC data array).** The symbol *intc* represents an array of CAMAC data words. Each element of *intc* has the same form as the CAMAC data word variable *int* (see 3.7). The length of *intc* is given by the value of the first element of *cb* (see 3.18) at the time the subroutine is executed.

3.17 *qa* **(Q responses).** The symbol *qa* represents an array of Q response values. Each element of *qa* has the same form and can have the same values as the parameter *q* (see 3.8). The length of *qa* is given by the value of the first element of *cb* (see 3.18) at the time the subroutine is executed.

3.18 *cb* **(control block).** The symbol *cb* represents an integer array having four elements. The contents of these elements are:

    element 1   Repeat Count

element 2   Tally
element 3   LAM identification
element 4   Channel identification

The repeat count specifies the number of individual CAMAC actions or the maximum number of data words to be transferred. Some multiple action and block transfer subroutines permit termination of the sequence upon a signal from the addressed module. In such cases the repeat count represents an upper limit. The tally is the number of actions actually performed or the number of CAMAC data words actually transferred. If the block transfer or multiple action is terminated by the controller due to exhaustion of the repeat count, the tally will be equal to the repeat count; otherwise it may be less. The LAM identification is an integer value having the same form and information content as the variable *lam* (see 3.10). The channel identification is an integer value which identifies system-dependent facilities which may be necessary to perform the block transfer or multiple action. This number, if it is required, has the same form and content as the parameter *chan* (see Appendix A, A6.2) and can be created by the subroutine CDCHN (see Appendix A, A3.1).

3.19 *extb* **(external addresses).** The symbol *extb* represents an array of integers containing external CAMAC addresses. The array has two elements: (1) The starting address for an Address Scan multiple action; (2) The final address which can be permitted to participate in the Address Scan sequence. Each element has the same form and information content as the parameter *ext* (see 3.1).

4. Bibliography

[1] ANSI/IEEE Std 583-1982, Standard Modular Instrumentation and Digital Interface System (CAMAC).

[2] ANSI/IEEE Std 595-1982, Standard Serial Highway Interface System (CAMAC).

[3] ANSI/IEEE Std 596-1982, Standard Parallel Highway Interface System (CAMAC).

[4] ANSI/IEEE Std 675-1982, Standard Multiple Controllers in a CAMAC Crate.

[5] ANSI/IEEE Std 683-1976 (Reaff 1981), Recommended Practice for Block Transfers in CAMAC Systems.

[6] US Energy Research and Development Report TID — 22615 CAMAC, The Definition of IML.

[7] ANSI/IEEE Std 726-1982, Real Time BASIC for CAMAC.

[8] ERDA US Energy Research and Development Administration Report TID-26614 CAMAC ANALOG Signals.

[9] ERDA US Energy Research and Development Administration Report TID-26615 IML (A CAMAC Language).

[10] ERDA US Energy Research and Development Administration Report TID-26618 CAMAC Tutorial Articles.

[11] CAMAC Instrumentation and Interface Standards, IEEE SH08482, Library of Congress Catalog No 8185060.

Appendixes

(These Appendixes are not a part of ANSI/IEEE Std 758-1979, IEEE Standard Subroutines for CAMAC.)

Appendix A
System-Dependent Subroutines

A1. Introduction

It is often useful, even necessary, to utilize subroutines which depend strongly on an individual computer or operating system architecture. These subroutines cannot be made completely transportable even in an implementation for a highly standardized language. The functions they must perform are usually very similar, however, and may be identical within restricted classes of systems. This appendix describes a number of such system-dependent subroutines. They are not required to be in any level of implementation, but any of them may be supplied in an implementation at the discretion of the implementor. The documentation for a given implementation must indicate which ones are supplied and must describe completely their system-dependent features and parameters.

A2. Access to Special Signals

These subroutines give access to features of CAMAC systems for which a standard access method is not defined or which are not available in all systems.

A2.1 Branch Initialize.
Name: CCINIT
Parameter:
   b (branch number, see 3.2)
Function: CCINIT causes the branch initialize signal (BZ) to be asserted on the branch identified by the parameter b. If b identifies a group of crates not interfaced via a parallel branch, the function may be simulated by causing crate initialize signals to be asserted in the crates. The user of a system should be warned that the response of many modules to the initialize signal is such as to place the system in undesirable states; thus the feature should be used with caution.

A2.2 Test Status of Preceding Action.
Name: CTSTAT
Parameter:
   k (status code, see A6.1)
Function: CTSTAT stores an integer status code into the parameter k. The status code stored reflects the results of the last action executed in the program; the code has a value given by the expression $4e+d$, where e and d are non-negative integers. The meaning of d is given by the following table:

Value	Meaning
0	Q=1, X=1
1	Q=0, X=1
2	Q=1, X=0
3	Q=0, X=0

If e is zero, there are no errors reported; if e is greater than zero, then some error or exception is indicated. The exact meanings of e are system-dependent and must be defined in the documentation for each implementation.

A3. Channel Identifier

It may be necessary in some computers or in some operating systems to utilize special I/O channel controllers or operating system features such as logical unit numbers and device codes to access CAMAC systems. These two subroutines provide a means to specify or interrogate such information.

A3.1 Declare Channel.
Name: CDCHN
Parameters:
   chan   (channel identifier, see A6.2)
           additional parameters, depending on implementation

Function: CDCHN encodes system-dependent parameters into a convenient form and returns the results as the value of *chan*. It specifies the computer hardware and software or both I/0 facilities to be used and possibly additional information relating to their use. The additional parameters must be completely described in the documentation for each implementation.

A3.2 Analyze Channel Declaration.
Name: CGCHN
Parameters:
   chan   (channel identifier, see A6.2)
           additional parameters, depending on implementation

Function: CGCHN extracts the system-dependent parameters from the channel identifier *chan*. It reverses the process performed by CDCHN.

A4. Short Data-Word Transfers

In many systems and applications it is inconvenient and inefficient to utilize the full twenty-four bit length of the CAMAC data word and sufficient to use a truncated data word. The appropriate degree of truncation is dependent on the system, but the computer word length is the most common choice. The leftmost bits of the CAMAC data word are discarded, while the rightmost bits are retained. The length of a truncated CAMAC data word must be specified in the documentation. These subroutines duplicate functions performed by subroutines described in Section 2 of the main body of this standard, except that truncated CAMAC data words are used.

A4.1 Perform Single CAMAC Action.
Name: CSSA
Parameters:
   f   (function code, see 3.6)
   ext   (external address, see 3.1)
   ints (truncated CAMAC data word, see A6.3)
   q   (Q response, see 3.8)

Function: CSSA performs the same function as CFSA (see 2.1.2) except that *ints* contains a truncated CAMAC data word.

A4.2 General Multiple Action.
Name: CSGA
Parameters:
   fa   (function codes, see 3.14)
   exta (external addresses, see 3.15)
   intt (truncated CAMAC data array, see A6.4)
   qa   (Q response, see 3.17)
   cb   (control block, see 3.18)

Function: CSGA performs the same function as CFGA (see 2.3.1) except that the elements of *intt* are truncated CAMAC data words.

A4.3 Address Scan.
Name: CSMAD
Parameters:
   f   (function code, see 3.6)
   extb (external addresses, see 3.19)
   intt (truncated CAMAC data array, see A6.4)
   cb   (control block, see 3.18)

Function: CSMAD performs the same function as CFMAD (see 2.3.2) except that the elements of *intt* are truncated CAMAC data words.

A4.4 Controller-Synchronized Block Transfer.
Name: CSUBC
Parameters:
   f   (function code, see 3.6)
   ext   (external address, see 3.1)
   intt (truncated CAMAC data array, see A6.4)
   cb   (control block, see 3.18)

Function: CSUBC performs the same function as CFUBC (see 2.3.3) except that the elements of *intt* are truncated CAMAC data words.

A4.5 LAM-Synchronized Block Transfer.
Name: CSUBL
Parameters:
   f   (function code, see 3.6)
   ext   (external address, see 3.1)
   intt (truncated CAMAC data array, see A6.4)
   cb   (control block, see 3.18)

Function: CSUBL performs the same function as CFUBL (see 2.3.4) except that the elements of *intt* are truncated CAMAC data words.

A4.6 Repeat Mode Block Transfer.
Name: CSUBR
Parameters:
   f   (function code, see 3.6)

SUBROUTINES FOR CAMAC

ext (external address, see 3.1)
intt (truncated CAMAC data array, see A6.4)
cb (control block, see 3.18)

Function: CSUBR performs the same function as CFUBR (see 2.3.5) except that the elements of *intt* are truncated CAMAC data words.

A5. Define Crate Identifier

Name: CDCRT
Parameters:
c (crate number, see 3.3)
intb (integer array, see A6.5)

Function: CDCRT defines the crate number *c* in terms of the system-dependent information contained in the array *intb*. It permits the meaning of the crate number *c* to be changed within a procedure execution.

A6. Definitions of Parameters

Although many of the parameters used in the subroutines of Appendix A are identical with those described in Section 3 of the main body of this standard, a few new ones have been introduced. The parameters referred to only in this appendix are described in this section.

A6.1 *k* **(status code).** The symbol *k* represents an integer used to provide status information with respect to a CAMAC action. The rightmost two bits contain the complements of the X and Q responses; the remaining bits in a word contain system-dependent error codes. The lowest-order bit contains the complement of the Q response; the next bit to the left (the 2's bit) contains the complement of X.

A6.2 *chan* **(channel identifier).** The symbol *chan* represents an integer which contains a channel identifier.

A6.3 *ints* **(truncated CAMAC data word).** The symbol *ints* represents a storage cell in computer memory used to hold bits 1 to *n* of a CAMAC data word, where *n* is defined for the implementation to be less than twenty-four. Higher bits in the CAMAC word are not used when the function is a READ; they are set to zeros if the function is a WRITE.

A6.4 *intt* **(truncated CAMAC data array).** The symbol *intt* represents an array in computer memory, each element of which contains a truncated CAMAC data word having the same form as *ints* (see A6.3).

A6.5 *intb* **(integer array).** The symbol *intb* represents an integer array, the length and contents of which are not defined in this standard. It is intended to contain system-dependent or implementation-dependent information associated with the definition of a crate number identifier. The information may include physical crate number, highway or computer interface identification, or operating system parameters necessary for access. The documentation for an implementation must describe the requirements for any parameters contained in this array.

Appendix B
FORTRAN Implementation

B1. General

This appendix recommends an implementation of the CAMAC standard subroutines for use with FORTRAN for the purpose of communicating with CAMAC-interfaced devices and synchronizing program execution with events associated with such devices.

B2. Description of Subroutines

B2.1 Primary Subroutines.
B2.1.1 Declare CAMAC Register.
Form: CALL CDREG (*ext,b,c,n,a*)
Function: CDREG combines the components *b*, *c*, *n*, and *a* into a single CAMAC register reference.

IMPLEMENTATION NOTES:
ext: The subroutine must return a value which uniquely defines the specified CAMAC register or crate for other subroutines in the implementation. The means by which this objective is accomplished are not specified. The most common procedure expected is to return a value in which the parameters *b*, *c*, *n*, and *a* have been packed into a single integer with the fields defined for the most efficient execution of CAMAC actions on the interface to which the implementation applies. Other possibilities are allowed, however. For example, the values of the parameters could be stored in a data structure and a pointer to the entry returned. If *n* and *a* are defined to be zero, then *ext* is defined to be a crate. If *n* and *a* are not zero, then *ext* is defined to be a register or other addressable entity.
Tables of Register References: It is advantageous in many applications to have tables of CAMAC register references, identical in form to the value returned by CDREG, available on a permanent basis to be loaded by programs or as arrays of constants within programs. Such tables not only save the time required to define the register references each time a program is executed but also can form parts of larger tables of logical device definitions. If such facilities can be supported by an implementation, the documentation should describe how to create these references in assembler code or in DATA statements in the compiler.

B2.1.2 Perform Single CAMAC Action.
Form: CALL CFSA (*f, ext, int, q*)
Function: This subroutine performs a single CAMAC function at a single CAMAC address. If the function *f* is a read or a write function, a 24-bit data word is transferred.

IMPLEMENTATION NOTES:
int: The internal reference *int* must be able to contain 24 bits. If the FORTRAN implementation supports integer variables of 24 bits or more, *int* can be an integer expression (for write) or an integer variable or array element (for read). Otherwise *int* must be an integer array (or a portion of one) with enough elements to contain 24 bits. When a CAMAC data word is stored in an integer variable, or when an integer constant is used as *int*, CAMAC bit 1 occupies the low-order bit position, and bit 24 occupies the twenty-fourth bit position to the left. (In a binary machine; in other radices an appropriate transformation to equivalent integer values must be made). If the FORTRAN implementation does not support integers with a length as great as 24 bits, then the bits of the CAMAC data word are stored in the array elements such that bit 1 is the lowest order bit in the highest numbered element. All the bits in the highest element are filled and the remaining bits are stored beginning in the lowest order bit of the next lowest numbered element, etc. For what is probably the most common example, 16-bit words, one could define an array of length 2, say CAMDAT. Then bits 1 to 16 are stored in CAMDAT (2) with bit 1 the lowest and bit 16 the highest. Bits 17 to 24 are stored in CAMDAT (1) with bit 17 in the lowest order position in the word. The

upper eight bits of CAMDAT (1) are cleared to zero.

Restrictions on Parameter Values: In multiple-user systems where CAMAC facilities are shared among several independent users, it may be necessary to restrict access to crate controllers and certain modules by not allowing crate controller actions to be issued via CFSA and by forbidding access via CFSA to certain other modules used for system control. In simpler systems, however, such restrictions are unnecessary and prevent the implementation of test and diagnostic codes. Each implementor must judge his own situation. Such restrictions can, of course, be controlled by setting mode switches either during operation or at system generation. Where such restrictions appear to be necessary, it is recommended that the implementation give the system manager the option of invoking them or not.

B2.2 Single Action Subroutines.

B2.2.1 Generate Dataway Initialize.
Form: CALL CCCZ (*ext*)
Function: This subroutine generates the crate initialize signal, Z, on the dataway of the addressed crate.

B2.2.2 Generate Crate Clear.
Form: CALL CCCC (*ext*)
Function: This subroutine generates the crate clear signal on the dataway of the addressed crate.

B2.2.3 Set or Clear Dataway Inhibit.
Form: CALL CCCI (*ext, l*)
Function: This subroutine sets the crate inhibit signal in the addressed crate if l is *true* and clears it if l is *false*.

B2.2.4 Test Dataway Inhibit.
Form: CALL CTCI (*ext,l*)
Function: This subroutine returns the value *true* in l if the inhibit signal in the addressed crate is set, *false* if not.

B2.2.5 Enable or Disable Crate Demands.
Form: CALL CCDD (*ext, l*)
Function: This subroutine enables (if l is true) or disables (if l is false) demands from the specified crate.

B2.2.6 Test Crate Demand Enabled.
Form: CALL CTCD (*ext,l*)
Function: This subroutine returns in l the value *true* if crate demands are enabled, *false* if not.

B2.2.7 Test Demand Present.
Form: CALL CTGL (*ext,l*)
Function: This subroutine returns in l the value *true* if any GL bit is present, *false* if not.

B2.2.8 Declare LAM.
Form: CALL CDLAM (*lam,b,c,n,m,inta*).
Function: CDLAM combines the components *b, c, n, m,* and *inta* into a single reference to the LAM.

IMPLEMENTATION NOTES:

lam: The subroutine must return a value which uniquely defines the specified LAM for other subroutines in the implementation. The means by which this objective is accomplished are not specified. If the parameters necessary to specify the address of a LAM and any necessary auxiliary information can be packed into a single integer variable, then the value returned may contain this information. Otherwise it may be necessary to supply an arbitrary index, or other code, which identifies the LAM to the other subroutines which access it.

Permanent Assignment of Logical LAMs: In many systems LAMs are not dynamic and can be known to the operating system or the CAMAC I/0 programs more or less permanently. In such cases the function CDLAM is not required. The LAMs can be defined via tables which are part of the CAMAC I/0 software or which may be loaded by a particular program to identify the LAMs with which it deals. If a particular implementation supports definition by such tables, the documentation should describe how to create LAM references using assembler code or compiler DATA statements.

B2.2.9 Enable or Disable LAM.
Form: CALL CCLM (*lam, l*)
Function: This subroutine enables (if l is true) or disables (if l is false) the module LAM identified by *lam*.

B2.2.10 Clear LAM.
Form: CALL CCLC (*lam*)
Function: This subroutine clears the module LAM identified by the argument.

B2.2.11 Test LAM.
Form: CALL CTLM (*lam, l*)
Function: This subroutine returns in l the value *true* if the addressed LAM is asserted, *false* if not.

B2.2.12 Link LAM to Service Procedure.
Form: CALL CCLNK (*lam, label*)
Function: Perform a run time association between a logical LAM and an interrupt service routine.

B2.3 Block Transfers, Multiple Actions, and Inverse Declarations.

B2.3.1 General Multiple Action.
Form: CALL CFGA (*fa, exta, intc, qa, cb*)

Function: This subroutine performs a sequence of CAMAC functions at a corresponding sequence of CAMAC addresses and returns a sequence of Q responses. The number of functions executed, addresses accessed, and Q responses returned is given by the value of the parameter $cb(1)$. CAMAC data words are transferred between the CAMAC system and the array *intc* whenever the specified function is read or write. Since the array *fa* can contain both read and write functions, data words may be written to and read from the same array by a single call to this subroutine. CAMAC-data-word positions in the array *intc* which correspond to functions in the array *fa* which are neither read nor write are not accessed by this subroutine.

IMPLEMENTATION NOTES:

fa: Note that in this subroutine a succession of possibly unrelated functions are executed at a succession of CAMAC addresses.

intc: The internal reference *intc*, unless all functions in the array *fa* are control or test functions which use no data, must be able to contain a CAMAC data word for each function executed. In order to maintain the positional correspondence between values in *fa*, *exta*, *intc*, and *qa*, a position in *intc* is skipped whenever a control or test function is executed. If the FORTRAN implementation supports integer array elements of 24 bits or more, then each execution of the function accesses a successive element of the array. The CAMAC data word occupies the low-order 24 bits of the integer array element; higher order bits are cleared by a read function and ignored by a write. If the FORTRAN implementation supports only integer arrays whose elements contain fewer than 24 bits, then each CAMAC data word is contained in an integral number of elements (a *block*) sufficiently large to contain 24 bits. The length of the array must be great enough to include a block for each CAMAC data word transferred. Each CAMAC data word is positioned within its block so that the element within the block with the highest index is filled with the lowest order bits of the data word, and any bits not required to contain data are at the highest order positions of the element with the lowest index. Thus in the typical 16-bit case, where a block consists of elements n and $n + l$, element n contains CAMAC bits 24 through 17 in its low-order eight bits and element $n + l$ contains bits 16 through 1.

qa: This array contains the Q responses for each function executed. The format of each element of the array *qa* is identical to the format of the variable *q* in subroutine CFSA, that is, a logical value is returned. No action based on these values is taken by the subroutine.

Restrictions on Parameter Values: In multiple-user systems where CAMAC facilities are shared among several independent users, it may be necessary to restrict access to crate controllers and certain modules by not allowing crate controller actions to be issued via CFGA and by forbidding access via CFGA to certain other modules used for system control. In simpler systems, however, such restrictions are unneccessary and prevent the implementation of test and diagnostic codes. Each implementor must judge his own situation. Such restrictions can, of course, be controlled by setting mode switches either during operation or at system generation. Where such restrictions appear to be necessary, it is recommended that the implementation give the system manager the option of invoking them or not.

Applicability: This subroutine is useful primarily in systems in which the overhead associated with initiating or specifying a CAMAC action is great compared with the time required to perform an individual CAMAC command. Systems which suffer from a high I/O overhead in the system software or where there is a remote intelligent processor with a relatively high communications overhead are among those in which this subroutine may be needed.

B2.3.2 Address-Scan.

Form: CALL CFMAD (f, *extb*, *intc*, *cb*)

Function: This subroutine executes a single CAMAC operation, f, at a succession of addresses computed using the Address Scan algorithm.

IMPLEMENTATION NOTES:

intc: See implementation note on *intc* in B2.3.1.

Q: The Q responses to the CAMAC functions are not saved. The Q response is used to control the addressing algorithm and cannot provide information to the calling program.

X: Special attention must be paid to the X response in the Address Scan mode. When a module responds with Q=0, it may at the time give either X=1 or X=0; consequently in this mode of access the X response is ignored when the Q response is 0.

Restrictions on Parameter Values: In multiple-user systems where CAMAC facilities are shared among several independent users, it may be necessary to restrict access to crate controllers and certain modules by not allowing crate controller actions to be issued via CFMAD and by forbidding access via CFMAD to certain other modules used for system control.

B2.3.3 Controller-Synchronized Block Transfer.

Form: CALL CFUBC (*f, ext, inc, cb*)
Function: This subroutine executes a single CAMAC function, *f*, at a single address, with the usage of the Q response defined as for the Stop Mode.

IMPLEMENTATION NOTES:

intc: See implementation note on *intc* in B2.3.1.
Q: The Q responses to the CAMAC functions are not saved. Since the Q response is used by the module to indicate end of block, it cannot carry information to the calling program.
Restrictions on Parameter Values: In multiple-user systems where CAMAC facilities are shared among several independent users, it may be necessary to restrict access to crate controllers and certain modules by not allowing crate controller actions to be issued via CFUBC and by forbidding access via CFUBC to certain other modules used for system control.

B2.3.4 LAM-Synchronized Block Transfer.

Form: CALL CFUBL (*f, ext, intc, cb*)
Function: This subroutine executes a single CAMAC function, *f*, at a single address, with the usage of the Q response and LAM signal for the ULS mode.

IMPLEMENTATION NOTES:

intc: See implementation note on *intc* in B2.3.1.
A: The Q responses to the CAMAC functions are not saved. Since the Q response is used by the module to indicate end of block, it cannot carry information to the calling program.
LAM Control: Modules designed to be used in this mode must clear the synchronizing LAM during the execution of a read or write command.
Restrictions on Parameter Values: In multiple-user systems where CAMAC facilities are shared among several independent users, it may be necessary to restrict access to crate controllers and certain modules by not allowing crate controller actions to be issued via CFUBL and by forbidding access via CFUBL to certain other modules used for system control.

B2.3.5 Repeat-Mode Block Transfer.

Form: CALL CFUBR (*f, ext, intc, cb*)
Function: This subroutine executes a single CAMAC function, *f*, in the Repeat Mode.

IMPLEMENTATION NOTES:

intc: See implementation note on *intc* in B2.3.1.
Q: The Q responses to the CAMAC functions are not saved. Since the Q response is used by the module to indicate data synchronization, it cannot carry information to the calling program.
System Integrity: A malfunction or erroneous operation may produce a situation in which a module will never respond with Q=1, causing an unending loop. To guard against this case a time out or maximum repetition count should be implemented in the controller hardware or software.
Restrictions on Parameter Values: In multiple-user systems where CAMAC facilities are shared among several independent users, it may be necessary to restrict access to crate controllers and certain modules by not allowing crate controller actions to be issued via CFUBR and by forbidding access via CFUBR to certain other modules used for system control.

B2.3.6 Analyze LAM Identifier.

Form: CALL CGLAM (*lam, b, c, n, m, inta*)
Function: CGLAM accepts a LAM identifier as input and returns the values of the hardware or operating system parameters which define it. It performs the inverse transformation of CDLAM.

B2.3.7 Analyze Register Identifier.

Form: CALL CGREG (*ext, b, c, n, a*)
Function: CGREG accepts a register identifier, *ext*, as input, analyzes it into its components, and stores them into *b, c, n,* and *a*.

B3. Parameter Types

The types of the subroutine parameters are given in this section. See Section 3 for the meaning of each parameter.

B3.1 Single Integers. If a subroutine returns a value in a parameter of this type, it must be either an integer variable or an integer array element. If no value is returned, it may be any integer expression. The parameters of this class are:

a, b, c, ext, f, lam, m, n.

B3.2 Single Logical Values. If a subroutine returns a value in a parameter of this type, it must be either a logical variable or a logical array element. If no value is returned, it may be any logical expression. The parameters of this class are:

l, q.

B3.3 Integer Arrays. These parameters must be integer arrays. The parameters of this class are:

cb, exta, extb, fa, inta.

B3.4 Logical Array. This parameter must be a logical array. The only parameter of this class is:

qa.

B3.5 CAMAC Data Word. This parameter must be capable of storing a CAMAC data word twenty-four bits in length. If the FORTRAN integer variable can contain twenty-four bits or more, then the type for CAMAC data word is the same as for a single integer (see B3.1). If the integer variable length is greater than the equivalent of twenty-four bits, the CAMAC data word is represented as an unsigned integer with bit 24 representing the highest order bit and bit 1 representing the lowest. If the FORTRAN integer variable cannot contain twenty-four bits, then the parameter must be an integer array sufficiently long to contain twenty-four bits. The CAMAC data word must be represented in the array as a multiple-precision integer with the lowest-order portion in the last array element and the highest-order portion in the first element. Bit 24 of the CAMAC word is taken to be the highest-order bit and bit 1 the lowest. The only parameter of this class is:

int.

B3.6 CAMAC Data Array. This parameter must be an array of elements each of the same type as the CAMAC data word (see B3.5). The only parameter of this class is:

intc.

B3.7 Label. This parameter cannot be defined or assigned a type within the bounds of strictly standard FORTRAN, since it labels a procedure which is intended to be executed out of sequence with respect to the calling program. It must be defined appropriately for each implementation. The only parameter of this class is:

label.

Appendix C

Function-Code Mnemonic Symbols

Symbol	Value	Definition	Symbol	Value	Definition
RD1	0	Read group 1 register	WT2	17	Write group 2 register
RD2	1	Read group 2 register	SS1	18	Selective set group 1 register
RC1	2	Read and clear group 1 register	SS2	19	Selective set group 2 register
RCM	3	Read complement of group 1 register	SC1	21	Selective clear group 1 register
TLM	8	Test LAM	SC2	23	Selective clear group 2 register
CL1	9	Clear group 1 register	DIS	24	Disable
CLM	10	Clear LAM	XEQ	25	Execute
CL2	11	Clear group 2 register	ENB	26	Enable
WT1	16	Write group 1 register	TST	27	Test